Mineralogy

Mineralogy

Edited by
Cyaden Ross

☰ Larsen & Keller
www.larsen-keller.com

Mineralogy
Edited by Cyaden Ross
ISBN: 978-1-63549-185-2 (Hardback)

© 2017 Larsen & Keller

☰ Larsen & Keller

Published by Larsen and Keller Education,
5 Penn Plaza,
19th Floor,
New York, NY 10001, USA

Cataloging-in-Publication Data

Mineralogy / edited by Cyaden Ross.
    p. cm.
Includes bibliographical references and index.
ISBN 978-1-63549-185-2
    1. Mineralogy. 2. Minerals. I. Ross, Cyaden.
QE363.2 .M56 2017
549--dc23

The publisher's policy is to use permanent paper from mills that operate a sustainable forestry policy. Furthermore, the publisher ensures that the text paper and cover boards used have met acceptable environmental accreditation standards.

Printed and bound in the United States of America.

For more information regarding Larsen and Keller Education and its products, please visit the publisher's website www.larsen-keller.com

Table of Contents

Preface

This book provides comprehensive insights into the field of mineralogy. It talks in detail about the various important methods and techniques used in this intricate field. Mineralogy refers to the study of minerals. It examines the physical, chemical and crystal structure properties of different minerals. This study is conducted to classify them, know about their origin, to get information about their geographical distribution and also their uses. Most of the topics introduced in this text cover new uses and the applications of mineralogy. The various subfields of the area along with technological progress that have future implications are glanced at in it. It will provide comprehensive knowledge to the readers. This textbook aims to serve as a resource guide for students and facilitate the study of the discipline.

Given below is the chapter wise description of the book:

Chapter 1- The study of minerals and mineralized compounds, their physical, chemical and crystalline structure is called mineralogy. It is a branch of geology that specifically studies the origin, formation, classification, geological distribution and the utilization of minerals. This chapter helps familiarize the reader to the discipline of mineralogy. A section traces the history of mineralogy as well.

Chapter 2- There are several branches of mineralogy that study various aspects- optical mineralogy which studies the optical properties of minerals and rocks to identify their origin and evolution; crystallography which is the experimental science that determines the arrangement of atoms in crystalline solids; amateur geology which the recreational study and hobby of collecting rocks and mineral specimens from the natural habitat; magnetic mineralogy which is the study of the magnetic properties of minerals and gemology which is the study of gems and their properties.

Chapter 3- Naturally occurring compounds, mostly crystalline in structure, are called minerals. There are over 5,300 species of known minerals whose diversity and distribution is controlled by the Earth's chemistry. They can be distinguished by their physical and chemical composition. This chapter acquaints the reader with the basic definition of a mineral and its defining characteristics.

Chapter 4- This chapter elucidates the crucial theories and concepts of minerals. Biomineralization refers to the process by which living organisms produce minerals often to stiffen and harden tissues. Biocrystallization is the formation of crystals from macromolecules by living organisms. The content explores the topics of biocrystallization, biomineralization, mineral alteration, crystal growth, evaporite etc.

Chapter 5- The chapter lists and studies the density and hardness of various minerals. It introduces the reader to the Rosiwal and Mohs scales of mineral hardness. The content also studies the tenacity, cleavage, fracture, crystal habit and lustre of minerals. These characteristics form the distinguishing characteristics that help classify minerals.

Chapter 6- Minerals differ in their response to magnetic fields. Based on their attraction or repulsion, minerals can be classified as possessing features like diamagnetism, paramagnetism, ferromagnetism, ferrimagnetism, antiferromagnetism and superparamagnetism. This chapter exclusively deals with the origins of magnetism and cites examples to deepen the understanding of each category of magnetic ability.

Chapter 7- Mineraloids are mineral like substances that do not exhibit crystalline nature. Most mineraloids are amorphous and do not possess an ordered atomic structure. The chapter introduces the reader to this category of minerals, lists the characteristic properties with suitable examples. Some examples explored include pearl, jet, obsidian, amber, ebonite etc.

Indeed, my job was extremely crucial and challenging as I had to ensure that every chapter is informative and structured in a student-friendly manner. I am thankful for the support provided by my family and colleagues during the completion of this book.

Editor

Introduction to Mineralogy

The study of minerals and mineralized compounds, their physical, chemical and crystalline structure is called mineralogy. It is a branch of geology that specifically studies the origin, formation, classification, geological distribution and the utilization of minerals. This chapter helps familiarize the reader to the discipline of mineralogy. A section traces the history of mineralogy as well.

Mineralogy

Mineralogy is a subject of geology specializing in the scientific study of chemistry, crystal structure, and physical (including optical) properties of minerals. Specific studies within mineralogy include the processes of mineral origin and formation, classification of minerals, their geographical distribution, as well as their utilization.

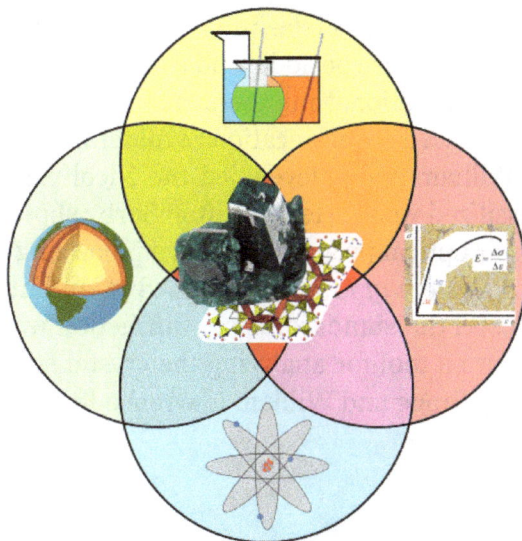

Mineralogy is a mixture of chemistry, materials science, physics and geology.

History

Early writing on mineralogy, especially on gemstones, comes from ancient Babylonia, the ancient Greco-Roman world, ancient and medieval China, and Sanskrit texts from ancient India and the ancient Islamic World. Books on the subject included the *Naturalis Historia* of Pliny the Elder, which not only described many different minerals but also explained many of their properties, and Kitab al Jawahir (Book of Precious Stones) by Persian scientist Al Biruni. The German Renaissance specialist Georgius Agricola wrote works such as *De re metallica* (*On Metals*, 1556) and *De Natura Fossilium* (*On the Nature of Rocks*, 1546) which began the scientific approach to the subject. Systematic scientific studies of minerals and rocks developed in post-Renaissance Europe. The modern study of mineralogy was founded on the principles of crystallography (the origins of geometric crystallography, itself, can be traced back to the mineralogy practiced in the eighteenth and nineteenth centuries) and to the microscopic study of rock sections with the invention of the microscope in the 17th century.

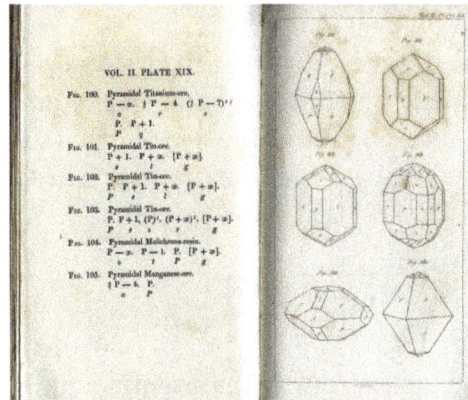

Page from *Treatise on mineralogy* by Friedrich Mohs (1825).

Nicholas Steno first observed the law of constancy of interfacial angles (also known as the first law of crystallography) in quartz crystals in 1669. This was later generalized and established experimentally by Jean-Baptiste L. Romé de l'Islee in 1783. René Just Haüy, the "father of modern crystallography", showed that crystals are periodic and established that the orientations of crystal faces can be expressed in terms of rational numbers, as later encoded in the Miller indices. In 1814, Jöns Jacob Berzelius introduced a classification of minerals based on their chemistry rather than their crystal structure. William Nicol developed the Nicol prism, which polarizes light, in 1827–1828 while studying fossilized wood; Henry Clifton Sorby showed that thin sections of minerals could be identified by their optical properties using a polarizing microscope. James D. Dana published his first edition of *A System of Mineralogy* in 1837, and in a later edition introduced a chemical classification that is still the standard. X-ray diffraction was demonstrated by Max von Laue in 1912, and developed into a tool for analyzing the crystal structure of minerals by the father/son team of William Henry Bragg and William Lawrence Bragg.

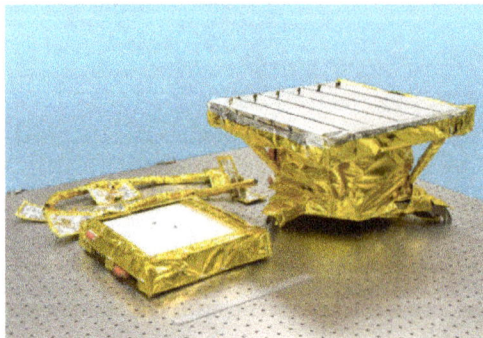

Left side of the Moon Mineralogy Mapper, a spectrometer that mapped the lunar surface.

More recently, driven by advances in experimental technique (such as neutron diffraction) and available computational power, the latter of which has enabled extremely accurate atomic-scale simulations of the behaviour of crystals, the science has branched out to consider more general problems in the fields of inorganic chemistry and solid-state physics. It, however, retains a focus on the crystal structures commonly encountered in rock-forming minerals (such as the perovskites, clay minerals and framework silicates). In particular, the field has made great advances in the understanding of the relationship between the atomic-scale structure of minerals and their function; in nature, prominent examples would be accurate measurement and prediction of the elastic properties of minerals, which has led to new insight into seismological behaviour of rocks and depth-related discontinuities in seismograms of the Earth's mantle. To this end, in their focus on the connection between atomic-scale phenomena and macroscopic properties, the *mineral sciences* (as they are now commonly known) display perhaps more of an overlap with materials science than any other discipline.

Physical Properties

An initial step in identifying a mineral is to examine its physical properties, many of which can be measured on a hand sample. These can be classified into density (often given as specific gravity); measures of mechanical cohesion (hardness, tenacity, cleavage, fracture, parting); macroscopic visual properties (luster, color, streak, luminescence, diaphaneity); magnetic and electric properties; radioactivity and solubility in hydrogen chloride (HCl).

If the mineral is well crystallized, it will also have a distinctive crystal habit (for example, hexagonal, columnar, botryoidal) that reflects the crystal structure or internal arrangement of atoms. It is also affected by crystal defects and twinning. Many crystals are polymorphic, having more than one possible crystal structure depending on factors such as pressure and temperature.

Crystal Structure

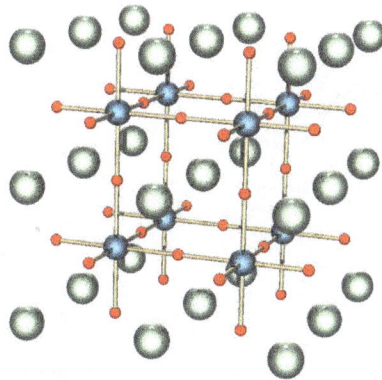

The perovskite crystal structure. The most abundant mineral in the Earth, bridgmanite, has this structure. Its chemical formula is $(Mg,Fe)SiO_3$; the red spheres are oxygen, the blue spheres silicon and the green spheres magnesium or iron.

The crystal structure is the arrangement of atoms in a crystal. It is represented by a lattice of points which repeats a basic pattern, called a unit cell, in three dimensions. The lattice can be characterized by its symmetries and by the dimensions of the unit cell. These dimensions are represented by three *Miller indices*. The lattice remains unchanged by certain symmetry operations about any given point in the lattice: reflection, rotation, inversion, and rotary inversion, a combination of rotation and reflection. Together, they make up a mathematical object called a *crystallographic point group* or *crystal class*. There are 32 possible crystal classes. In addition, there are operations that displace all the points: translation, screw axis, and glide plane. In combination with the point symmetries, they form 230 possible space groups.

Most geology departments have X-ray powder diffraction equipment to analyze the crystal structures of minerals. X-rays have wavelengths that are the same order of magnitude as the distances between atoms. Diffraction, the constructive and destructive interference between waves scattered at different atoms, leads to distinctive patterns of high and low intensity that depend on the geometry of the crystal. In a sample that is ground to a powder, the X-rays sample a random distribution of all crystal orientations. Powder diffraction can distinguish between minerals that may appear the same in a hand sample, for example quartz and its polymorphs tridymite and cristobalite.

isomorphous minerals of different compositions have similar powder diffraction patterns, the main difference being in spacing and intensity of lines. For example, the NaCl (halite) crystal structure is space group *Fm3m*; this structure is shared by sylvite (KCl), periclase (MgO), bunsenite (NiO), galena (PbS), alabandite (MnS), chlorargyrite (AgCl), and osbornite (TiN).

Chemical

A few minerals are chemical elements, including sulfur, copper, silver, and gold, but the vast majority are compounds. Before about 1947, the main method for identifying composition was *wet chemical analysis*, which involved dissolving a mineral in an acid such as hydrochloric acid (HCl). The elements in solution were then identified using colorimetry, volumetric analysis or gravimetric analysis. A variation on the wet methods is atomic absorption spectroscopy, which also requires the dissolution of the sample but is much faster and cheaper than the above methods. The solution is vaporized and its absorption spectrum is measured in the visible and ultraviolet range. Other techniques are X-ray fluorescence, electron microprobe analysis and optical emission spectrography.

Optical

Photomicrograph of olivine adcumulate, Archaean Komatiite, Agnew, Western Australia.

In addition to macroscopic properties such as color or lustre, minerals have properties that require a polarizing microscope to observe.

Transmitted Light

When light passes from air or a vacuum into a transparent crystal, some of it is reflected at the surface and some refracted. The latter is a bending of the light path that occurs because the speed of light changes as it goes into the crystal; Snell's law relates the bending angle to the Refractive index, the ratio of speed in a vacuum to speed in the crystal. Crystals whose point symmetry group falls in the cubic system are *isotropic*: the index does not depend on direction. All other crystals are *anisotropic*: light passing through them is broken up into two plane polarized rays that travel at different speeds and refract at different angles. A polarizing microscope is similar to an ordinary microscope, but it has two plane-polarized filters, a (*polarizer*) below the sample and an analyzer above it, polarized perpendicular to each other. Light passes successively through the polarizer, the sample and the analyzer. If there is no sample, the analyzer blocks all the light from the polarizer. However, an anisotropic sample will generally change the polarization so some of the light can pass through. Thin sections and powders can be used as samples.

When an isotropic crystal is viewed, it appears dark because it does not change the polarization of the light. However, when it is immersed in a calibrated liquid with a lower index of refraction and the microscope is thrown out of focus, a bright line called a *Becke line* appears around the perimeter of the crystal. By observing the presence or absence of such lines in liquids with different indices, the index of the crystal can be estimated, usually to within ± 0.003.

Systematic

Hanksite, $Na_{22}K(SO_4)_9(CO_3)_2Cl$, one of the few minerals that is considered a carbonate and a sulfate

Systematic mineralogy is the identification and classification of minerals by their properties. Historically, mineralogy was heavily concerned with taxonomy of the rock-forming minerals. In 1959, the International Mineralogical Association formed the Commission of New Minerals and Mineral Names to rationalize the nomenclature and regulate the introduction of new names. In July 2006, it was merged with the Commission on Classification of Minerals to form the Commission on New Minerals, Nomenclature, and Classification. There are over 6,000 named and unnamed minerals, and about 100 are discovered each year. The *Manual of Mineralogy* places minerals in the following classes: native elements, sulfides, sulfosalts, oxides and hydroxides, halides, carbonates, nitrates and borates, sulfates, chromates, molybdates and tungstates, phosphates, arsenates and vanadates, and silicates.

Formation Environments

The environments of mineral formation and growth are highly varied, ranging from slow crystallization at the high temperatures and pressures of igneous melts deep within the Earth's crust to the low temperature precipitation from a saline brine at the Earth's surface.

Various possible methods of formation include:

- sublimation from volcanic gases

- deposition from aqueous solutions and hydrothermal brines

- crystallization from an igneous magma or lava

- recrystallization due to metamorphic processes and metasomatism

- crystallization during diagenesis of sediments

- formation by oxidation and weathering of rocks exposed to the atmosphere or within the soil environment.

Biomineralogy

Biomineralogy is a cross-over field between mineralogy, paleontology and biology. It is the study of how plants and animals stabilize minerals under biological control, and the sequencing of mineral replacement of those minerals after deposition. It uses techniques from chemical mineralogy, especially isotopic studies, to determine such things as growth forms in living plants and animals as well as things like the original mineral content of fossils.

A new approach to mineralogy called "mineral evolution" explores the co-evolution of the geosphere and biosphere, including the role of minerals in the origin of life and processes as mineral-catalyzed organic synthesis and the selective adsorption of organic molecules on mineral surfaces.

Uses

A color chart of some raw forms of commercially valuable metals.

Minerals are essential to various needs within human society, such as minerals used as ores for essential components of metal products used in various commodities and machinery, essential components to building materials such as limestone, marble, granite, gravel, glass, plaster, cement, etc. Minerals are also used in fertilizers to enrich the growth of agricultural crops.

Collecting

Mineral collecting is also a recreational study and collection hobby, with clubs and societies representing the field. Museums, such as the Smithsonian National Museum of Natural History Hall of Geology, Gems, and Minerals, the Natural History Museum of Los Angeles County, the Natural History Museum, London, and the private Mim Mineral Museum in Beirut, Lebanon, have popular collections of mineral specimens on permanent display.

History of Mineralogy

Early writing on mineralogy, especially on gemstones, comes from ancient Babylonia, the ancient Greco-Roman world, ancient and medieval China, and Sanskrit texts from ancient India. Books on the subject included the Naturalis Historia of Pliny the Elder which not only described many different minerals but also explained many of their properties. The German Renaissance specialist Georgius Agricola wrote works such as *De re metallica* (*On Metals*, 1556) and *De Natura Fossilium* (*On the Nature of Rocks*, 1546) which began the scientific approach to the subject. Systematic scientific studies of minerals and rocks developed in post-Renaissance Europe. The modern study of mineralogy was founded on the principles of crystallography and microscopic study of rock sections with the invention of the microscope in the 17th century.

Europe and The Middle East

Theophrastus

The ancient Greek writers Aristotle (384–322 BC) and Theophrastus (370–285 BC) were the first in the Western tradition to write of minerals and their properties, as well as metaphysical explanations for them. The Greek philosopher Aristotle wrote his *Meteorologica*, and in it theorized that all the known substances were composed of water, air, earth, and fire, with the properties of dryness, dampness, heat, and cold. The Greek philosopher and botanist Theophrastus wrote his *De Mineralibus*, which accepted Aristotle's view, and divided minerals into two categories: those affected by heat and those affected by dampness.

The metaphysical emanation and exhalation (*anathumiaseis*) theory of the Greek philosopher Aristotle included early speculation on earth sciences including mineralogy. According to his theory, while metals were supposed to be congealed by means of moist exhalation, dry gaseous exhalation (*pneumatodestera*) was the efficient material cause of minerals found in the Earth's soil. He postulated these ideas by using the examples of moisture on the surface of the earth (a moist vapor 'potentially like water'), while the other was from the earth itself, pertaining to the attributes of hot, dry, smoky, and highly combustible ('potentially like fire'). Aristotle's metaphysical theory from times of antiquity had wide-ranging influence on similar theory found in later medieval Europe, as the historian Berthelot notes:

The theory of exhalations was the point of departure for later ideas on the generation of metals in the earth, which we meet with Proclus, and which reigned throughout the middle ages.

Fibrous asbestos on muscovite

Ancient Greek terminology of minerals has also stuck through the ages with widespread usage in modern times. For example, the Greek word asbestos (meaning 'inextinguishable', or 'unquenchable'), for the unusual mineral known today containing fibrous structure. The ancient historians Strabo (63 BC–19 AD) and Pliny the Elder (23–79 AD) both wrote of asbestos, its qualities, and its origins, with the Hellenistic belief that it was of a type of vegetable. Pliny the Elder listed it as a mineral common in India, while the historian Yu Huan (239–265 AD) of China listed this 'fireproof cloth' as a product of ancient Rome or Arabia (Chinese: Daqin). Although documentation of these minerals in ancient times does not fit the manner of modern scientific classification, there was nonetheless extensive written work on early mineralogy.

Pliny The Elder

For example, Pliny devoted five entire volumes of his work Naturalis Historia (77 AD) to the classification of "earths, metals, stones, and gems". He not only describes many minerals not known to Theophrastus, but discusses their applications and properties. He is the first to correctly recognise the origin of amber for example, as the fossilized remnant of tree resin from the observation of insects trapped in some samples. He laid the basis of crystallography by discussing crystal habit, especially the octahedral shape of diamond. His discussion of mining methods is unrivalled in the ancient world, and includes, for example, an eye-witness account of gold mining in northern Spain, an account which is fully confirmed by modern research.

octahedral shape of diamond.

Baltic amber necklace with trapped insects

However, before the more definitive foundational works on mineralogy in the 16th century, the ancients recognized no more than roughly 350 minerals to list and describe.

Jabir and Avicenna

With philosophers such as Proclus, the theory of Neoplatonism also spread to the Islamic world during the Middle Ages, providing a basis for metaphyiscal ideas on mineralogy in the medieval Middle East as well. The medieval Islamic scientists expanded upon this as well, including the Persian scientist Ibn Sina (ابوعلی سینا/بورسینا) (980-1037 AD), also known as *Avicenna*, who rejected alchemy and the earlier notion of Greek metaphysics that metallic and other elements could be transformed into one another. However, what was largely accurate of the ancient Greek and medieval metaphysical ideas on mineralogy was the slow chemical change in composition of the Earth's crust. There was also the Islamic alchemist and scientist Jābir ibn Hayyān (721-815 AD), who was the first to bring the experimental method into alchemy. Aided by Greek mathematics and Islamic mathematics, he discovered the syntheses for hydrochloric acid, nitric acid, distillation and crystallization (the latter two being essential for the understanding of modern mineralogy).

Georgius Agricola, 'Father of Mineralogy'

In the early 16th century AD, the writings of the German scientist Georg Bauer, pen-name Georgius Agricola (1494-1555 AD), in his *Bermannus, sive de re metallica dialogus* (1530) is considered to be the official establishment of mineralogy in the modern sense of its study. He wrote the treatise while working as a town physician and making observations in Joachimsthal, which was then a center for mining and metallurgic smelting industries. In 1544, he published his written work *De ortu et causis subterraneorum*, which is considered to be the foundational work of modern physical geology. In it (much like Ibn Sina) he heavily criticized the theories laid out by the ancient Greeks such as Aristotle. His work on mineralogy and metallurgy continued with the publication of *De veteribus et novis metallis* in 1546, and culminated in his best known works, the *De re metallica* of 1556. It was an impressive work outlining applications of mining, refining, and smelting metals, alongside discussions on geology of ore bodies, surveying, mine construction, and ventilation. He praises Pliny the Elder for his pioneering work Naturalis Historia and makes extensive references to his discussion of minerals and mining methods. For the next two centuries this written work remained the authoritative text on mining in Europe.

Agricola, author of De re metallica

Agricola had many various theories on mineralogy based on empirical observation, including understanding of the concept of ore channels that were formed by the circulation of ground waters ('succi') in fissures subsequent to the deposition of the surrounding rocks. As will be noted below, the medieval Chinese previously had conceptions of this as well.

For his works, Agricola is posthumously known as the "Father of Mineralogy".

After the foundational work written by Agricola, it is widely agreed by the scientific community that the *Gemmarum et Lapidum Historia* of Anselmus de Boodt (1550–1632) of Bruges is the first definitive work of modern mineralogy. The German mining chemist J.F. Henckel wrote his *Flora Saturnisans* of 1760, which was the first treatise in Europe to deal with geobotanical minerals, although the Chinese had mentioned this in earlier treatises of 1421 and 1664. In addition, the Chinese writer Du Wan made clear references to weathering and erosion processes in his *Yun Lin Shi Pu* of 1133, long before Agricola's work of 1546.

China and The Far East

In ancient China, the oldest literary listing of minerals dates back to at least the 4th century BC, with the *Ji Ni Zi* book listing twenty four of them. Chinese ideas of metaphysical mineralogy span back to at least the ancient Han Dynasty (202 BC–220 AD). From the 2nd century BC text of the

Huai Nan Zi, the Chinese used ideological Taoist terms to describe meteorology, precipitation, different types of minerals, metallurgy, and alchemy. Although the understanding of these concepts in Han times was Taoist in nature, the theories proposed were similar to the Aristotelian theory of mineralogical exhalations (noted above). By 122 BC, the Chinese had thus formulated the theory for metamorphosis of minerals, although it is noted by historians such as Dubs that the tradition of alchemical-mineralogical Chinese doctrine stems back to the School of Naturalists headed by the philosopher Zou Yan (305 BC–240 BC). Within the broad categories of rocks and stones (shi) and metals and alloys (jin), by Han times the Chinese had hundreds (if not thousands) of listed types of stones and minerals, along with theories for how they were formed.

In the 5th century AD, Prince Qian Ping Wang of the Liu Song Dynasty wrote in the encyclopedia *Tai-ping Yu Lan* (circa 444 AD, from the lost book *Dian Shu*, or *Management of all Techniques*):

The most precious things in the world are stored in the innermost regions of all. For example, there is orpiment. After a thousand years it changes into realgar. After another thousand years the realgar becomes transformed into yellow gold.

In ancient and medieval China, mineralogy became firmly tied to empirical observations in pharmaceutics and medicine. For example, the famous horologist and mechanical engineer Su Song (1020–1101 AD) of the Song Dynasty (960–1279 AD) wrote of mineralogy and pharmacology in his *Ben Cao Tu Jing* of 1070. In it he created a systematic approach to listing various different minerals and their use in medicinal concoctions, such as all the variously known forms of mica that could be used to cure various ills through digestion. Su Song also wrote of the subconchoidal fracture of native cinnabar, signs of ore beds, and provided description on crystal form. Similar to the ore channels formed by circulation of ground water mentioned above with the German scientist Agricola, Su Song made similar statements concerning copper carbonate, as did the earlier *Ri Hua Ben Cao* of 970 AD with copper sulfate.

The Yuan Dynasty scientist Zhang Si-xiao (died 1332 AD) provided a groundbreaking treatise on the conception of ore beds from the circulation of ground waters and rock fissures, two centuries before Georgius Agricola would come to similar conclusions. In his *Suo-Nan Wen Ji*, he applies this theory in describing the deposition of minerals by evaporation of (or precipitation from) ground waters in ore channels.

In addition to alchemical theory posed above, later Chinese writers such as the Ming Dynasty physician Li Shizhen (1518–1593 AD) wrote of mineralogy in similar terms of Aristotle's metaphysical theory, as the latter wrote in his pharmaceutical treatise *Běncǎo Gāngmù* (本草綱目, *Compendium of Materia Medica*, 1596). Another figure from the Ming era, the famous geographer Xu Xiake (1587–1641) wrote of mineral beds and mica schists in his treatise. However, while European literature on mineralogy became wide and varied, the writers of the Ming and Qing dynasties wrote little of the subject (even compared to Chinese of the earlier Song era). The only other works from these two eras worth mentioning were the *Shi Pin* (Hierarchy of Stones) of Yu Jun in 1617, the *Guai Shi Lu* (Strange Rocks) of Song Luo in 1665, and the *Guan Shi Lu* (On Looking at Stones) in 1668. However, one figure from the Song era that is worth mentioning above all is Shen Kuo.

Theories of Shen Kuo

The medieval Chinese Song Dynasty statesman and scientist Shen Kuo (1031-1095 AD) wrote of his land formation theory involving concepts of mineralogy. In his *Meng Xi Bi Tan* (梦溪笔谈; *Dream Pool Essays*, 1088), Shen formulated a hypothesis for the process of land formation (geomorphology); based on his observation of marine fossil shells in a geological stratum in the Taihang Mountains hundreds of miles from the Pacific Ocean. He inferred that the land was formed by erosion of the mountains and by deposition of silt, and described soil erosion, sedimentation and uplift. In an earlier work of his (circa 1080), he wrote of a curious fossil of a sea-orientated creature found far inland. It is also of interest to note that the contemporary author of the *Xi Chi Cong Yu* attributed the idea of particular places under the sea where serpents and crabs were petrified to one Wang Jinchen. With Shen Kuo's writing of the discovery of fossils, he formulated a hypothesis for the shifting of geographical climates throughout time. This was due to hundreds of petrified bamboos found underground in the dry climate of northern China, once an enormous landslide upon the bank of a river revealed them. Shen theorized that in pre-historic times, the climate of Yanzhou must have been very rainy and humid like southern China, where bamboos are suitable to grow.

Shen Kuo (沈括) (1031-1095))

In a similar way, the historian Joseph Needham likened Shen's account with the Scottish scientist Roderick Murchison (1792–1871), who was inspired to become a geologist after observing a providential landslide. In addition, Shen's description of sedimentary deposition predated that of James Hutton, who wrote his groundbreaking work in 1802 (considered the foundation of modern geology). The influential philosopher Zhu Xi (1130–1200) wrote of this curious natural phenomena of fossils as well, and was known to have read the works of Shen Kuo. In comparison, the first mentioning of fossils found in the West was made nearly two centuries later with Louis IX of France in 1253 AD, who discovered fossils of marine animals (as recorded in Joinville's records of 1309 AD).

America

Perhaps the most influential mineralogy text in the 19th and 20th centuries was the *Manual of Mineralogy* by James Dwight Dana, Harvard professor, first published in 1848. The fourth edition was entitled *Manual of Mineralogy and Lithology* (ed. 4, 1887). It became a standard college text, and has been continuously revised and updated by a succession of editors including W. E. Ford (13th-14th eds., 1912–1929), Cornelius S. Hurlbut (15th-21st eds., 1941–1999), and beginning with the 22nd by Cornelis Klein. The 23rd edition is now in print under the title *Manual of Mineral Science (Manual of Mineralogy)* (2007), revised by Cornelis Klein and Barbara Dutrow.

Equally influential was Dana's *System of Mineralogy*, first published in 1837, which has consistently been updated and revised. The 6th edition (1892) being edited by his son Edward Salisbury Dana. A 7th edition was published in 1944, and the 8th edition was published in 1997 under the title *Dana's New Mineralogy: The System of Mineralogy of James Dwight Dana and Edward Salisbury Dana*, edited by R. V. Gaines *et al.*

References

- Gribble, C.D.; Hall, A.J. (1993). Optical Mineralogy: Principles And Practice. London: CRC Press. ISBN 9780203498705.

- Tisljar, S.K. Haldar, Josip (2013). Introduction to mineralogy and petrology. Burlington: Elsevier Science. ISBN 9780124167100.

- Chan, Alan Kam-leung and Gregory K. Clancey, Hui-Chieh Loy (2002). Historical Perspectives on East Asian Science, Technology and Medicine. Singapore: Singapore University Press ISBN 9971-69-259-7

Branches of Mineralogy

There are several branches of mineralogy that study various aspects- optical mineralogy which studies the optical properties of minerals and rocks to identify their origin and evolution; crystallography which is the experimental science that determines the arrangement of atoms in crystalline solids; amateur geology which the recreational study and hobby of collecting rocks and mineral specimens from the natural habitat; magnetic mineralogy which is the study of the magnetic properties of minerals and gemology which is the study of gems and their properties.

Optical Mineralogy

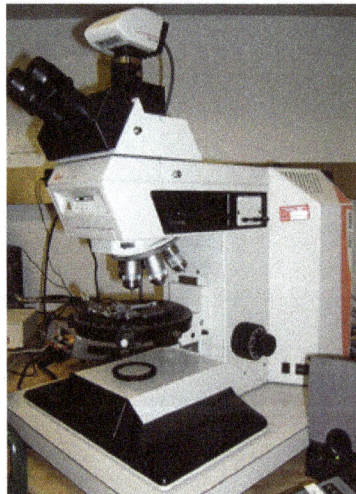

A petrographic microscope, which is an optical microscope fitted with cross-polarizing lenses, a conoscopic lens, and compensators (plates of anisotropic materials; gypsum plates and quartz wedges are common), for crystallographic analysis.

Optical mineralogy is the study of minerals and rocks by measuring their optical properties. Most commonly, rock and mineral samples are prepared as thin sections or grain mounts for study in the laboratory with a petrographic microscope. Optical mineralogy is used to identify the mineralogical composition of geological materials in order to help reveal their origin and evolution.

Some of the properties and techniques used include:

- Refractive index
- Birefringence
- Michel-Lévy Interference colour chart
- Pleochroism

- Extinction angle

- Conoscopic interference pattern (Interference figure)

- Becke line test

- Optical relief

- Sign of elongation (Length fast vs. length slow)

- Wave plate

History

William Nicol, whose name is associated with the creation of the Nicol prism, seems to have been the first to prepare thin slices of mineral substances, and his methods were applied by Henry Thronton Maire Witham (1831) to the study of plant petrifactions. This method, of such far-reaching importance in petrology, was not at once made use of for the systematic investigation of rocks, and it was not until 1858 that Henry Clifton Sorby pointed out its value. Meanwhile, the optical study of sections of crystals had been advanced by Sir David Brewster and other physicists and mineralogists and it only remained to apply their methods to the minerals visible in rock sections.

Sections

A rock-section should be about one-thousandth of an inch (30 micrometres) in thickness, and is relatively easy to make. A thin splinter of the rock, about 1 centimetre may be taken; it should be as fresh as possible and free from obvious cracks. By grinding it on a plate of planed steel or cast iron with a little fine carborundum it is soon rendered flat on one side and is then transferred to a sheet of plate glass and smoothed with the very finest emery till all minute pits and roughnesses are removed and the surface is a uniform plane. The rock-chip is then washed, and placed on a copper or iron plate which is heated by a spirit or gas lamp. A microscopic glass slip is also warmed on this plate with a drop of viscous natural Canada balsam on its surface. The more volatile ingredients of the balsam are dispelled by the heat, and when that is accomplished the smooth, dry, warm rock is pressed firmly into contact with the glass plate so that the film of balsam intervening may be as thin as possible and free from air-bubbles. The preparation is allowed to cool and then the rock chip is again ground down as before, first with carborundum and, when it becomes transparent, with fine emery till the desired thickness is obtained. It is then cleaned, again heated with a little more balsam, and covered with a cover glass. The labor of grinding the first surface may be avoided by cutting off a smooth slice with an iron disk armed with crushed diamond powder. A second application of the slitter after the first face is smoothed and cemented to the glass will in expert hands leave a rock-section so thin as to be already transparent. In this way the preparation of a section may require only twenty minutes.

Microscope

The microscope employed is usually one which is provided with a rotating stage beneath which there is a polarizer, while above the objective or the eyepiece an analyzer is mounted; alternatively the stage may be fixed and the polarizing and analyzing prisms may be capable of simultaneous rotation by means of toothed wheels and a connecting-rod. If ordinary light and not polarized light is desired, both prisms may be withdrawn from the axis of the instrument; if the polarizer only is inserted the light transmitted is plane polarized; with both prisms in position the slide is viewed in cross-polarized light, also known as "crossed nicols." A microscopic rock-section in ordinary light, if a suitable magnification (say 30) be employed, is seen to consist of grains or crystals varying in color, size and shape.

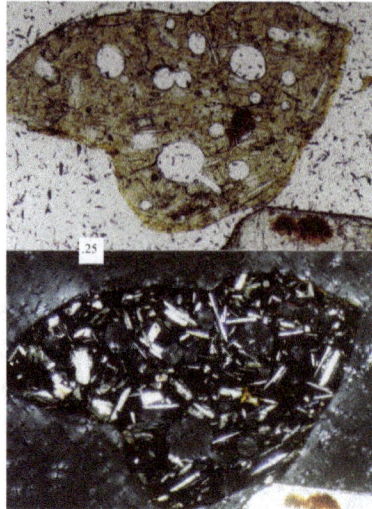

Photomicrograph of a volcanic lithic fragment (sand grain); upper picture is plane-polarized light, bottom picture is cross-polarized light, scale box at left-center is 0.25 millimeter.

Characters of Minerals

Some minerals are colorless and transparent (quartz, calcite, feldspar, muscovite, etc.), others are yellow or brown (rutile, tourmaline, biotite), green (diopside, hornblende, chlorite), blue (glaucophane), pink (garnet), etc. The same mineral may present a variety of colors, in the same or different rocks, and these colors may be arranged in zones parallel to the surfaces of the crystals. Thus tourmaline may be brown, yellow, pink, blue, green, violet, grey, or colorless, but every mineral has one or more characteristic, most common tints. The shapes of the crystals determine in a general way the outlines of the sections of them presented on the slides. If the mineral has one or more good cleavages they will be indicated by systems of cracks. The refractive index is also clearly shown by the appearance of the section, which are rough, with well-defined borders if they have a much stronger refraction than the medium in which they are mounted. Some minerals decompose readily and become turbid and semi-transparent (e.g. feldspar); others remain always perfectly fresh and clear (e.g. quartz), others yield characteristic secondary products (such as green chlorite after biotite). The inclusions in the crystals (both solid and fluid) are of great interest; one mineral may enclose another, or may contain spaces occupied by glass, by fluids or by gases.

Microstructure

Lastly the structure of the rock, that is to say, the relation of its components to one another, is usually clearly indicated, whether it be fragmented or massive; the presence of glassy matter in contradistinction to a completely crystalline or "holo-crystalline" condition; the nature and origin of organic fragments; banding, foliation or lamination; the pumiceous or porous structure of many lavas; these and many other characters, though often not visible in the hand specimens of a rock, are rendered obvious by the examination of a microscopic section. Many refined methods of observation may be introduced, such as the measurement of the size of the elements of the rock by the help of micrometers; their relative proportions by means of a glass plate ruled in small squares; the angles between cleavages or faces seen in section by the use of the rotating graduated stage, and the estimation of the refractive index of the mineral by comparison with those of different mounting media.

Pleochroism

The light vibrates now only in one plane, and in passing through doubly refracting crystals in the slide, is, speaking generally, broken up into rays, which vibrate at right angles to one another. In many colored minerals such as biotite, hornblende, tourmaline, chlorite, these two rays have different colors, and when a section containing any of these minerals is rotated the change of color is often very striking. This property, known as "pleochroism" is of great value in the determination of rock-making minerals.

Pleochroism is often especially intense in small spots which surround minute enclosures of other minerals, such as zircon and epidote, these are known as "pleochroic halos."

Double Refraction

If the analyzer is now inserted in such a position that it is crossed relatively to the polarizer, the field of view will be dark where there are no minerals or where the light passes through isotropic substances such as glass, liquids and cubic crystals. All other crystalline bodies, being doubly refracting, will appear bright in some position as the stage is rotated. The only exception to this rule is provided by sections which are perpendicular to the optic axes of birefringent crystals; these remain dark or nearly dark during a whole rotation, and as will be seen later, their investigation is of special importance.

Extinction

The doubly refracting mineral sections, however, will in all cases appear black in certain positions as the stage is rotated. They are said to go "extinct" when this takes place. If we note these positions we may measure the angle between them and any cleavages, faces or other structures of the crystal by means of the rotating stage. These angles are characteristic of the system to which the mineral belongs and often of the mineral species itself. To facilitate measurement of extinction angles various kinds of eyepieces have been devised, some having a stereoscopic calcite plate, others with two or four plates of quartz cemented together; these are often found to give more exact results than are obtained by observing merely the position in which the mineral section is most completely dark between crossed nicols.

The mineral sections when not extinguished are not only bright but are colored and the colors they show depend on several factors, the most important of which is the strength of the double refraction. If all the sections are of the same thickness as is nearly true of well-made slides, the minerals with strongest double refraction yield the highest polarization colors. The order in which the colors are arranged in what is known as Newton's scale, the lowest being dark grey, then grey, white, yellow, orange, red, purple, blue and so on. The difference between the refractive indexes of the ordinary and the extraordinary ray in quartz is .009, and in a rock-section about 1/500 of an inch thick this mineral gives grey and white polarization colours; nepheline with weaker double refraction gives dark grey; augite on the other hand will give red and blue, while calcite with the stronger double refraction will appear pinkish or greenish white. All sections of the same mineral, however, will not have the same color; it was stated above that sections perpendicular to an optic axis will be nearly black, and, in general, the more nearly any section approaches this direction the lower its polarization colors will be. By taking the average, or the highest color given by any mineral, the relative value of its double refraction can be estimated; or if the thickness of the section be precisely known the difference between the two refractive indexes can be ascertained. If the slides be thick the colors will be on the whole higher than in thin slides.

It is often important to find out whether of the two axes of elasticity (or vibration traces) in the section is that of greater elasticity (or lesser refractive index). The quartz wedge or selenite plate enables us to do this. Suppose a doubly refracting mineral section so placed that it is "extinguished"; if now is rotated through 45 degrees it will be brightly illuminated. If the quartz wedge be passed across it so that the long axis of the wedge is parallel to the axis of elasticity in the section the polarization colors will rise or fall. If they rise the axes of greater elasticity in the two minerals are parallel; if they sink the axis of greater elasticity in the one is parallel to that of lesser elasticity in the other. In the latter case by pushing the wedge sufficiently far complete darkness or compensation will result. Selenite wedges, selenite plates, mica wedges and mica plates are also used for this purpose. A quartz wedge also may be calibrated by determining the amount of double refraction in all parts of its length. If now it be used to produce compensation or complete extinction in any doubly refracting mineral section, we can ascertain what is the strength of the double refraction of the section because it is obviously equal and opposite to that of a known part of the quartz wedge.

A further refinement of microscopic methods consists of the use of strongly convergent polarized light (konoscopic methods). This is obtained by a wide angled achromatic condenser above the polarizer, and a high power microscopic objective. Those sections are most useful which are perpendicular to an optic axis, and consequently remain dark on rotation. If they belong to uniaxial crystals they show a dark cross or convergent light between crossed nicols, the bars of which remain parallel to the wires in the field of the eyepiece. Sections perpendicular to an optic axis of a biaxial mineral under the same conditions show a dark bar which on rotation becomes curved to a hyperbolic shape. If the section is perpendicular to a "bisectrix" a black cross is seen which on rotation opens out to form two hyperbolas, the apices of which are turned towards one another. The optic axes emerge at the apices of the hyperbolas and may be surrounded by colored rings, though owing to the thinness of minerals in rock sections these are only seen when the double refraction of the mineral is strong. The distance between the axes as seen in the field of the microscope depends partly on the axial angle of the crystal and partly on the numerical aperture of the objective. If it is measured by means of eye-piece micrometer, the optic axial angle of the mineral can be found by a simple calculation. The quartz wedge, quarter mica plate or selenite plate permit the determination of the positive or negative character of the crystal by the changes in the color or shape of the figures observed in the field. These operations are precisely similar to those employed by the mineralogist in the examination of plates cut from crystals. It is sufficient to point out that the petrological microscope in its modern development is an optical instrument of great precision, enabling us to determine physical constants of crystallized substances as well as serving to produce magnified images like the ordinary microscope. A great variety of accessory apparatus has been devised to fit it for these special uses.

Examination of Rock Powders

Although rocks are now studied principally in microscopic sections the investigation of fine crushed rock powders, which was the first branch of microscopic petrology to receive attention, is by no means discontinued. The modern optical methods are perfectly applicable to transparent mineral fragments of any kind. Minerals are almost as easily determined in powder as in section, but it is otherwise with rocks, as the structure or relation of the components to one another, which is an element of great importance in the study of the history and classification of rocks, is almost completely destroyed by grinding them to powder.

Crystallography

A crystalline solid: atomic resolution image of strontium titanate. Brighter atoms are strontium and darker ones are titanium.

Crystallography is the experimental science of determining the arrangement of atoms in the crystalline solids. The word "crystallography" derives from the Greek words *crystallon* "cold drop, frozen drop", with its meaning extending to all solids with some degree of transparency, and *grapho* "I write". In July 2012, the United Nations recognised the importance of the science of crystallography by proclaiming that 2014 would be the International Year of Crystallography. X-ray crystallography is used to determine the structure of large biomolecules such as proteins.

Before the development of X-ray diffraction crystallography, the study of crystals was based on physical measurements of their geometry. This involved measuring the angles of crystal faces relative to each other and to theoretical reference axes (crystallographic axes), and establishing the symmetry of the crystal in question. This physical measurement is carried out using a goniometer. The position in 3D space of each crystal face is plotted on a stereographic net such as a Wulff net or Lambert net. The pole to each face is plotted on the net. Each point is labelled with its Miller index. The final plot allows the symmetry of the crystal to be established.

Crystallographic methods now depend on analysis of the diffraction patterns of a sample targeted by a beam of some type. X-rays are most commonly used; other beams used include electrons or neutrons. This is facilitated by the wave properties of the particles. Crystallographers often explicitly state the type of beam used, as in the terms *X-ray crystallography, neutron diffraction* and *electron diffraction*. These three types of radiation interact with the specimen in different ways.

X-rays interact with the spatial distribution of electrons in the sample.

Electrons are charged particles and therefore interact with the total charge distribution of both the atomic nuclei and the electrons of the sample.

Neutrons are scattered by the atomic nuclei through the strong nuclear forces, but in addition, the magnetic moment of neutrons is non-zero. They are therefore also scattered by magnetic fields. When neutrons are scattered from hydrogen-containing materials, they produce diffraction patterns with high noise levels. However, the material can sometimes be treated to substitute deuterium for hydrogen.

Because of these different forms of interaction, the three types of radiation are suitable for different crystallographic studies.

Theory

An image of a small object is made using a lens to focus the beam, similar to a lens in a microscope. However, the wavelength of visible light (about 4000 to 7000 ångström) is three orders of magnitude longer than the length of typical atomic bonds and atoms themselves (about 1 to 2 Å). Therefore, obtaining information about the spatial arrangement of atoms requires the use of radiation with shorter wavelengths, such as X-ray or neutron beams. Employing shorter wavelengths implied abandoning microscopy and true imaging, however, because there exists no material from which a lens capable of focusing this type of radiation can be created.(Nevertheless, scientists have had some success focusing X-rays with microscopic Fresnel zone plates made from gold, and by critical-angle reflection inside long tapered capillaries.) Diffracted X-ray or neutron beams cannot be focused to produce images, so the sample structure must be reconstructed from the diffraction pattern. Sharp features in the diffraction pattern arise from periodic, repeating structure in the sample, which are often very strong due to coherent reflection of many photons from many regularly spaced instances of similar structure, while non-periodic components of the structure result in diffuse (and usually weak) diffraction features - areas with a higher density and repetition of atom order tend to reflect more light toward one point in space when compared to those areas with fewer atoms and less repetition.

Because of their highly ordered and repetitive structure, crystals give diffraction patterns of sharp Bragg reflection spots, and are ideal for analyzing the structure of solids.

Notation

Coordinates in *square brackets* such as [100] denote a direction vector (in real space).

Coordinates in *angle brackets* or *chevrons* such as <100> denote a *family* of directions which are related by symmetry operations. In the cubic crystal system for example, <100> would mean [100], [010], [001] or the negative of any of those directions.

Miller indices in *parentheses* such as (100) denote a plane of the crystal structure, and regular repetitions of that plane with a particular spacing. In the cubic system, the normal to the (hkl) plane is the direction [hkl], but in lower-symmetry cases, the normal to (hkl) is not parallel to [hkl].

Indices in *curly brackets* or *braces* such as {100} denote a family of planes and their normals which are equivalent in cubic materials due to symmetry operations, much the way angle brackets denote a family of directions. In non-cubic materials, <hkl> is not necessarily perpendicular to {hkl}.

Techniques

Some materials that have been analyzed crystallographically, such as proteins, do not occur naturally as crystals. Typically, such molecules are placed in solution and allowed to slowly crystallize through vapor diffusion. A drop of solution containing the molecule, buffer, and precipitants is sealed in a container with a reservoir containing a hygroscopic solution. Water in the drop diffuses to the reservoir, slowly increasing the concentration and allowing a crystal to form. If the concentration were to rise more quickly, the molecule would simply precipitate out of solution, resulting in disorderly granules rather than an orderly and hence usable crystal.

Once a crystal is obtained, data can be collected using a beam of radiation. Although many universities that engage in crystallographic research have their own X-ray producing equipment, synchrotrons are often used as X-ray sources, because of the purer and more complete patterns such sources can generate. Synchrotron sources also have a much higher intensity of X-ray beams, so data collection takes a fraction of the time normally necessary at weaker sources. Complementary neutron crystallography techniques are used to identify the positions of hydrogen atoms, since X-rays only interact very weakly with light elements such as hydrogen.

Producing an image from a diffraction pattern requires sophisticated mathematics and often an iterative process of modelling and refinement. In this process, the mathematically predicted diffraction patterns of an hypothesized or "model" structure are compared to the actual pattern generated by the crystalline sample. Ideally, researchers make several initial guesses, which through refinement all converge on the same answer. Models are refined until their predicted patterns match to as great a degree as can be achieved without radical revision of the model. This is a painstaking process, made much easier today by computers.

The mathematical methods for the analysis of diffraction data only apply to *patterns,* which in turn result only when waves diffract from orderly arrays. Hence crystallography applies for the most part only to crystals, or to molecules which can be coaxed to crystallize for the sake of measurement. In spite of this, a certain amount of molecular information can be deduced from patterns that are generated by fibers and powders, which while not as perfect as a solid crystal, may exhibit a degree of order. This level of order can be sufficient to deduce the structure of simple molecules, or to determine the coarse features of more complicated molecules. For example, the double-helical structure of DNA was deduced from an X-ray diffraction pattern that had been generated by a fibrous sample.

In Materials Science

Crystallography is used by materials scientists to characterize different materials. In single crystals, the effects of the crystalline arrangement of atoms is often easy to see macroscopically, because the natural shapes of crystals reflect the atomic structure. In addition, physical properties are often controlled by crystalline defects. The understanding of crystal structures is an important prerequisite for understanding crystallographic defects. Mostly, materials do not occur as a single crystal, but in poly-crystalline form (i.e., as an aggregate of small crystals with different orientations). Because of this, the powder diffraction method, which takes diffraction patterns of polycrystalline samples with a large number of crystals, plays an important role in structural determination.

Other physical properties are also linked to crystallography. For example, the minerals in clay form small, flat, platelike structures. Clay can be easily deformed because the platelike particles can slip along each other in the plane of the plates, yet remain strongly connected in the direction perpendicular to the plates. Such mechanisms can be studied by crystallographic texture measurements.

In another example, iron transforms from a body-centered cubic (bcc) structure to a face-centered cubic (fcc) structure called austenite when it is heated. The fcc structure is a close-packed structure unlike the bcc structure; thus the volume of the iron decreases when this transformation occurs.

Crystallography is useful in phase identification. When manufacturing or using a material, it is generally desirable to know what compounds and what phases are present in the material, as their composition, structure and proportions will influence the material's properties. Each phase has a characteristic arrangement of atoms. X-ray or neutron diffraction can be used to identify which patterns are present in the material, and thus which compounds are present. Crystallography covers the enumeration of the symmetry patterns which can be formed by atoms in a crystal and for this reason is related to group theory and geometry.

Biology

X-ray crystallography is the primary method for determining the molecular conformations of biological macromolecules, particularly protein and nucleic acids such as DNA and RNA. In fact, the double-helical structure of DNA was deduced from crystallographic data. The first crystal structure of a macromolecule was solved in 1958, a three-dimensional model of the myoglobin molecule obtained by X-ray analysis. The Protein Data Bank (PDB) is a freely accessible repository for the structures of proteins and other biological macromolecules. Computer programs such as RasMol or Pymol can be used to visualize biological molecular structures. Neutron crystallography is often used to help refine structures obtained by X-ray methods or to solve a specific bond; the methods are often viewed as complementary, as X-rays are sensitive to electron positions and scatter most strongly off heavy atoms, while neutrons are sensitive to nucleus positions and scatter strongly even off many light isotopes, including hydrogen and deuterium. Electron crystallography has been used to determine some protein structures, most notably membrane proteins and viral capsids.

Scientists of Note

- William Astbury
- William Barlow
- C. Arnold Beevers
- John Desmond Bernal
- William Henry Bragg
- William Lawrence Bragg
- Auguste Bravais
- Glenn H. Brown
- Martin Julian Buerger

- Francis Crick
- Pierre Curie
- Peter Debye
- Johann Deisenhofer
- Boris Delone
- Gautam R. Desiraju
- Jack Dunitz
- David Eisenberg
- Paul Peter Ewald
- Evgraf Stepanovich Fedorov
- Rosalind Franklin
- Georges Friedel
- Paul Heinrich von Groth
- René Just Haüy
- Wayne Hendrickson
- Carl Hermann
- Johann Friedrich Christian Hessel
- Dorothy Crowfoot Hodgkin
- Robert Huber
- Isabella Karle
- Jerome Karle
- Aaron Klug
- Max von Laue
- Otto Lehmann
- Michael Levitt
- Henry Lipson
- Kathleen Lonsdale
- Ernest-François Mallard
- Charles-Victor Mauguin
- William Hallowes Miller

- Friedrich Mohs
- Paul Niggli
- Louis Pasteur
- Arthur Lindo Patterson
- Max Perutz
- Friedrich Reinitzer
- Hugo Rietveld
- Jean-Baptiste L. Romé de l'Isle
- Michael Rossmann
- Paul Scherrer
- Arthur Moritz Schönflies
- Dan Shechtman
- George M. Sheldrick
- Tej P. Singh
- Nicolas Steno
- Constance Tipper
- Daniel Vorländer
- Christian Samuel Weiss
- Don Craig Wiley
- Ralph Walter Graystone Wyckoff
- Ada Yonath
- Dorothy Hodgkin

Amateur Geology

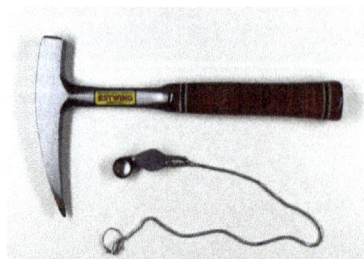

A rockhound's tools; a geologist's hammer and loupe

Amateur geology (known as rockhounding in the United States and Canada, and regionally known as rock hunting in southern Oregon) is the recreational study and hobby of collecting rocks and mineral specimens from their natural environment.

Collecting

The first amateur geologists were prospectors looking for valuable minerals and gemstones for commercial purposes. Eventually, however, more and more people have been drawn to amateur geology for recreational purposes, mainly for the beauty that rocks and minerals provide.

One reason for the rise in popularity of amateur geology is that a collection can begin by simply picking up a rock. There are also many clubs and groups that search for specimens and compare them in groups as a hobby. Information on where to find such groups can be found at libraries, bookstores, and "gem and mineral shows". Tourist information centers and small-town chambers of commerce can also supply valuable local information. The Internet can also be a useful search tool as it can help find other amateur geologists.

The amateur geologist's principal piece of equipment is the geologist's hammer. This is a small tool with a pick-like point on one end, and a flat hammer on the other. The hammer end is for breaking rocks, and the pick end is mainly used for prying and digging into crevices. The pick end of most rock hammers can dull quickly if struck onto bare rock. Rock collectors may also bring a sledgehammer to break hard rocks. Good places for a collector to look are quarries, road cuts, rocky hills and mountains, and streams.

There are many different laws in place regarding the collection of rocks and minerals from public areas, so it is advisable to read up on local laws before prospecting. Rock and mineral collecting is prohibited in most if not all national parks in the United States.

Related Fields

Avid rock collectors often use their specimens to learn about petrology, mineralogy and geology as well as skills in the identification and classifying of specimen rocks, and preparing them for display. The hobby can lead naturally into lapidary projects, and also the cutting, polishing, and mounting of gemstones and minerals. The equipment needed to do this includes rock saws and polishers. Many beautiful crystal varieties are typically found in very small samples which requires a good microscope for working with and photographing the specimen. The hobby can be as simple as finding pretty rocks for a windowsill or develop into a detailed and comprehensive museum quality display.

Magnetic Mineralogy

Magnetic mineralogy is the study of the magnetic properties of minerals. The contribution of a mineral to the total magnetism of a rock depends strongly on the type of magnetic order or disorder. Magnetically disordered minerals (diamagnets and paramagnets) contribute a weak magnetism and have no remanence. The more important minerals for rock magnetism are the minerals that can be magnetically ordered, at least at some temperatures. These are the ferromagnets, ferrimagnets and certain kinds of antiferromagnets. These minerals have a much stronger response to the field and can have a remanence.

Weakly Magnetic Minerals

Non-Iron-Bearing Minerals

Most minerals with no iron content are diamagnetic. Some such minerals may have a significant positive magnetic susceptibility, for example serpentine, but this is because the minerals have inclusions containing strongly magnetic minerals such as magnetite. The susceptibility of such minerals is negative and small (Table 1).

Table 1. Susceptibilities of non-iron-bearing minerals	
Mineral	Volume susceptibility at room temperature (SI)
graphite	-80 to -200
calcite	-7.5 to -39
anhydrite	-14 to -60
gypsum	-13 to -29
ice	-9
orthoclase	-13 to -17
magnesite	-15
forsterite	-12
halite	-10 to -16
galena	-33
quartz	-13 to -17
celestine	-16 to -18
sphalerite	-31 to -750

Iron-Bearing Paramagnetic Minerals

Reddish crystals: biotite.

Most iron-bearing carbonates and silicates are paramagnetic at all temperatures. Some sulfides are paramagnetic, but some are strongly magnetic. In addition, many of the strongly magnetic minerals discussed below are paramagnetic above a critical temperature (the Curie temperature or Néel temperature). In Table 2 are given susceptibilities for some iron-bearing minerals. The susceptibilities are positive and an order of magnitude or more larger than diamagnetic susceptibilities.

Table 2. Susceptibilities of some paramagnetic minerals	
Mineral	Volume susceptibility (SI)
garnet	2,700
illite	410
montmorillonite	330-350
biotite	1,500-2,900
siderite	1,300-11,000
chromite	3,000-120,000
orthopyroxene	1,500-1,800
fayalite	5,500
olivine	1,600
jacobsite	25,000
franklinite	450,000

Strongly Magnetic Minerals

Iron-Titanium Oxides

Magnetite-bearing lodestone displaying strong magnetic properties.

Many of the most important magnetic minerals on Earth are oxides of iron and titanium. Their compositions are conveniently represented on a ternary plot with axes corresponding to the proportions of Ti^{4+}, Fe^{2+}, and Fe^{3+}. Important regions on the diagram include the *titanomagnetites*, which form a line of compositions $Fe_{3-x}Ti_xO_4$ for x between 0 and 1. At the $x=0$ end is magnetite, while the $x=1$ composition is ulvöspinel. The titanomagnetites have an inverse spinel crystal structure and at high temperatures are a solid solution series. Crystals formed from titanomagnetites by cation-deficient oxidation are called *titanomaghemites*, an important example of which is maghemite. Another series, the *titanohematites*, have hematite and ilmenite as their end members, and so are also called *hemoilmenites*. The crystal structure of hematite is trigonal-hexagonal. It has the same composition as maghemite; to distinguish between them, their chemical formulae are generally given as γFe_2O_3 for hematite and αFe_2O_3 for maghemite.

Iron Sulfides

The other important class of strongly magnetic minerals is the iron sulfides, particularly greigite and pyrrhotite.

Iron Alloys

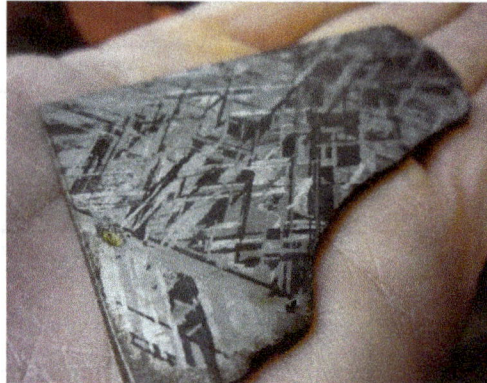

Meteorite slice with intergrowth of kamacite and taenite.

Extraterrestrial environments being low in oxygen, minerals tend to have very little Fe^{3+}. The primary magnetic phase on the Moon is ferrite, the body-centered cubic (bcc) phase of iron. As the proportion of iron decreases, the crystal structure changes from bcc to face centered cubic (fcc). Nickel iron mixtures tend to exsolve into a mixture of iron-rich kamacite and iron-poor taenite.

Gemology

A selection of gemstone pebbles made by tumbling rough rock with abrasive grit in a rotating drum. The biggest pebble here is 40 millimetres (1.6 in) long.

Gemology or gemmology is the science dealing with natural and artificial gemstone materials. It is considered a geoscience and a branch of mineralogy. Some jewelers are academically trained gemologists and are qualified to identify and evaluate gems.

Background

Rudimentary education in gemology for jewelers and gemologists began in the nineteenth century, but the first qualifications were instigated after the National Association of Goldsmiths of Great Britain (NAG) set up a Gemmological Committee for this purpose in 1908. This committee matured into the Gemmological Association of Great Britain (also known as Gem-A), now an educational charity and accredited awarding body with its courses taught worldwide. The first US graduate of Gem-A's Diploma Course, in 1929, was Robert Shipley, who later established both the Gemological Institute of America and the American Gem Society. There are now several professional schools and associations of gemologists and certification programs around the world.

The first gemological laboratory serving the jewelry trade was established in London in 1925, prompted by the influx of the newly developed "cultured pearl" and advances in the synthesis of rubies and sapphires. There are now numerous gem labs around the world requiring ever more advanced equipment and experience to identify the new challenges - such as treatments to gems, new synthetics, and other new materials.

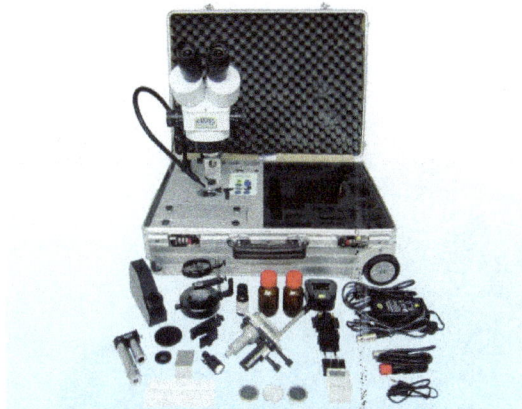

Gemmological travel lab KA52KRS

It is often difficult to obtain an expert judgement from a neutral laboratory. Analysis and estimation in the gemstone trade usually have to take place on site. Professional gemologists and gemstone buyers use mobile laboratories, which pool all necessary instruments in a travel case. Such so-called travel labs even have their own current supply, which makes them independent from infrastructure. They are also suitable for gemological expeditions.

Gemstones are basically categorized based on their crystal structure, specific gravity, refractive index, and other optical properties, such as pleochroism. The physical property of "hardness" is defined by the non-linear Mohs scale of mineral hardness.

Gemologists study these factors while valuing or appraising cut and polished gemstones. Gemological microscopic study of the internal structure is used to determine whether a gem is synthetic or natural by revealing natural fluid inclusions or partially melted exogenous crystals that are evidence of heat treatment to enhance color.

The spectroscopic analysis of cut gemstones also allows a gemologist to understand the atomic structure and identify its origin, which is a major factor in valuing a gemstone. For example, a ruby from Burma will have definite internal and optical activity variance from a Thai ruby.

When the gemstones are in a rough state, the gemologist studies the external structure; the host rock and mineral association; and natural and polished color. Initially, the stone is identified by its color, refractive index, optical character, specific gravity, and examination of internal characteristics under magnification.

General Identification of Gems

Gem identification is basically a process of elimination. Gemstones of similar color undergo non-destructive optical testing until there is only one possible identity. Any single test is indicative, only. For example, the specific gravity of ruby is 4.00, glass is 3.15–4.20, and cubic zirconia is 5.6–5.9. So one can easily tell the difference between cubic zirconia and the other two; however, there is overlap between ruby and glass.

And, as with all naturally occurring materials, no two gems are identical. The geological environment they are created in influences the overall process so that although the basics can be identified, the presence of chemical "impurities" and substitutions along with structural imperfections create "individuals".

Identification by Refractive Index

Traditional handheld refractometer

One test to determine the gem's identity is to measure the refraction of light in the gem. Every material has a critical angle, above which point light is reflected back internally. This can be measured and thus used to determine the gem's identity. Typically this is measured using a refractometer, although it is possible to measure it using a microscope.

Identification by Specific Gravity

Specific gravity, also known as relative density, varies depending upon the chemical composition and crystal structure type. Heavy liquids with a known specific gravity are used to test loose gemstones.

Specific gravity is measured by comparing the weight of the gem in air with the weight of the gem suspended in water.

Identification by Spectroscopy

This method uses a similar principle to how a prism works to separate white light into its component colors. A gemological spectroscope is employed to analyze the selective absorption of light in the gem material. Essentially, when light passes from one medium to another, it bends. Blue light bends more than red light. How much the light bends will vary depending on the gem material. Coloring agents or chromophores show bands in the spectroscope and indicate which element is responsible for the gem's color.

References

- Dunlop, David J.; Özdemir, Özden (1997). Rock magnetism: Fundamentals and Frontiers. Cambridge Univ. Press. ISBN 0-521-32514-5.

Mineral: An Overview

Naturally occurring compounds, mostly crystalline in structure, are called minerals. There are over 5,300 species of known minerals whose diversity and distribution is controlled by the Earth's chemistry. They can be distinguished by their physical and chemical composition. This chapter acquaints the reader with the basic definition of a mineral and its defining characteristics.

Mineral

A mineral is a naturally occurring chemical compound. Most often, they are crystalline and abiogenic in origin. A mineral is different from a rock, which can be an aggregate of minerals or non-minerals and does not have one specific chemical composition, as a mineral does. The exact definition of a mineral is under debate, especially with respect to the requirement that a valid species be abiogenic, and to a lesser extent with regard to it having an ordered atomic structure.

Amethyst, a variety of quartz

The study of minerals is called mineralogy. There are over 5,300 known mineral species; over 5,070 of these have been approved by the International Mineralogical Association (IMA). The silicate minerals compose over 90% of the Earth's crust. The diversity and abundance of mineral species is controlled by the Earth's chemistry. Silicon and oxygen constitute approximately 75% of the Earth's crust, which translates directly into the predominance of silicate minerals.

Minerals are distinguished by various chemical and physical properties. Differences in chemical composition and crystal structure distinguish the various species, which were determined by the mineral's geological environment when formed. Changes in the temperature, pressure, or bulk composition of a rock mass cause changes in its minerals.

Minerals can be described by their various physical properties, which are related to their chemical structure and composition. Common distinguishing characteristics include crystal structure and habit, hardness, lustre, diaphaneity, colour, streak, tenacity, cleavage, fracture, parting, and specific gravity. More specific tests for describing minerals include magnetism, taste or smell, radioactivity and reaction to acid.

Minerals are classified by key chemical constituents; the two dominant systems are the Dana classification and the Strunz classification. The silicate class of minerals is subdivided into six subclasses by the degree of polymerization in the chemical structure. All silicate minerals have a base unit of a $[SiO_4]^{4-}$ silica tetrahedron—that is, a silicon cation coordinated by four oxygen anions, which gives the shape of a tetrahedron. These tetrahedra can be polymerized to give the subclasses: orthosilicates (no polymerization, thus single tetrahedra), disilicates (two tetrahedra bonded together), cyclosilicates (rings of tetrahedra), inosilicates (chains of tetrahedra), phyllosilicates (sheets of tetrahedra), and tectosilicates (three-dimensional network of tetrahedra). Other important mineral groups include the native elements, sulfides, oxides, halides, carbonates, sulfates, and phosphates.

Definition
Basic Definition

The general definition of a mineral encompasses the following criteria:

- Naturally occurring

- Stable at room temperature

- Represented by a chemical formula

- Usually abiogenic (not resulting from the activity of living organisms)

- Ordered atomic arrangement

The first three general characteristics are less debated than the last two. The first criterion means that a mineral has to form by a natural process, which excludes anthropogenic compounds. Stability at room temperature, in the simplest sense, is synonymous to the mineral being solid. More specifically, a compound has to be stable or metastable at 25 °C. Classical examples of exceptions to this rule include native mercury, which crystallizes at −39 °C, and water ice, which is solid only below 0 °C; as these two minerals were described prior to 1959, they were grandfathered by the International Mineralogical Association (IMA). Modern advances have included extensive study of liquid crystals, which also extensively involve mineralogy. Minerals are chemical compounds, and as such they can be described by fixed or a variable formula. Many mineral groups and species are composed of a solid solution; pure substances are not usually found because of contamination or chemical substitution. For example, the olivine group is described by the variable formula $(Mg, Fe)_2SiO_4$, which is a solid solution of two end-member species, magnesium-rich forsterite and iron-rich fayalite, which are described by a fixed chemical formula. Mineral species themselves could have a variable compositions, such as the sulfide mackinawite, $(Fe, Ni)_9S_8$, which is mostly a ferrous sulfide, but has a very significant nickel impurity that is reflected in its formula.

The requirement that a valid mineral species be abiogenic has also been described as similar to being inorganic; however, this criterion is imprecise and organic compounds have been assigned a separate classification branch. Finally, the requirement of an ordered atomic arrangement is usually synonymous with crystallinity; however, crystals are also periodic, so the broader criterion is used instead. An ordered atomic arrangement gives rise to a variety of macroscopic physical properties, such as crystal form, hardness, and cleavage. There have been several recent proposals to amend the definition to consider biogenic or amorphous substances as minerals. The formal definition of a mineral approved by the IMA in 1995:

"A mineral is an element or chemical compound that is normally crystalline and that has been formed as a result of geological processes."

In addition, biogenic substances were explicitly excluded:

"Biogenic substances are chemical compounds produced entirely by biological processes without a geological component (e.g., urinary calculi, oxalate crystals in plant tissues, shells of marine molluscs, etc.) and are not regarded as minerals. However, if geological processes were involved in the genesis of the compound, then the product can be accepted as a mineral."

Recent Advances

Mineral classification schemes and their definitions are evolving to match recent advances in mineral science. Recent changes have included the addition of an organic class, in both the new Dana and the Strunz classification schemes. The organic class includes a very rare group of minerals with hydrocarbons. The IMA Commission on New Minerals and Mineral Names adopted in 2009 a hierarchical scheme for the naming and classification of mineral groups and group names and established seven commissions and four working groups to review and classify minerals into an official listing of their published names. According to these new rules, "mineral species can be grouped in a number of different ways, on the basis of chemistry, crystal structure, occurrence, association, genetic history, or resource, for example, depending on the purpose to be served by the classification."

The Nickel (1995) exclusion of biogenic substances was not universally adhered to. For example, Lowenstam (1981) stated that "organisms are capable of forming a diverse array of minerals, some of which cannot be formed inorganically in the biosphere." The distinction is a matter of classification and less to do with the constituents of the minerals themselves. Skinner (2005) views all solids as potential minerals and includes biominerals in the mineral kingdom, which are those that are created by the metabolic activities of organisms. Skinner expanded the previous definition of a mineral to classify "element or compound, amorphous or crystalline, formed through *biogeochemical* processes," as a mineral.

Recent advances in high-resolution genetics and X-ray absorption spectroscopy are providing revelations on the biogeochemical relations between microorganisms and minerals that may make Nickel's (1995) biogenic mineral exclusion obsolete and Skinner's (2005) biogenic mineral inclusion a necessity. For example, the IMA commissioned "Environmental Mineralogy and Geochemistry Working Group" deals with minerals in the hydrosphere, atmosphere, and biosphere. The group's scope includes mineral-forming microorganisms, which exist on nearly every rock, soil, and particle surface spanning the globe to depths of at least 1600 metres below the sea floor and

70 kilometres into the stratosphere (possibly entering the mesosphere). Biogeochemical cycles have contributed to the formation of minerals for billions of years. Microorganisms can precipitate metals from solution, contributing to the formation of ore deposits. They can also catalyze the dissolution of minerals.

Prior to the International Mineralogical Association's listing, over 60 biominerals had been discovered, named, and published. These minerals (a sub-set tabulated in Lowenstam (1981)) are considered minerals proper according to the Skinner (2005) definition. These biominerals are not listed in the International Mineral Association official list of mineral names, however, many of these biomineral representatives are distributed amongst the 78 mineral classes listed in the Dana classification scheme. Another rare class of minerals (primarily biological in origin) include the mineral liquid crystals that have properties of both liquids and crystals. To date over 80,000 liquid crystalline compounds have been identified.

The Skinner (2005) definition of a mineral takes this matter into account by stating that a mineral can be crystalline or amorphous, the latter group including liquid crystals. Although biominerals and liquid mineral crystals, are not the most common form of minerals, they help to define the limits of what constitutes a mineral proper. The formal Nickel (1995) definition explicitly mentioned crystallinity as a key to defining a substance as a mineral. A 2011 article defined icosahedrite, an aluminium-iron-copper alloy as mineral; named for its unique natural icosahedral symmetry, it is a quasicrystal. Unlike a true crystal, quasicrystals are ordered but not periodic.

Rocks, Ores, and Gems

Schist is a metamorphic rock characterized by an abundance of platy minerals. In this example, the rock has prominent sillimanite porphyroblasts as large as 3 cm (1.2 in).

Minerals are not equivalent to rocks. A rock is either an aggregate of one or more minerals, or not composed of minerals at all. Rocks like limestone or quartzite are composed primarily of one mineral—calcite or aragonite in the case of limestone, and quartz in the latter case. Other rocks can be defined by relative abundances of key (essential) minerals; a granite is defined by proportions of quartz, alkali feldspar, and plagioclase feldspar. The other minerals in the rock are termed accessory, and do not greatly affect the bulk composition of the rock. Rocks can also be composed entirely of non-mineral material; coal is a sedimentary rock composed primarily of organically derived carbon.

In rocks, some mineral species and groups are much more abundant than others; these are termed the rock-forming minerals. The major examples of these are quartz, the feldspars, the micas, the amphiboles, the pyroxenes, the olivines, and calcite; except the last one, all of the minerals are silicates. Overall, around 150 minerals are considered particularly important, whether in terms of their abundance or aesthetic value in terms of collecting.

Commercially valuable minerals and rocks are referred to as industrial minerals. For example, muscovite, a white mica, can be used for windows (sometimes referred to as isinglass), as a filler, or as an insulator. Ores are minerals that have a high concentration of a certain element, typically a metal. Examples are cinnabar (HgS), an ore of mercury, sphalerite (ZnS), an ore of zinc, or cassiterite (SnO_2), an ore of tin. Gems are minerals with an ornamental value, and are distinguished from non-gems by their beauty, durability, and usually, rarity. There are about 20 mineral species that qualify as gem minerals, which constitute about 35 of the most common gemstones. Gem minerals are often present in several varieties, and so one mineral can account for several different gemstones; for example, ruby and sapphire are both corundum, Al_2O_3.

Nomenclature and Classification

Minerals are classified by variety, species, series and group, in order of increasing generality. The basic level of definition is that of mineral species, which is distinguished from other species by specific and unique chemical and physical properties. For example, quartz is defined by its formula, SiO_2, and a specific crystalline structure that distinguishes it from other minerals with the same chemical formula (termed polymorphs). When there exists a range of composition between two minerals species, a mineral series is defined. For example, the biotite series is represented by variable amounts of the endmembers phlogopite, siderophyllite, annite, and eastonite. In contrast, a mineral group is a grouping of mineral species with some common chemical properties that share a ccrystal structure. The pyroxene group has a common formula of $XY(Si,Al)_2O_6$, where X and Y are both cations, with X typically bigger than Y; the pyroxenes are single-chain silicates that crystallize in either the orthorhombic or monoclinic crystal systems. Finally, a mineral variety is a specific type of mineral species that differs by some physical characteristic, such as colour or crystal habit. An example is amethyst, which is a purple variety of quartz.

Two common classifications, Dana and Strunz, are used for minerals; both rely on composition, specifically with regards to important chemical groups, and structure. James Dwight Dana, a leading geologist of his time, first published his *System of Mineralogy* in 1837; as of 1997, it is in its eighth edition. The Dana classification assigns a four-part number to a mineral species. Its class number is based on important compositional groups; the type gives the ratio of cations to anions in the mineral; and the last two numbers group minerals by structural similarity within a given type or class. The less commonly used Strunz classification, named for German mineralogist Karl Hugo Strunz, is based on the Dana system, but combines both chemical and structural criteria, the latter with regards to distribution of chemical bonds.

As of January 2016, 5,090 mineral species are approved by the IMA. They are most commonly named after a person (45%), followed by discovery location (23%); names based on chemical composition (14%) and physical properties (8%) are the two other major groups of mineral name etymologies.

Fake Minerals

The fake minerals are minerals (or gems, that is to say outstanding minerals) not natural, man-made. This can be a natural mineral transformed by man into another, or an entirely artificial mineral. Also referred to as false to the synonyms. In short, counterfeiting may be partial (sample processed) or total (sample created by humans) or cover the name given to the sample. They have always existed, their marketing is growing very rapidly. There are now numerous examples of fakes in mineralogy and gemology. If there are criteria for authentication of minerals, it is sometimes difficult to distinguish between fake and genuine ones. Fake samples are sold for real deception, but when the infringement is announced or found, it may have a financial interest, decorative or teaching for the buyer.

Mineral Chemistry

Hübnerite, the manganese-rich end-member of the wolframite series, with minor quartz in the background

The abundance and diversity of minerals is controlled directly by their chemistry, in turn dependent on elemental abundances in the Earth. The majority of minerals observed are derived from the Earth's crust. Eight elements account for most of the key components of minerals, due to their abundance in the crust. These eight elements, summing to over 98% of the crust by weight, are, in order of decreasing abundance: oxygen, silicon, aluminium, iron, magnesium, calcium, sodium and potassium. Oxygen and silicon are by far the two most important — oxygen composes 46.6% of the crust by weight, and silicon accounts for 27.7%.

The minerals that form are directly controlled by the bulk chemistry of the parent body. For example, a magma rich in iron and magnesium will form mafic minerals, such as olivine and the pyroxenes; in contrast, a more silica-rich magma will crystallize to form minerals that incorporate more SiO_2, such as the feldspars and quartz. In a limestone, calcite or aragonite (both $CaCO_3$) form because the rock is rich in calcium and carbonate. A corollary is that a mineral will not be found in a rock whose bulk chemistry does not resemble the bulk chemistry of a given mineral with the exception of trace minerals. For example, kyanite, Al_2SiO_5 forms from the metamorphism of aluminium-rich shales; it would not likely occur in aluminium-poor rock, such quartzite.

The chemical composition may vary between end member species of a solid solution series. For example, the plagioclase feldspars comprise a continuous series from sodium-rich end member albite ($NaAlSi_3O_8$) to calcium-rich anorthite ($CaAl_2Si_2O_8$) with four recognized intermediate varieties between them (given in order from sodium- to calcium-rich): oligoclase, andesine, labradorite, and bytownite. Other examples of series include the olivine series of magnesium-rich forsterite and iron-rich fayalite, and the wolframite series of manganese-rich hübnerite and iron-rich ferberite.

Chemical substitution and coordination polyhedra explain this common feature of minerals. In nature, minerals are not pure substances, and are contaminated by whatever other elements are present in the given chemical system. As a result, it is possible for one element to be substituted for another. Chemical substitution will occur between ions of a similar size and charge; for example, K^+ will not substitute for Si^{4+} because of chemical and structural incompatibilities caused by a big difference in size and charge. A common example of chemical substitution is that of Si^{4+} by Al^{3+}, which are close in charge, size, and abundance in the crust. In the example of plagioclase, there are three cases of substitution. Feldspars are all framework silicates, which have a silicon-oxygen ratio of 2:1, and the space for other elements is given by the substitution of Si^{4+} by Al^{3+} to give a base unit of $[AlSi_3O_8]^-$; without the substitution, the formula would be charge-balanced as SiO_2, giving quartz. The significance of this structural property will be explained further by coordination polyhedra. The second substitution occurs between Na^+ and Ca^{2+}; however, the difference in charge has to accounted for by making a second substitution of Si^{4+} by Al^{3+}.

Coordination polyhedra are geometric representation of how a cation is surrounded by an anion. In mineralogy, due its abundance in the crust, coordination polyhedra are usually considered in terms of oxygen. The base unit of silicate minerals is the silica tetrahedron — one Si^{4+} surrounded by four O^{2-}. An alternate way of describing the coordination of the silicate is by a number: in the case of the silica tetrahedron, the silicon is said to have a coordination number of 4. Various cations have a specific range of possible coordination numbers; for silicon, it is almost always 4, except for very high-pressure minerals where compound is compressed such that silicon is in six-fold (octahedral) coordination by oxygen. Bigger cations have a bigger coordination number because of the increase in relative size as compared to oxygen (the last orbital subshell of heavier atoms is different too). Changes in coordination numbers between leads to physical and mineralogical differences; for example, at high pressure such as in the mantle, many minerals, especially silicates such as olivine and garnet will change to a perovskite structure, where silicon is in octahedral coordination. Another example are the aluminosilicates kyanite, andalusite, and sillimanite (polymorphs, as they share the formula Al_2SiO_5), which differ by the coordination number of the Al^{3+}; these minerals transition from one another as a response to changes in pressure and temperature. In the case of silicate materials, the substitution of Si^{4+} by Al^{3+} allows for a variety of minerals because of the need to balance charges.

When minerals react, the products will sometimes assume the shape of the reagent; the product mineral is termed to be a pseudomorph of (or after) the reagent. Illustrated here is a pseudomorph of kaolinite after orthoclase. Here, the pseudomorph preserved the Carlsbad twinning common in orthoclase.

Changes in temperature and pressure, and composition alter the mineralogy of a rock sample. Changes in composition can be caused by processes such as weathering or metasomatism (hydrothermal alteration). Changes in temperature and pressure occur when the host rock undergoes tectonic or magmatic movement into differing physical regimes. Changes in thermodynamic conditions make it favourable for mineral assemblages to react with each other to produce new minerals; as such, it is possible for two rocks to have an identical or a very similar bulk rock chemistry without having a similar mineralogy. This process of mineralogical alteration is related to the rock cycle. An example of a series of mineral reactions is illustrated as follows.

Orthoclase feldspar ($KAlSi_3O_8$) is a mineral commonly found in granite, a plutonic igneous rock. When exposed to weathering, it reacts to form kaolinite ($Al_2Si_2O_5(OH)_4$, a sedimentary mineral, and silicic acid):

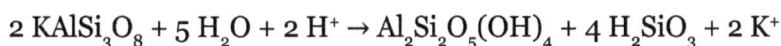

$$2\ KAlSi_3O_8 + 5\ H_2O + 2\ H^+ \rightarrow Al_2Si_2O_5(OH)_4 + 4\ H_2SiO_3 + 2\ K^+$$

Under low-grade metamorphic conditions, kaolinite reacts with quartz to form pyrophyllite ($Al_2Si_4O_{10}(OH)_2$):

$$Al_2Si_2O_5(OH)_4 + SiO_2 \rightarrow Al_2Si_4O_{10}(OH)_2 + H_2O$$

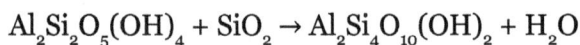

As metamorphic grade increases, the pyrophyllite reacts to form kyanite and quartz:

$$Al_2Si_4O_{10}(OH)_2 \rightarrow Al_2SiO_5 + 3\ SiO_2 + H_2O$$

Alternatively, a mineral may change its crystal structure as a consequence of changes in temperature and pressure without reacting. For example, quartz will change into a variety of its SiO_2 polymorphs, such as tridymite and cristobalite at high temperatures, and coesite at high pressures.

Physical Properties of Minerals

Classifying minerals ranges from simple to difficult. A mineral can be identified by several physical properties, some of them being sufficient for full identification without equivocation. In other cases, minerals can only be classified by more complex optical, chemical or X-ray diffraction analysis; these methods, however, can be costly and time-consuming. Physical properties applied for classification include crystal structure and habit, hardness, lustre, diaphaneity, colour, streak, cleavage and fracture, and specific gravity. Other less general tests include fluorescence, phosphorescence, magnetism, radioactivity, tenacity (response to mechanical induced changes of shape or form), piezoelectricity and reactivity to dilute acids.

Crystal Structure and Habit

Crystal structure results from the orderly geometric spatial arrangement of atoms in the internal structure of a mineral. This crystal structure is based on regular internal atomic or ionic arrangement that is often expressed in the geometric form that the crystal takes. Even when the mineral grains are too small to see or are irregularly shaped, the underlying crystal structure is always periodic and can be determined by X-ray diffraction. Minerals are typically described by their symmetry content. Crystals are restricted to 32 point groups, which differ by their symmetry. These groups are classified in turn into more broad categories, the most encompassing of these being the six crystal families.

Topaz has a characteristic orthorhombic elongated crystal shape.

These families can be described by the relative lengths of the three crystallographic axes, and the angles between them; these relationships correspond to the symmetry operations that define the narrower point groups. They are summarized below; a, b, and c represent the axes, and α, β, γ represent the angle opposite the respective crystallographic axis (e.g. α is the angle opposite the a-axis, viz. the angle between the b and c axes):

Crystal family	Lengths	Angles	Common examples
Isometric	a=b=c	$\alpha=\beta=\gamma=90°$	Garnet, halite, pyrite
Tetragonal	a=b≠c	$\alpha=\beta=\gamma=90°$	Rutile, zircon, andalusite
Orthorhombic	a≠b≠c	$\alpha=\beta=\gamma=90°$	Olivine, aragonite, orthopyroxenes
Hexagonal	a=b≠c	$\alpha=\beta=90°, \gamma=120°$	Quartz, calcite, tourmaline
Monoclinic	a≠b≠c	$\alpha=\gamma=90°, \beta≠90°$	Clinopyroxenes, orthoclase, gypsum
Triclinic	a≠b≠c	$\alpha≠\beta≠\gamma≠90°$	Anorthite, albite, kyanite

The hexagonal crystal family is also split into two crystal *systems* — the trigonal, which has a three-fold axis of symmetry, and the hexagonal, which has a six-fold axis of symmetry.

Chemistry and crystal structure together define a mineral. With a restriction to 32 point groups, minerals of different chemistry may have identical crystal structure. For example, halite (NaCl), galena (PbS), and periclase (MgO) all belong to the hexaoctahedral point group (isometric family), as they have a similar stoichiometry between their different constituent elements. In contrast, polymorphs are groupings of minerals that share a chemical formula but have a different structure. For example, pyrite and marcasite, both iron sulfides, have the formula FeS_2; however, the former is isometric while the latter is orthorhombic. This polymorphism extends to other sulfides with the generic AX_2 formula; these two groups are collectively known as the pyrite and marcasite groups.

Polymorphism can extend beyond pure symmetry content. The aluminosilicates are a group of three minerals — kyanite, andalusite, and sillimanite — which share the chemical formula Al_2SiO_5. Kyanite is triclinic, while andalusite and sillimanite are both orthorhombic and belong to the dipyramidal point group. These difference arise correspond to how aluminium is coordinated within the crystal structure. In all minerals, one aluminium ion is always in six-fold coordination by oxygen; the silicon, as a general rule is in four-fold coordination in all minerals; an exception is a case like stishovite (SiO_2, an ultra-high pressure quartz polymorph with rutile structure). In kyanite, the second aluminium is in six-fold coordination; its chemical formula can be expressed as $AlAlSiO_5$, to reflect its crystal structure. Andalusite has the second aluminium in five-fold coordination ($AlAlSiO_5$) and sillimanite has it in four-fold coordination ($AlAlSiO_5$).

Differences in crystal structure and chemistry greatly influence other physical properties of the mineral. The carbon allotropes diamond and graphite have vastly different properties; diamond is the hardest natural substance, has an adamantine lustre, and belongs to the isometric crystal family, whereas as graphite is very soft, has a greasy lustre, and crystallises in the hexagonal family. This difference is accounted by differences in bonding. In diamond, the carbons are in sp³ hybrid orbitals, which means they form a framework where each carbon is covalently bonded to four neighbours in a tetrahedral fashion; on the other hand, graphite is composed of sheets of carbons in sp² hybrid orbitals, where each carbon is bonded covalently to only three others. These sheets are held together by much weaker van der Waals forces, and this discrepancy translates to big macroscopic differences.

Contact twins, as seen in spinel

Twinning is the intergrowth of two or more crystal of a single mineral species. The geometry of the twinning is controlled by the mineral's symmetry. As a result, there are several types of twins, including contact twins, reticulated twins, geniculated twins, penetration twins, cyclic twins, and polysynthetic twins. Contact, or simple twins, consist of two crystals joined at a plane; this type of twinning is common in spinel. Reticulated twins, common in rutile, are interlocking crystals resembling netting. Geniculated twins have a bend in the middle that is caused by start of the twin. Penetration twins consist of two single crystals that have grown into each other; examples of this twinning include cross-shaped staurolite twins and Carlsbad twinning in orthoclase. Cyclic twins are caused by repeated twinning around a rotation axis. It occurs around three, four, five, six, or eight-fold axes, and the corresponding patterns are called threelings, fourlings, fivelings, sixlings, and eightlings. Sixlings are common in aragonite. Polysynthetic twins are similar to cyclic twinning by the presence of repetitive twinning; however, instead of occurring around a rotational axis, it occurs along parallel planes, usually on a microscopic scale.

Crystal habit refers to the overall shape of crystal. Several terms are used to describe this property. Common habits include acicular, which described needlelike crystals like in natrolite, bladed, dendritic (tree-pattern, common in native copper), equant, which is typical of garnet, prismatic (elongated in one direction), and tabular, which differs from bladed habit in that the former is platy whereas the latter has a defined elongation. Related to crystal form, the quality of crystal faces is diagnostic of some minerals, especially with a petrographic microscope. Euhedral crystals have a defined external shape, while anhedral crystals do not; those intermediate forms are termed subhedral.

Hardness

The hardness of a mineral defines how much it can resist scratching. This physical property is controlled by the chemical composition and crystalline structure of a mineral. A mineral's hardness is not necessarily constant for all sides, which is a function of its structure; crystallographic weakness renders some directions softer than others. An example of this property exists in kyanite, which has a Mohs hardness of 5½ parallel to but 7 parallel to .

The most common scale of measurement is the ordinal Mohs hardness scale. Defined by ten indicators, a mineral with a higher index scratches those below it. The scale ranges from talc, a phyllosilicate, to diamond, a carbon polymorph that is the hardest natural material. The scale is provided below:

Mohs hardness	Mineral	Chemical formula
1	Talc	$Mg_3Si_4O_{10}(OH)_2$
2	Gypsum	$CaSO_4 \cdot 2H_2O$
3	Calcite	$CaCO_3$
4	Fluorite	CaF_2
5	Apatite	$Ca_5(PO_4)_3(OH,Cl,F)$
6	Orthoclase	$KAlSi_3O_8$
7	Quartz	SiO_2
8	Topaz	$Al_2SiO_4(OH,F)_2$
9	Corundum	Al_2O_3
10	Diamond	C

Lustre and Diaphaneity

Pyrite has a metallic lustre.

Lustre indicates how light reflects from the mineral's surface, with regards to its quality and intensity. There are numerous qualitative terms used to describe this property, which are split into metallic and non-metallic categories. Metallic and sub-metallic minerals have high reflectivity like metal; examples of minerals with this lustre are galena and pyrite. Non-metallic lustres include: adamantine, such as in diamond; vitreous, which is a glassy lustre very common in silicate minerals; pearly, such as in talc and apophyllite, resinous, such as members of the garnet group, silky which common in fibrous minerals such as asbestiform chrysotile.

The diaphaneity of a mineral describes the ability of light to pass through it. Transparent minerals do not diminish the intensity of light passing through it. An example of such a mineral is muscovite (potassium mica); some varieties are sufficiently clear to have been used for windows. Translucent minerals allow some light to pass, but less than those that are transparent. Jadeite and nephrite (mineral forms of jade are examples of minerals with this property). Minerals that do not allow light to pass are called opaque.

The diaphaneity of a mineral depends on thickness of the sample. When a mineral is sufficiently thin (e.g., in a thin section for petrography), it may become transparent even if that property is not seen in hand sample. In contrast, some minerals, such as hematite or pyrite are opaque even in thin-section.

Colour and Streak

Colour is typically not a diagnostic property of minerals. Shown are green uvarovite (left) and red-pink grossular (right), both garnets. The diagnostic features would include dodecahedral crystals, resinous lustre, and hardness around 7.

Colour is the most obvious property of a mineral, but it is often non-diagnostic. It is caused by electromagnetic radiation interacting with electrons (except in the case of incandescence, which does not apply to minerals). Two broad classes of elements are defined with regards to their contribution to a mineral's colour. Idiochromatic elements are essential to a mineral's composition; their contribution to a mineral's colour is diagnostic. Examples of such minerals are malachite (green) and azurite (blue). In contrast, allochromatic elements in minerals are present in trace amounts as impurities. An example of such a mineral would be the ruby and sapphire varieties of the mineral corundum. The colours of pseudochromatic minerals are the result of interference of light waves. Examples include labradorite and bornite.

In addition to simple body colour, minerals can have various other distinctive optical properties, such as play of colours, asterism, chatoyancy, iridescence, tarnish, and pleochroism. Several of these properties involve variability in colour. Play of colour, such as in opal, results in the sample reflecting different colours as it is turned, while pleochroism describes the change in colour as light passes through a mineral in a different orientation. Iridescence is a variety of the play of colours where light scatters off a coating on the surface of crystal, cleavage planes, or off layers having minor gradations in chemistry. In contrast, the play of colours in opal is caused by light refracting from ordered microscopic silica spheres within its physical structure. Chatoyancy ("cat's eye") is the wavy banding of colour that is observed as the sample is rotated; asterism, a variety of chatoyancy, gives the appearance of a star on the mineral grain. The latter property is particularly common in gem-quality corundum.

The streak of a mineral refers to the colour of a mineral in powdered form, which may or may not be identical to its body colour. The most common way of testing this property is done with a streak plate, which is made out of porcelain and coloured either white or black. The streak of a mineral is independent of trace elements or any weathering surface. A common example of this property is illustrated with hematite, which is coloured black, silver, or red in hand sample, but has a cherry-red to reddish-brown streak. Streak is more often distinctive for metallic minerals, in contrast to non-metallic minerals whose body colour is created by allochromatic elements. Streak testing is constrained by the hardness of the mineral, as those harder than 7 powder the *streak plate* instead.

Cleavage, Parting, Fracture, and Tenacity

Perfect basal cleavage as seen in biotite (black), and good cleavage seen in the matrix (pink orthoclase).

By definition, minerals have a characteristic atomic arrangement. Weakness in this crystalline structure causes planes of weakness, and the breakage of a mineral along such planes is termed cleavage. The quality of cleavage can be described based on how cleanly and easily the mineral breaks; common descriptors, in order of decreasing quality, are "perfect", "good", "distinct", and "poor". In particularly transparent mineral, or in thin-section, cleavage can be seen a series of parallel lines marking the planar surfaces when viewed at a side. Cleavage is not a universal property among minerals; for example, quartz, consisting of extensively interconnected silica tetrahedra, does not have a crystallographic weakness which would allow it to cleave. In contrast, micas, which have perfect basal cleavage, consist of sheets of silica tetrahedra which are very weakly held together.

As cleavage is a function of crystallography, there are a variety of cleavage types. Cleavage occurs typically in either one, two, three, four, or six directions. Basal cleavage in one direction is a distinctive property of the micas. Two-directional cleavage is described as prismatic, and occurs in minerals such as the amphiboles and pyroxenes. Minerals such as galena or halite have cubic (or isometric) cleavage in three directions, at 90°; when three directions of cleavage are present, but not at 90°, such as in calcite or rhodochrosite, it is termed rhombohedral cleavage. Octahedral cleavage (four directions) is present in fluorite and diamond, and sphalerite has six-directional dodecahedral cleavage.

Minerals with many cleavages might not break equally well in all of the directions; for example, calcite has good cleavage in three direction, but gypsum has perfect cleavage in one direction, and poor cleavage in two other directions. Angles between cleavage planes vary between minerals. For example, as the amphiboles are double-chain silicates and the pyroxenes are single-chain silicates, the angle between their cleavage planes is different. The pyroxenes cleave in two directions at approximately 90°, whereas the amphiboles distinctively cleave in two directions separated by approximately 120° and 60°. The cleavage angles can be measured with a contact goniometer, which is similar to a protractor.

Parting, sometimes called "false cleavage", is similar in appearance to cleavage but is instead produced by structural defects in the mineral as opposed to systematic weakness. Parting varies from crystal to crystal of a mineral, whereas all crystals of a given mineral will cleave if the atomic structure allows for that property. In general, parting is caused by some stress applied to a crystal. The sources of the stresses include deformation (e.g. an increase in pressure), exsolution, or twinning. Minerals that often display parting include the pyroxenes, hematite, magnetite, and corundum.

When a mineral is broken in a direction that does not correspond to a plane of cleavage, it is termed to have been fractured. There are several types of uneven fracture. The classic example is conchoidal fracture, like that of quartz; rounded surfaces are created, which are marked by smooth curved lines. This type of fracture occurs only in very homogeneous minerals. Other types of fracture are fibrous, splintery, and hackly. The latter describes a break along a rough, jagged surface; an example of this property is found in native copper.

Tenacity is related to both cleavage and fracture. Whereas fracture and cleavage describes the surfaces that are created when a mineral is broken, tenacity describes how resistant a mineral is to such breaking. Minerals can be described as brittle, ductile, malleable, sectile, flexible, or elastic.

Specific Gravity

Galena, PbS, is a mineral with a high specific gravity.

Specific gravity numerically describes the density of a mineral. The dimensions of density are mass divided by volume with units: kg/m^3 or g/cm^3. Specific gravity measures how much water a mineral sample displaces. Defined as the quotient of the mass of the sample and difference between the weight of the sample in air and its corresponding weight in water, specific gravity is a unitless ratio. Among most minerals, this property is not diagnostic. Rock forming minerals — typically silicates or occasionally carbonates — have a specific gravity of 2.5–3.5.

High specific gravity is a diagnostic property of a mineral. A variation in chemistry (and consequently, mineral class) correlates to a change in specific gravity. Among more common minerals, oxides and sulfides tend to have a higher specific gravity as they include elements with higher atomic mass. A generalization is that minerals with metallic or adamantine lustre tend to have higher specific gravities than those having a non-metallic to dull lustre. For example, hematite, Fe_2O_3, has a specific gravity of 5.26 while galena, PbS, has a specific gravity of 7.2–7.6, which is a result of their high iron and lead content, respectively. A very high specific gravity becomes very pronounced in native metals; kamacite, an iron-nickel alloy common in iron meteorites has a specific gravity of 7.9, and gold has an observed specific gravity between 15 and 19.3.

Other Properties

Carnotite (yellow) is a radioactive uranium-bearing mineral.

Other properties can be used to diagnose minerals. These are less general, and apply to specific minerals.

Dropping dilute acid (often 10% HCl) aids in distinguishing carbonates from other mineral classes. The acid reacts with the carbonate ($[CO_3]^{2-}$) group, which causes the affected area to effervesce, giving off carbon dioxide gas. This test can be further expanded to test the mineral in its original crystal form or powdered. An example of this test is done when distinguish calcite from dolomite, especially within rocks (limestone and dolostone respectively). Calcite immediately effervesces in acid, whereas acid must be applied to powdered dolomite (often to a scratched surface in a rock), for it to effervesce. Zeolite minerals will not effervesce in acid; instead, they become frosted after 5–10 minutes, and if left in acid for a day, they dissolve or become a silica gel.

When tested, magnetism is a very conspicuous property of minerals. Among common minerals, magnetite exhibits this property strongly, and it is also present, albeit not as strongly, in pyrrhotite and ilmenite.

Minerals can also be tested for taste or smell. Halite, NaCl, is table salt; its potassium-bearing counterpart, sylvite, has a pronounced bitter taste. Sulfides have a characteristic smell, especially as samples are fractured, reacting, or powdered.

Radioactivity is a rare property; minerals may be composed of radioactive elements. They could be a defining constituent, such as uranium in uraninite, autunite, and carnotite, or as trace impurities. In the latter case, the decay of a radioactive element damages the mineral crystal; the result, termed a *radioactive halo* or *pleochroic halo*, is observable by various techniques, such as thin-section petrography.

Mineral Classes

As the composition of the Earth's crust is dominated by silicon and oxygen, silicate elements are by far the most important class of minerals in terms of rock formation and diversity. However, non-silicate minerals are of great economic importance, especially as ores.

Non-silicate minerals are subdivided into several other classes by their dominant chemistry, which included native elements, sulfides, halides, oxides and hydroxides, carbonates and nitrates, borates, sulfates, phosphates, and organic compounds. The majority of non-silicate mineral species are extremely rare (constituting in total 8% of the Earth's crust), although some are relative common, such as calcite, pyrite, magnetite, and hematite. There are two major structural styles observed in non-silicates: close-packing and silicate-like linked tetrahedra. The close-packed structures, which is a way to densely pack atoms while minimizing interstitial space. Hexagonal close-packing involves stacking layers where every other layer is the same ("ababab"), whereas cubic close-packing involves stacking groups of three layers ("abcabcabc"). Analogues to linked silica tetrahedra include SO_4 (sulfate), PO_4 (phosphate), AsO_4 (arsenate), and VO_4 (vanadate). The non-silicates have great economic importance, as they concentrate elements more than the silicate minerals do.

The largest grouping of minerals by far are the silicates; most rocks are composed of greater than 95% silicate minerals, and over 90% of the Earth's crust is composed of these minerals. The two main constituents of silicates are silicon and oxygen, which are the two most abundant elements in the Earth's crust. Other common elements in silicate minerals correspond to other common elements in the Earth's crust, such aluminium, magnesium, iron, calcium, sodium, and potassium. Some important rock-forming silicates include the feldspars, quartz, olivines, pyroxenes, amphiboles, garnets, and micas.

Silicates

Aegirine, an iron-sodium clinopyroxene, is part of the inosilicate subclass.

The base of unit of a silicate mineral is the $[SiO_4]^{4-}$ tetrahedron. In the vast majority of cases, silicon is in four-fold or tetrahedral coordination with oxygen. In very high-pressure situations, silicon will be six-fold or octahedral coordination, such as in the perovskite structure or the quartz polymorph stishovite (SiO_2). In the latter case, the mineral no longer has a silicate structure, but that of rutile (TiO_2), and its associated group, which are simple oxides. These silica tetrahedra are then polymerized to some degree to create various structures, such as one-dimensional chains, two-dimensional sheets, and three-dimensional frameworks. The basic silicate mineral where no polymerization of the tetrahedra has occurred requires other elements to balance out the base 4-charge. In other silicate structures, different combinations of elements are required to balance out the resultant negative charge. It is common for the Si^{4+} to be substituted by Al^{3+} because of similarity in ionic radius and charge; in those case, the $[AlO_4]^{5-}$ tetrahedra form the same structures as do the unsubstituted tetrahedra, but their charge-balancing requirements are different.

The degree of polymerization can be described by both the structure formed and how many tetrahedral corners (or coordinating oxygens) are shared (for aluminium and silicon in tetrahedral sites). Orthosilicates (or nesosilicates) have no linking of polyhedra, thus tetrahedra share no corners. Disilicates (or sorosilicates) have two tetrahedra sharing one oxygen atom. Inosilicates are chain silicates; single-chain silicates have two shared corners, whereas double-chain silicates have two or three shared corners. In phyllosilicates, a sheet structure is formed which requires three shared oxygens; in the case of double-chain silicates, some tetrahedra must share two corners instead of three as otherwise a sheet structure would result. Framework silicates, or tectosilicates, have tetrahedra that share all four corners. The ring silicates, or cyclosilicates, only need tetrahedra to share two corners to form the cyclical structure.

The silicate subclasses are described below in order of decreasing polymerization.

Tectosilicates

Natrolite is a mineral series in the zeolite group; this sample has a very prominent acicular crystal habit.

Tectosilicates, also known as framework silicates, have the highest degree of polymerization. With all corners of a tetrahedra shared, the silicon:oxygen ratio becomes 1:2. Examples are quartz, the feldspars, feldspathoids, and the zeolites. Framework silicates tend to be particularly chemically stable as a result of strong covalent bonds.

Forming 12% of the Earth's crust, quartz (SiO_2) is the most abundant mineral species. It is characterized by its high chemical and physical resistivity. Quartz has several polymorphs, including tridymite and cristobalite at high temperatures, high-pressure coesite, and ultra-high pressure stishovite. The latter mineral can only be formed on Earth by meteorite impacts, and its structure has been composed so much that it had changed from a silicate structure to that of rutile (TiO_2). The silica polymorph that is most stable at the Earth's surface is α-quartz. Its counterpart, β-quartz, is present only at high temperatures and pressures (changes to α-quartz below 573 °C at 1 bar). These two polymorphs differ by a "kinking" of bonds; this change in structure gives β-quartz greater symmetry than α-quartz, and they are thus also called high quartz (β) and low quartz (α).

Feldspars are the most abundant group in the Earth's crust, at about 50%. In the feldspars, Al^{3+} substitutes for Si^{4+}, which creates a charge imbalance that must be accounted for by the addition of cations. The base structure becomes either $[AlSi_3O_8]^-$ or $[Al_2Si_2O_8]^{2-}$ There are 22 mineral species of feldspars, subdivided into two major subgroups—alkali and plagioclase—and two less common groups—celsian and banalsite. The alkali feldspars are most commonly in a series between potassium-rich orthoclase and sodium-rich albite; in the case of plagioclase, the most common series ranges from albite to calcium-rich anorthite. Crystal twinning is common in feldspars, especially polysynthetic twins in plagioclase and Carlsbad twins in alkali feldspars. If the latter subgroup cools slowly from a melt, it forms exsolution lamellae because the two components—orthoclase and albite—are unstable in solid solution. Exsolution can be on a scale from microscopic to readily observable in hand-sample; perthitic texture forms when Na-rich feldspar exsolve in a K-rich host. The opposite texture (antiperthitic), where K-rich feldspar exsolves in a Na-rich host, is very rare.

Feldspathoids are structurally similar to feldspar, but differ in that they form in Si-deficient conditions which allows for further substitution by Al^{3+}. As a result, feldsapthoids cannot be associated with quartz. A common example of a feldsapthoid is nepheline ((Na, K)$AlSiO_4$); compared to alkali feldspar, nepheline has an Al_2O_3:SiO_2 ratio of 1:2, as opposed to 1:6 in the feldspar. Zeolites often have distinctive crystal habits, occurring in needles, plates, or blocky masses. They form in the presence of water at low temperatures and pressures, and have channels and voids in their structure. Zeolites have several industrial applications, especially in waste water treatment.

Phyllosilicates

Muscovite, a mineral species in the mica group, within the phyllosilicate subclass

Phyllosilicates consist of sheets of polymerized tetrahedra. They are bound at three oxygen sites, which gives a characteristic silicon:oxygen ratio of 2:5. Important examples include the mica, chlorite, and the kaolinite-serpentine groups. The sheets are weakly bound by van der Waals forces or hydrogen bonds, which causes a crystallographic weakness, in turn leading to a prominent basal cleavage among the phyllosilicates. In addition to the tetrahedra, phyllosilicates have a sheet of octahedra (elements in six-fold coordination by oxygen) that balanced out the basic tetrahedra, which have a negative charge (e.g. $[Si_4O_{10}]^{4-}$) These tetrahedra (T) and octahedra (O) sheets are stacked in a variety of combinations to create phyllosilicate groups. Within an octahedral sheet, there are three octahedral sites in a unit structure; however, not all of the sites may be occupied. In that case, the mineral is termed dioctahedral, whereas in other case it is termed trioctahedral.

The kaolinite-serpentine group consists of T-O stacks (the 1:1 clay minerals); their hardness ranges from 2 to 4, as the sheets are held by hydrogen bonds. The 2:1 clay minerals (pyrophyllite-talc) consist of T-O-T stacks, but they are softer (hardness from 1 to 2), as they are instead held together by van der Waals forces. These two groups of minerals are subgrouped by octahedral occupation; specifically, kaolinite and pyrophyllite are dioctahedral whereas serpentine and talc trioctahedral.

Micas are also T-O-T-stacked phyllosilicates, but differ from the other T-O-T and T-O-stacked subclass members in that they incorporate aluminium into the tetrahedral sheets (clay minerals have Al^{3+} in octahedral sites). Common examples of micas are muscovite, and the biotite series. The chlorite group is related to mica group, but a brucite-like ($Mg(OH)_2$) layer between the T-O-T stacks.

Because of their chemical structure, phyllosilicates typically have flexible, elastic, transparent layers that are electrical insulators and can be split into very thin flakes. Micas can be used in electronics as insulators, in construction, as optical filler, or even cosmetics. Chrysotile, a species of serpentine, is the most common mineral species in industrial asbestos, as it is less dangerous in terms of health than the amphibole asbestos.

Inosilicates

Asbestiform tremolite, part of the amphibole group in the inosilicate subclass

Inosilicates consist of tetrahedra repeatedly bonded in chains. These chains can be single, where a tetrahedron is bound to two others to form a continuous chain; alternatively, two chains can be merged to create double-chain silicates. Single-chain silicates have a silicon:oxygen ratio of 1:3 (e.g. $[Si_2O_6]^{4-}$), whereas the double-chain variety has a ratio of 4:11, e.g. $[Si_8O_{22}]^{12-}$. Inosilicates contain two important rock-forming mineral groups; single-chain silicates are most commonly pyroxenes, while double-chain silicates are often amphiboles. Higher-order chains exist (e.g. three-member, four-member, five-member chains, etc.) but they are rare.

The pyroxene group consists of 21 mineral species. Pyroxenes have a general structure formula of $XY(Si_2O_6)$, where X is an octahedral site, while Y can vary in coordination number from six to eight.

Most varieties of pyroxene consist of permutations of Ca^{2+}, Fe^{2+} and Mg^{2+} to balance the negative charge on the backbone. Pyroxenes are common in the Earth's crust (about 10%) and are a key constituent of mafic igneous rocks.

Amphiboles have great variability in chemistry, described variously as a "mineralogical garbage can" or a "mineralogical shark swimming a sea of elements". The backbone of the amphiboles is the $[Si_8O_{22}]^{12-}$; it is balanced by cations in three possible positions, although the third position is not always used, and one element can occupy both remaining ones. Finally, the amphiboles are usually hydrated, that is, they have a hydroxyl group ($[OH]^-$), although it can be replaced by a fluoride, a chloride, or an oxide ion. Because of the variable chemistry, there are over 80 species of amphibole, although variations, as in the pyroxenes, most commonly involve mixtures of Ca^{2+}, Fe^{2+} and Mg^{2+}. Several amphibole mineral species can have an asbestiform crystal habit. These asbestos minerals form long, thin, flexible, and strong fibres, which are electrical insulators, chemically inert and heat-resistant; as such, they have several applications, especially in construction materials. However, asbestos are known carcinogens, and cause various other illnesses, such as asbestosis; amphibole asbestos (anthophyllite, tremolite, actinolite, grunerite, and riebeckite) are considered more dangerous than chrysotile serpentine asbestos.

Cyclosilicates

An example of elbaite, a species of tourmaline, with distinctive colour banding.

Cyclosilicates, or ring silicates, have a ratio of silicon to oxygen of 1:3. Six-member rings are most common, with a base structure of $[Si_6O_{18}]^{12-}$; examples include the tourmaline group and beryl. Other ring structures exist, with 3, 4, 8, 9, 12 having been described. Cyclosilicates tend to be strong, with elongated, striated crystals.

Tourmalines have a very complex chemistry that can be described by a general formula $XY_3Z_6(BO_3)_3T_6O_{18}V_3W$. The T_6O_{18} is the basic ring structure, where T is usually Si^{4+}, but substitutable by Al_{3+} or B^{3+}. Tourmalines can be subgrouped by the occupancy of the X site, and from there further subdivided by the chemistry of the W site. The Y and Z sites can accommodate a variety of cations, especially various transition metals; this variability in structural transition metal content gives the tourmaline group greater variability in colour. Other cyclosilicates include beryl, $Al_2Be_3Si_6O_{18}$, whose varieties include the gemstones emerald (green) and aquamarine (bluish). Cordierite is structurally similar to beryl, and is a common metamorphic mineral.

Sorosilicates

Sorosilicates, also termed disilicates, have tetrahedron-tetrahedron bonding at one oxygen, which results in a 2:7 ratio of silicon to oxygen. The resultant common structural element is the $[Si_2O_7]^{6-}$ group. The most common disilicates by far are members of the epidote group. Epidotes are found in variety of geologic settings, ranging from mid-ocean ridge to granites to metapelites. Epidotes are built around the structure $[(SiO_4)(Si_2O_7)]^{10-}$ structure; for example, the mineral *species* epidote has calcium, aluminium, and ferric iron to charge balance: $Ca_2Al_2(Fe^{3+}, Al)(SiO_4)(Si_2O_7)O(OH)$. The presence of iron as Fe^{3+} and Fe^{2+} helps understand oxygen fugacity, which in turn is a significant factor in petrogenesis.

Other examples of sorosilicates include lawsonite, a metamorphic mineral forming in the blueschist facies (subduction zone setting with low temperature and high pressure), vesuvianite, which takes up a significant amount of calcium in its chemical structure.

Orthosilicates

Black andradite, an end-member of the orthosilicate garnet group.

Orthosilicates consist of isolated tetrahedra that are charge-balanced by other cations. Also termed nesosilicates, this type of silicate has a silicon:oxygen ratio of 1:4 (e.g. SiO_4). Typical orthosilicates tend to form blocky equant crystals, and are fairly hard. Several rock-forming minerals are part of this subclass, such as the aluminosilicates, the olivine group, and the garnet group.

The aluminosilicates—kyanite, andalusite, and sillimanite, all Al_2SiO_5—are structurally composed of one $[SiO_4]^{4-}$ tetrahedron, and one Al^{3+} in octahedral coordination. The remaining Al^{3+} can be in six-fold coordination (kyanite), five-fold (andalusite) or four-fold (sillimanite); which mineral forms in a given environment is depend on pressure and temperature conditions. In the olivine structure, the main olivine series of $(Mg, Fe)_2SiO_4$ consist of magnesium-rich forsterite and iron-rich fayalite. Both iron and magnesium are in octahedral by oxygen. Other mineral species having this structure exist, such as tephroite, Mn_2SiO_4. The garnet group has a general formula of $X_3Y_2(SiO_4)_3$, where X is a large eight-fold coordinated cation, and Y is a smaller six-fold coordinated cation. There are six ideal endmembers of garnet, split into two group. The pyralspite garnets have Al^{3+} in the Y position: pyrope $(Mg_3Al_2(SiO_4)_3)$, almandine $(Fe_3Al_2(SiO_4)_3)$, and spessartine $(Mn_3Al_2(SiO_4)_3)$. The ugrandite garnets have Ca^{2+} in the X position: uvarovite $(Ca_3Cr_2(SiO_4)_3)$, grossular $(Ca_3Al_2(SiO_4)_3)$ and andradite $(Ca_3Fe_2(SiO_4)_3)$. While there are two subgroups of garnet, solid solutions exist between all six end-members.

Other orthosilicates include zircon, staurolite, and topaz. Zircon ($ZrSiO_4$) is useful in geochronology as the Zr^{4+} can be substituted by U^{6+}; furthermore, because of its very resistant structure, it is difficult to reset it as a chronometer. Staurolite is a common metamorphic intermediate-grade index mineral. It has a particularly complicated crystal structure that was only fully described in 1986. Topaz ($Al_2SiO_4(F, OH)_2$, often found in granitic pegmatites associated with tourmaline, is a common gemstone mineral.

Non-Silicates

Native Elements

Native gold. Rare specimen of stout crystals growing off of a central stalk, size 3.7 x 1.1 x 0.4 cm, from Venezuela.

Native elements are those that are not chemically bonded to other elements. This mineral group includes native metals, semi-metals, and non-metals, and various alloys and solid solutions. The metals are held together by metallic bonding, which confers distinctive physical properties such as their shiny metallic lustre, ductility and malleability, and electrical conductivity. Native elements are subdivided into groups by their structure or chemical attributes.

The gold group, with a cubic close-packed structure, includes metals such as gold, silver, and copper. The platinum group is similar in structure to the gold group. The iron-nickel group is characterized by several iron-nickel alloy species. Two examples are kamacite and taenite, which are found in iron meteorites; these species differ by the amount of Ni in the alloy; kamacite has less than 5–7% nickel and is a variety of native iron, whereas the nickel content of taenite ranges from 7–37%. Arsenic group minerals consist of semi-metals, which have only some metallic; for example, they lack the malleability of metals. Native carbon occurs in two allotropes, graphite and diamond; the latter forms at very high pressure in the mantle, which gives it a much stronger structure than graphite.

Sulfides

The sulfide minerals are chemical compounds of one or more metals or semimetals with a sulfur; tellurium, arsenic, or selenium can substitute for the sulfur. Sulfides tend to be soft, brittle minerals with a high specific gravity. Many powdered sulfides, such as pyrite, have a sulfurous smell when powdered. Sulfides are susceptible to weathering, and many readily dissolve in water; these dissolved minerals can be later redeposited, which creates enriched secondary ore deposits. Sulfides are classified by the ratio of the metal or semimetal to the sulfur, such as M:S equal to 2:1, or

1:1. Many sulfide minerals are economically important as metal ores; examples include sphalerite (ZnS), an ore of zinc, galena (PbS), an ore of lead, cinnabar (HgS), an ore of mercury, and molybdenite (MoS_2, an ore of molybdenum. Pyrite (FeS_2), is the most commonly occurring sulfide, and can be found in most geological environments. It is not, however, an ore of iron, but can be instead oxidized to produce sulfuric acid. Related to the sulfides are the rare sulfosalts, in which a metallic element is bonded to sulfur and a semimetal such as antimony, arsenic, or bismuth. Like the sulfides, sulfosalts are typically soft, heavy, and brittle minerals.

Oxides

Oxide minerals are divided into three categories: simple oxides, hydroxides, and multiple oxides. Simple oxides are characterized by O^{2-} as the main anion and primarily ionic bonding. They can be further subdivided by the ratio of oxygen to the cations. The periclase group consists of minerals with a 1:1 ratio. Oxides with a 2:1 ratio include cuprite (Cu_2O) and water ice. Corundum group minerals have a 2:3 ratio, and includes minerals such as corundum (Al_2O_3), and hematite (Fe_2O_3). Rutile group minerals have a ratio of 1:2; the eponymous species, rutile (TiO_2) is the chief ore of titanium; other examples include cassiterite (SnO_2; ore of tin), and pyrolusite (MnO_2; ore of manganese). In hydroxides, the dominant anion is the hydroxyl ion, OH^-. Bauxites are the chief aluminium ore, and are a heterogeneous mixture of the hydroxide minerals diaspore, gibbsite, and bohmite; they form in areas with a very high rate of chemical weathering (mainly tropical conditions). Finally, multiple oxides are compounds of two metals with oxygen. A major group within this class are the spinels, with a general formula of $X^{2+}Y^{3+}_2O_4$. Examples of species include spinel ($MgAl_2O_4$), chromite ($FeCr_2O_4$), and magnetite (Fe_3O_4). The latter is readily distinguishable by its strong magnetism, which occurs as it has iron in two oxidation states ($Fe^{2+}Fe^{3+}_2O_4$), which makes it a multiple oxide instead of a single oxide.

Halides

Pink cubic halite ($NaCl$; halide class) crystals on a nahcolite matrix ($NaHCO_3$; a carbonate, and mineral form of sodium bicarbonate, used as baking soda).

The halide minerals are compounds where a halogen (fluorine, chlorine, iodine, and bromine) is the main anion. These minerals tend to be soft, weak, brittle, and water-soluble. Common examples of halides include halite (NaCl, table salt), sylvite (KCl), fluorite (CaF_2). Halite and sylvite commonly form as evaporites, and can be dominant minerals in chemical sedimentary rocks. Cryolite, Na_3AlF_6, is a key mineral in the extraction of aluminium from bauxites; however, as the only significant occurrence at Ivittuut, Greenland, in a granitic pegmatite, was depleted, synthetic cryolite can be made from fluorite.

Carbonates

The carbonate minerals are those were the main anionic group is carbonate, $[CO_3]^{2-}$. Carbonates tend to be brittle, many have rhombohedral cleavage, and all react with acid. Due to the last characteristic, field geologists often carry dilute hydrochloric acid to distinguish carbonates from non-carbonates. The reaction of acid with carbonates, most commonly found as the polymorph calcite and aragonite ($CaCO_3$), relates to the dissolution and precipitation of the mineral, which is a key in the formation of limestone caves, features within them such as stalactite and stalagmites, and karst landforms. Carbonates are most often formed as biogenic or chemical sediments in marine environments. The carbonate group is structurally a triangle, where a central C^{4+} cation is surrounded by three O^{2-} anions; different groups of minerals form from different arrangements of these triangles. The most common carbonate mineral is calcite, and is the primary constituent of sedimentary limestone and metamorphic marble. Calcite, $CaCO_3$, can have a high magnesium impurity; under high-Mg conditions, its polymorph aragonite will form instead; the marine geochemistry in this regard can be described as an aragonite or calcite sea, depending on which mineral preferentially forms. Dolomite is a double carbonate, with the formula $CaMg(CO_3)_2$. Secondary dolomitization of limestone is common, where calcite or aragonite are converted to dolomite; this reaction increases pore space (the unit cell volume of dolomite is 88% that of calcite), which can create a reservoir for oil and gas. These two minerals species are members of eponymous mineral groups: the calcite group includes carbonates with the general formula XCO_3, and the dolomite group constitutes minerals with general formula $XY(CO_3)_2$.

Sulfates

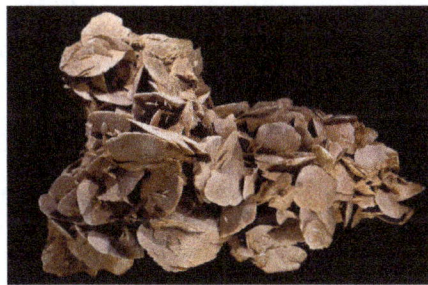

Gypsum desert rose

The sulfate minerals all contain the sulfate anion, $[SO_4]^{2-}$. They tend to be transparent to translucent, soft, and many are fragile. Sulfate minerals commonly form as evaporites, where they precipitate out of evaporating saline waters; alternative, sulfates can also be found in hydrothermal vein systems associated with sulfides, or as oxidation products of sulfides. Sulfates can be subdivided into anhydrous and hydrous minerals. The most common hydrous sulfate by far is gypsum, $CaSO_4 \cdot 2H_2O$. It forms as an evaporite, and is associated with other evaporites such as calcite and halite; if it incorporates sand grains as it crystallizes, gypsum can form desert roses. Gypsum has very low thermal conductivity and maintains a low temperature when heated as it loses that heat by dehydrating; as such, gypsum is used as an insulator in materials such as plaster and drywall. The anhydrous equivalent of gypsum is anhydrite; it can form directly from seawater in highly arid conditions. The barite group has the general formula XSO_4, where the X is a large 12-coordinated cation. Examples include barite ($BaSO_4$), celestine ($SrSO_4$), and anglesite ($PbSO_4$); anhydrite is not part of the barite group, as the smaller Ca^{2+} is only in eight-fold coordination.

Phosphates

The phosphate minerals are characterized by the tetrahedral $[PO_4]^{3-}$ unit, although the structure can be generalized, and phosphorus is replaced by antimony, arsenic, or vanadium. The most common phosphate is the apatite group; common species within this group are fluorapatite ($Ca_5(PO_4)_3F$), chlorapatite ($Ca_5(PO_4)_3Cl$) and hydroxylapatite ($Ca_5(PO_4)_3(OH)$). Minerals in this group are the main crystalline constituents of teeth and bones in vertebrates. The relatively abundant monazite group has a general structure of ATO_4, where T is phosphorus or arsenic, and A is often a rare-earth element (REE). Monazite is important in two ways: first, as a REE "sink", it can sufficiently concentrate these elements to become an ore; secondly, monazite group elements can incorporate relatively large amounts of uranium and thorium, which can be used to date the rock based on the decay of the U and Th to lead.

Organic Minerals

The Strunz classification includes a class for organic minerals. These rare compounds contain organic carbon, but can be formed by a geologic process. For example, whewellite, $CaC_2O_4 \cdot H_2O$ is an oxalate that can be deposited in hydrothermal ore veins. While hydrated calcium oxalate can be found in coal seams and other sedimentary deposits involving organic matter, the hydrothermal occurrence is not considered to be related to biological activity.

Astrobiology

It has been suggested that biominerals could be important indicators of extraterrestrial life and thus could play an important role in the search for past or present life on the planet Mars. Furthermore, organic components (biosignatures) that are often associated with biominerals are believed to play crucial roles in both pre-biotic and biotic reactions.

On January 24, 2014, NASA reported that current studies by the *Curiosity* and *Opportunity* rovers on Mars will now be searching for evidence of ancient life, including a biosphere based on autotrophic, chemotrophic and/or chemolithoautotrophic microorganisms, as well as ancient water, including fluvio-lacustrine environments (plains related to ancient rivers or lakes) that may have been habitable. The search for evidence of habitability, taphonomy (related to fossils), and organic carbon on the planet Mars is now a primary NASA objective.

Silicate Minerals

Silicate minerals are rock-forming minerals made up of silicate groups. They are the largest and most important class of rock-forming minerals and make up approximately 90 percent of the Earth's crust. They are classified based on the structure of their silicate groups, which contain different ratios of silicon and oxygen.

CNesosilicates or Orthosilicates

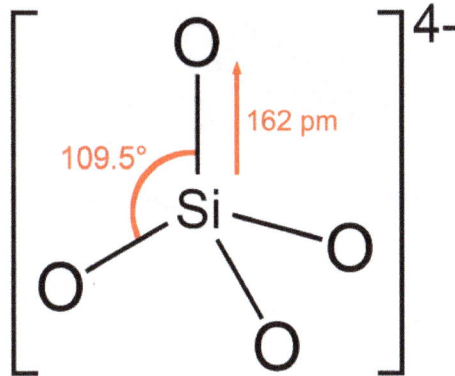

Basic (ortho-)silicate anion structure

Nesosilicates or orthosilicates, have the orthosilicate ion, which constitute isolated (insular) $[SiO_4]^{4-}$ tetrahedra that are connected only by interstitial cations. Nickel–Strunz classification: 09.A

- Phenakite group
    - Phenakite – Be_2SiO_4
    - Willemite – Zn_2SiO_4
- Olivine group
    - Forsterite – Mg_2SiO_4
    - Fayalite – Fe_2SiO_4
    - Tephroite – Mn_2SiO_4
- Garnet group
    - Pyrope – $Mg_3Al_2(SiO_4)_3$
    - Almandine – $Fe_3Al_2(SiO_4)_3$
    - Spessartine – $Mn_3Al_2(SiO_4)_3$
    - Grossular – $Ca_3Al_2(SiO_4)_3$
    - Andradite – $Ca_3Fe_2(SiO_4)_3$
    - Uvarovite – $Ca_3Cr_2(SiO_4)_3$
    - Hydrogrossular – $Ca_3Al_2Si_2O_8(SiO_4)_{3-m}(OH)_{4m}$
- Zircon group
    - Zircon – $ZrSiO_4$
    - Thorite – $(Th,U)SiO_4$

Kyanite crystals (unknown scale)

- Al_2SiO_5 group
    - Andalusite – Al_2SiO_5
    - Kyanite – Al_2SiO_5
    - Sillimanite – Al_2SiO_5
    - Dumortierite – $Al_{6.5-7}BO_3(SiO_4)_3(O,OH)_3$
    - Topaz – $Al_2SiO_4(F,OH)_2$
    - Staurolite – $Fe_2Al_9(SiO_4)_4(O,OH)_2$
- Humite group – $(Mg,Fe)_7(SiO_4)_3(F,OH)_2$
    - Norbergite – $Mg_3(SiO_4)(F,OH)_2$
    - Chondrodite – $Mg_5(SiO_4)_2(F,OH)_2$
    - Humite – $Mg_7(SiO_4)_3(F,OH)_2$
    - Clinohumite – $Mg_9(SiO_4)_4(F,OH)_2$
- Datolite – $CaBSiO_4(OH)$
- Titanite – $CaTiSiO_5$
- Chloritoid – $(Fe,Mg,Mn)_2Al_4Si_2O_{10}(OH)_4$
- Mullite (aka Porcelainite) – $Al_6Si_2O_{13}$

Sorosilicates

Sorosilicate exhibit at Museum of Geology in South Dakota

Sorosilicates have isolated double tetrahedra groups with $(Si_2O_7)^{6-}$ or a ratio of 2:7. Nickel–Strunz classification: 09.B

- Hemimorphite (calamine) – $Zn_4(Si_2O_7)(OH)_2 \cdot H_2O$
- Lawsonite – $CaAl_2(Si_2O_7)(OH)_2 \cdot H_2O$
- Ilvaite – $CaFe^{II}_2Fe^{III}O(Si_2O_7)(OH)$
- Epidote group (has both $(SiO_4)^{4-}$ and $(Si_2O_7)^{6-}$ groups)
    - Epidote – $Ca_2(Al,Fe)_3O(SiO_4)(Si_2O_7)(OH)$
    - Zoisite – $Ca_2Al_3O(SiO_4)(Si_2O_7)(OH)$
    - Clinozoisite – $Ca_2Al_3O(SiO_4)(Si_2O_7)(OH)$
    - Tanzanite – $Ca_2Al_3O(SiO_4)(Si_2O_7)(OH)$
    - Allanite – $Ca(Ce,La,Y,Ca)Al_2(Fe^{II},Fe^{III})O(SiO_4)(Si_2O_7)(OH)$
    - Dollaseite-(Ce) – $CaCeMg_2AlSi_3O_{11}F(OH)$
- Vesuvianite (idocrase) – $Ca_{10}(Mg,Fe)_2Al_4(SiO_4)_5(Si_2O_7)_2(OH)_4$

Cyclosilicates

Cyclosilicate specimens at the Museum of Geology, South Dakota

Cyclosilicates, or ring silicates, have linked tetrahedra with $(T_xO_{3x})^{2x-}$ or a ratio of 1:3. These exist as 3-member $(T_3O_9)^{6-}$ and 6-member $(T_6O_{18})^{12-}$ rings, where T stands for a tetrahedrally coordinated cation. Nickel–Strunz classification: 09.C

- 3-member ring
    - Benitoite – $BaTi(Si_3O_9)$
- 6-member ring
    - Axinite – $(Ca,Fe,Mn)_3Al_2(BO_3)(Si_4O_{12})(OH)$
    - Beryl/Emerald – $Be_3Al_2(Si_6O_{18})$
    - Sugilite – $KNa_2(Fe,Mn,Al)_2Li_3Si_{12}O_{30}$
    - Cordierite – $(Mg,Fe)_2Al_3(Si_5AlO_{18})$
    - Tourmaline – $(Na,Ca)(Al,Li,Mg)_{3-}(Al,Fe,Mn)_6(Si_6O_{18}(BO_3)_3(OH)_4$

Note that the ring in axinite contains two B and four Si tetrahedra and is highly distorted compared to the other 6-member ring cyclosilicates.

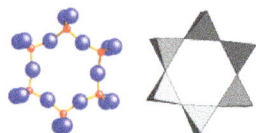

- Cyclosilicate, $[Si_6O_{18}]$ – 6-membered single rings, beryl (red: Si, blue: O)

- Cyclosilicate, $[Si_3O_9]$ – 3-membered single ring, benitoite

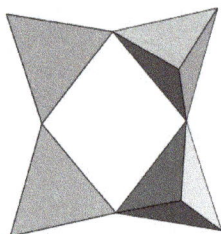

- Cyclosilicate, $[Si_4O_{12}]$ – 4-membered single ring, papagoite

- Cyclosilicate, $[Si_9O_{27}]$ – 9-membered ring, eudialyte

- Cyclosilicate, $[Si_6O_{18}]$ – 6-membered double ring, milarite

Inosilicates

Inosilicates, or chain silicates, have interlocking chains of silicate tetrahedra with either SiO_3, 1:3 ratio, for single chains or Si_4O_{11}, 4:11 ratio, for double chains. Nickel–Strunz classification: 09.D

Single Chain Inosilicates

- Pyroxene group

- Enstatite – orthoferrosilite series
  - Enstatite – $MgSiO_3$
  - Ferrosilite – $FeSiO_3$
- Pigeonite – $Ca_{0.25}(Mg,Fe)_{1.75}Si_2O_6$
- Diopside – hedenbergite series
  - Diopside – $CaMgSi_2O_6$
  - Hedenbergite – $CaFeSi_2O_6$
  - Augite – $(Ca,Na)(Mg,Fe,Al)(Si,Al)_2O_6$
- Sodium pyroxene series
  - Jadeite – $NaAlSi_2O_6$
  - Aegirine (Acmite) – $NaFe^{III}Si_2O_6$
- Spodumene – $LiAlSi_2O_6$
- Pyroxenoid group
  - Wollastonite – $CaSiO_3$
  - Rhodonite – $MnSiO_3$
  - Pectolite – $NaCa_2(Si_3O_8)(OH)$

Double Chain Inosilicates

- Amphibole group
  - Anthophyllite – $(Mg,Fe)_7Si_8O_{22}(OH)_2$
  - Cumingtonite series
    - Cummingtonite – $Fe_2Mg_5Si_8O_{22}(OH)_2$
    - Grunerite – $Fe_7Si_8O_{22}(OH)_2$
  - Tremolite series
    - Tremolite – $Ca_2Mg_5Si_8O_{22}(OH)_2$
    - Actinolite – $Ca_2(Mg,Fe)_5Si_8O_{22}(OH)_2$
  - Hornblende – $(Ca,Na)_{2-3}(Mg,Fe,Al)_5Si_6(Al,Si)_2O_{22}(OH)_2$
  - Sodium amphibole group
    - Glaucophane – $Na_2Mg_3Al_2Si_8O_{22}(OH)_2$
    - Riebeckite (asbestos) – $Na_2Fe^{II}_3Fe^{III}_2Si_8O_{22}(OH)_2$
    - Arfvedsonite – $Na_3(Fe,Mg)_4FeSi_8O_{22}(OH)_2$

- Inosilicate, pyroxene family, with 2-periodic single chain (Si_2O_6), diopside

- Inosilicate, clinoamphibole, with 2-periodic double chains (Si_4O_{11}), tremolite

- Inosilicate, unbranched 3-periodic single chain of wollastonite

- Inosilicate with 5-periodic single chain, rhodonite

Phyllosilicates

Phyllosilicates, or sheet silicates, form parallel sheets of silicate tetrahedra with Si_2O_5 or a 2:5 ratio. Nickel–Strunz classification: 09.E. All phyllosilicate minerals are hydrated, with either water or hydroxyl groups attached.

Kaolin

- Serpentine subgroup
    - Antigorite – $Mg_3Si_2O_5(OH)_4$
    - Chrysotile – $Mg_3Si_2O_5(OH)_4$
    - Lizardite – $Mg_3Si_2O_5(OH)_4$
- Clay minerals group
    - Halloysite – $Al_2Si_2O_5(OH)_4$
    - Kaolinite – $Al_2Si_2O_5(OH)_4$
    - Illite – $(K,H_3O)(Al,Mg,Fe)_2(Si,Al)_4O_{10}[(OH)_2,(H_2O)]$
    - Montmorillonite – $(Na,Ca)_{0.33}(Al,Mg)_2Si_4O_{10}(OH)_2 \cdot nH_2O$
    - Vermiculite – $(MgFe,Al)_3(Al,Si)_4O_{10}(OH)_2 \cdot 4H_2O$
    - Talc – $Mg_3Si_4O_{10}(OH)_2$
    - Sepiolite – $Mg_4Si_6O_{15}(OH)_2 \cdot 6H_2O$
    - Palygorskite (or attapulgite) – $(Mg,Al)_2Si_4O_{10}(OH) \cdot 4(H_2O)$
    - Pyrophyllite – $Al_2Si_4O_{10}(OH)_2$
- Mica group
    - Biotite – $K(Mg,Fe)_3(AlSi_3)O_{10}(OH)_2$
    - Muscovite – $KAl_2(AlSi_3)O_{10}(OH)_2$
    - Phlogopite – $KMg_3(AlSi_3)O_{10}(OH)_2$
    - Lepidolite – $K(Li,Al)_{2-3}(AlSi_3)O_{10}(OH)_2$
    - Margarite – $CaAl_2(Al_2Si_2)O_{10}(OH)_2$
    - Glauconite – $(K,Na)(Al,Mg,Fe)_2(Si,Al)_4O_{10}(OH)_2$
- Chlorite group
    - Chlorite – $(Mg,Fe)_3(Si,Al)_4O_{10}(OH)_2 \cdot (Mg,Fe)_3(OH)_6$

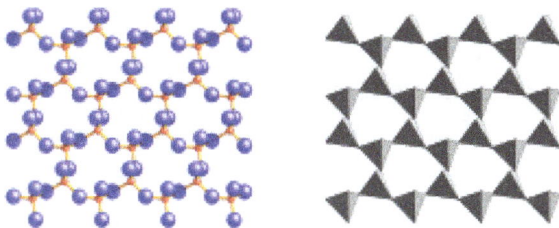

- Phyllosilicate, mica group, muscovite (red: Si, blue: O)

- Phyllosilicate, single net of tetrahedra with 4-membered rings, apophyllite-(KF)-apophyllite-(KOH) series

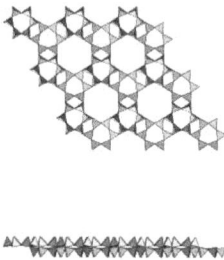

- Phyllosilicate, single tetrahedral nets of 6-membered rings, pyrosmalite-(Fe)-pyrosmalite-(Mn) series

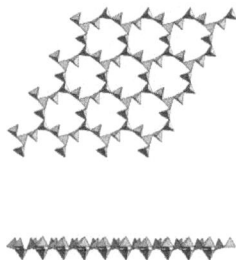

- Phyllosilicate, single tetrahedral nets of 6-membered rings, zeophyllite

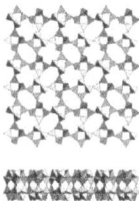

- Phyllosilicate, double nets with 4- and 6-membered rings, carletonite

Tectosilicates

Tectosilicates, or "framework silicates," have a three-dimensional framework of silicate tetrahedra with SiO_2 or a 1:2 ratio. This group comprises nearly 75% of the crust of the Earth. Tectosilicates, with the exception of the quartz group, are aluminosilicates. Nickel–Strunz classification: 09.F and 09.G, 04.DA (Quartz/ silica family)

Lunar Ferroan Anorthosite no. 60025 (Plagioclase Feldspar). Collected by Apollo 16 from the Lunar Highlands near Descartes Crater.

- Quartz group
  - Quartz – SiO_2
  - Tridymite – SiO_2
  - Cristobalite – SiO_2
  - Coesite – SiO_2
  - Stishovite – SiO_2
- Feldspar family
  - Alkali feldspars (potassium feldspars)
    - Microcline – $KAlSi_3O_8$
    - Orthoclase – $KAlSi_3O_8$
    - Anorthoclase – $(Na,K)AlSi_3O_8$
    - Sanidine – $KAlSi_3O_8$
    - Albite – $NaAlSi_3O_8$
  - Plagioclase feldspars
    - Albite – $NaAlSi_3O_8$
    - Oligoclase – $(Na,Ca)(Si,Al)_4O_8$ (Na:Ca 4:1)
    - Andesine – $(Na,Ca)(Si,Al)_4O_8$ (Na:Ca 3:2)
    - Labradorite – $(Ca,Na)(Si,Al)_4O_8$ (Na:Ca 2:3)
    - Bytownite – $(Ca,Na)(Si,Al)_4O_8$ (Na:Ca 1:4)
    - Anorthite – $CaAl_2Si_2O_8$
- Feldspathoid family

- Nosean – $Na_8Al_6Si_6O_{24}(SO_4)$
- Cancrinite – $Na_6Ca_2(CO_3,Al_6Si_6O_{24}).2H_2O$
- Leucite – $KAlSi_2O_6$
- Nepheline – $(Na,K)AlSiO_4$
- Sodalite – $Na_8(AlSiO_4)_6Cl_2$
  - Hauyne – $(Na,Ca)_{4-8}Al_6Si_6(O,S)24(SO_4,Cl)_{1-2}$
  - Lazurite – $(Na,Ca)_8(AlSiO_4)_6(SO_4,S,Cl)_2$
- Petalite – $LiAlSi_4O_{10}$
- Scapolite group
  - Marialite – $Na_4(AlSi_3O_8)_3(Cl_2,CO_3,SO_4)$
  - Meionite – $Ca_4(Al_2Si_2O_8)_3(Cl_2CO_3,SO_4)$
- Analcime – $NaAlSi_2O_6 \cdot H_2O$
- Zeolite family
  - Natrolite – $Na_2Al_2Si_3O_{10} \cdot 2H_2O$
  - Erionite – $(Na_2,K_2,Ca)_2Al_4Si_{14}O_{36} \cdot 15H_2O$
  - Chabazite – $CaAl_2Si_4O_{12} \cdot 6H_2O$
  - Heulandite – $CaAl_2Si_7O_{18} \cdot 6H_2O$
  - Stilbite – $NaCa_2Al_5Si_{13}O_{36} \cdot 17H_2O$
  - Scolecite – $CaAl_2Si_3O_{10} \cdot 3H_2O$
  - Mordenite – $(Ca,Na_2,K_2)Al_2Si_{10}O_{24} \cdot 7H_2O$

Classification of Silicate Minerals

This list gives an overview of the classification of minerals (silicates) and includes mostly IMA recognized minerals and its groupings. This list complements the alphabetical list on List of minerals (complete) and List of minerals. Rocks, ores, mineral mixtures, non-IMA approved minerals and non-named minerals are mostly excluded. Mostly major groups only, or groupings used by *New Dana Classification* and *Mindat*.

Classification of Minerals

Introduction

The grouping of the New Dana Classification and of the mindat.org is similar only, and so this classification is an overview only. Consistency is missing too on the group name endings (group, sub-group, series) between New Dana Classification and mindat.org. Category, class and supergroup name endings are used as layout tools in the list as well.

- Abbreviations:
  - "*" - mineral not IMA Approved.
  - "^{Q}" - doubtful/ questionable.
  - Rn - renaming.
  - Rd - redefinition.
  - "REE" - rare earth element (Sc, Y, La, Ce, Pr, Nd, Pm, Sm, Eu, Gd, Tb, Dy, Ho, Er, Tm, Yb, Lu)
  - "PGE" - platinum group element (Ru, Rh, Pd, Os, Ir, Pt)
  - "s.p." - special procedure.

Category '9': Silicate Minerals

- Unclassified silicates

Subclass '9.A': Nesosilicates

- Zircon group: $MSiO_4$ (a group of simple tetragonal silicates where M = tetravalent Zr, Th, Hf)
  - Hafnon $HfSiO_4$, Stetindite $Ce[SiO_4]$, Thorite $(Th,U)SiO_4$, Zircon $ZrSiO_4$, Coffinite $U(SiO_4)_{1-x}(OH)_{4x}$, Thorogummite $Th(SiO_4)_{1-x}(OH)_{4x}$, IMA2008-035 $CeSiO_4$
- Olivine group
  - Calcio-Olivine Ca_2SiO_4, Fayalite $(Fe^{2+})_2[SiO_4]$, Forsterite $Mg_2[SiO_4]$, Laihunite $Fe^{2+}(Fe^{3+})_2(SiO_4)_2$, Liebenbergite $(Ni,Mg)_2[SiO_4]$, Olivine $(Mg,Fe^{2+})_2[SiO_4]$, Tephroite $(Mn^{2+})_2[SiO_4]$
- Phenakite group
  - Phenakite Be_2SiO_4, Willemite Zn_2SiO_4, Eucryptite $LiAlSiO_4$
- $Al_2(SiO_4)O$
  - Sillimanite subgroup
    - Sillimanite $AlAlOSiO_4$, Mullite $Al_{(4+2x)}Si_{(2-2x)}O_{(10-x)}$ (x =0.17 to 0.59), Boromullite $Al_{4.5}SiB_{0.5}O_{9.5}$
  - Andalusite subgroup
    - Andalusite $AlAlOSiO_4$, Kanonaite $(Mn^{3+},Al)AlSiO_5$, Yoderite $Mg_2(Al,Fe^{3+})_6 Si_4O_{18}(OH)_2$

- Kyanite $AlAlOSiO_4$

- Titanite group

  - Titanite $CaTiSiO_5$, Malayaite $CaSnOSiO_4$, Vanadomalayaite $CaVOSiO_4$

- Cerite Group

  - Cerite-(Ce) $(Ce^{3+})_9Fe^{3+}(SiO_4)_6[(SiO_3)(OH)](OH)_3$, Cerite-(La) $(La,Ce,Ca)_9(Mg,Fe^{3+})(SiO_4)_6[SiO_3(OH)](OH)_3$, Aluminocerite-(Ce) $(Ce,REE,Ca)_9(Al,Fe^{3+})(SiO_4)_3[SiO_3(OH)]_4(OH)_3$

- Silicate apatites

  - Ellestadite* $Ca_5(SiO_4,PO_4,SO_4)_3(F,OH,Cl)$, Britholite-(Ce) $(Ce,Ca,Th,La,Nd)_5(SiO_4,PO_4)_3(OH,F)$, Britholite-(Y) $(Y,Ca)_5(SiO_4,PO_4)_3(OH,F)$, Ellestadite-(F) $Ca_5(SiO_4,PO_4,SO_4)_3(F,OH,Cl)$, Ellestadite-(OH) $Ca_5(SiO_4,SO_4)_3(OH,Cl,F)$, Ellestadite-(Cl) $Ca_5(SiO_4,PO_4,SO_4)_3(Cl,OH,F)$, Mattheddleite $Pb_2o(SiO_4)_7(SO_4)_4Cl_4$, Karnasurtite-(Ce) $(Ce,La,Th)(Ti,Nb)(Al,Fe^{3+})(Si,P)2O_7(OH)_4 \cdot 3(H_2O)$ (?), Fluorbritholite-(Ce) $(Ca,Ce,La,Na)_5(SiO_4,PO_4)_3(OH,F)$, Fluorcalciobritholite $(Ca,REE)_5[(Si,P)O_4]_3F$

- Uranophane group

  - Kasolite $Pb(UO_2)SiO_4 \cdot H_2O$, Uranophane $Ca(UO_2)_2SiO_3(OH) \cdot 5H_2O$, Sklodowskite $(H_3O)_2Mg(UO_2)_2(SiO_4)_2 \cdot 4H_2O$, Cuprosklodowskite $Cu[(UO_2)(SiO_2OH)]_2 \cdot 6H_2O$, Boltwoodite $HK(UO_2)(SiO_4) \cdot 1.5H_2O$, Natroboltwoodite $(H_3O)(Na,K)(UO_2)SiO_4 \cdot H_2O$, Oursinite $(Co,Mg)(H_3O)_2[(UO_2)SiO_4]_2 \cdot 3H_2O$, Swamboite $U^{7+}H_6(UO_2)_6(SiO_4)_6 \cdot 3oH_2O$, Uranophane-beta $Ca[(UO_2)SiO_3(OH)]_2 \cdot H_2O$

- Datolite group

  - Datolite series

    - Datolite $CaBSiO_4(OH)$, Hingganite-(Ce) $(Ce,Ca)_2([\ ],Fe)Be_2Si_2O_8[(OH),O]_2$, Hingganite-(Y) $Y_2([\ ])Be_2Si_2O_8(OH)_2$, Hingganite-(Yb) $(Yb,Y)_2([\ ])Be_2Si_2O_8(OH)_2$, Calcybeborosilite-(Y) $(REE,Ca)_2[\ ](B,Be)_2(SiO_4)_2(OH,O)_2$

- Homilite series

    - Bakerite $Ca_4B_4(BO_4)(SiO_4)_3(OH)_3 \cdot H_2O$, Gadolinite-(Ce) $(Ce,La,Nd,Y)_2Fe^{2+}Be_2Si_2O_{10}$, Gadolinite-(Y) $Y2Fe^{2+}Be_2Si_2O_{10}$, Calciogadolinite? $CaREE(Fe^{3+})Be_2Si_2O_{10}$, Homilite $Ca_2(Fe^{2+},Mg)B_2Si_2O_{10}$, Minasgeraisite-(Y) $CaY_2Be_2Si_2O_{10}$

- Hellandite group

  - Hellandite-(Y) $(Ca,REE)_4(Y,Ce)_2(Al,[\ ])_2[Si_4B_4O_{22}](OH)_2$

  - Tadzhikite-(Y) $Ca_4(Y,Ce)_2(Ti,Al,Fe^{3+},[\ ])_2[Si_4B_4O_{22}](OH)_2$

  - Tadzhikite-(Ce) $Ca_4(Ce,Y)_2(Ti,Al,Fe^{3+},[\ ])_2[Si_4B_4O_{22}](OH)_2$

  - Hellandite-(Ce) $(Ca_3REE)_4Ce_2Al[\ ]_2[Si_4B_4O_{22}](OH)_2$

  - Mottanaite-(Ce) $Ca4(Ce,Ca)2AlBe2[Si_4B_4O_{22}]O_2$

- Ciprianiite $Ca_4[(Th,U)(REE)]_2(Al,[\ ])_2[Si_4B_4O_{22}](OH,F)_2$
- Piergorite-(Ce) $Ca_8Ce_2(Al_{0.5}(Fe^{3+})_{0.5})([\ ],Li,Be)_2Si_6B_8O_{36}(OH,F)_2$
- Vicanite group
  - Vicanite-(Ce) $(Ca,Ce,La,Th)_{15}As^{5+}(As^{3+}_{0.5},Na_{0.5})Fe^{3+}Si_6B_4O_9OF_7$
  - Hundholmenite-(Y) $(Y,REE,Ca,Na)_{15}(Al,Fe^{3+})Cax(As^{3+})_{1-x}(Si,As^{5+})Si_6B_3(O,F)_{48}$
  - Proshchenkoite-(Y) $(Y,REE,Ca,Na,Mn)_{15}Fe^{2+}Ca(P,Si)Si_6B_3(O,F)_{48}$

"Garnet" Supergroup

- Nesosilicate insular SiO_4 groups only with cations in and > coordination
- Category:Garnet group, $X_3Z_2(TO_4)_3$ (X = Ca, Fe, etc., Z = Al, Cr, etc., T = Si, As, V, etc.)
  - Pyralspite series
    - Pyrope $Mg_3Al_2(SiO_4)_3$, Almandine $(Fe^{2+})_3Al_2(SiO_4)_3$, Spessartine $(Mn^{2+})_3Al_2(SiO_4)_3$, Knorringite $Mg_3Cr_2(SiO_4)_3$, Majorite $Mg_3(Fe,Al,Si)_2(SiO_4)_3$, Calderite $(Mn^{2+},-Ca)_3(Fe^{3+},Al)_2(SiO_4)_3$
  - Ugrandite series
    - Andradite $Ca_3(Fe^{3+})_2(SiO_4)_3$, Grossular $Ca_3Al_2(SiO_4)_3$, Uvarovite $Ca_3Cr_2(SiO_4)_3$, Goldmanite $Ca_3(V,Al,Fe^{3+})_2(SiO_4)_3$, Yamatoite? $(Mn^{2+},Ca)_3(V^{3+},Al)_2(SiO_4)_3$
  - Schorlomite - Kimzeyite series
    - Schorlomite $Ca_3(Ti,Fe^{3+},Al)_2[(Si,Fe^{3+},Fe^{2+})O_4]_3$, Kimzeyite $Ca_3(Z-r,Ti)_2(Si,Al,Fe^{3+})_3-O_{12}$, Morimotoite $Ca_3TiFe^{2+}Si_3O_{12}$
  - Hydrogarnet
    - Hibschite $Ca_3Al_2(SiO_4)_{3-x}(OH)_{4x}$ (x=0.2-1.5), Katoite $Ca_3Al_2(SiO_4)_{3-x}(OH)_{4x}$ (x=1.5-3)
  - Tetragonal Hydrogarnet
    - Henritermierite $Ca_3(Mn,Al)_2(SiO_4)_2(OH)_4$, Holtstamite $Ca_3(Al,Mn^{3+})_2(-SiO_4)_2(OH)_4$
- Bredigite $Ca_7Mg(SiO_4)_4$, Merwinite $Ca_3Mg(SiO_4)_2$, Wadalite $Ca_6Al_5Si_2O_{16}Cl_3$, Rondorfite $Ca_8Mg(SiO_4)4Cl_2$

"Humite" Supergroup

- Nesosilicate insular SiO_4 groups and O, OH, F, and H_2O with cations in coordination only
- Topaz group
  - Topaz $Al_2SiO_4(F,OH)_2$, Krieselite $(Al,Ga)_2(Ge,C)O_4(OH)_2$
- Humite, general formula $A_n(SiO_4)_m(F,OH)_2$

- Chondrodite series

  - Alleghanyite $(Mn^{2+})_5(SiO_4)_2(OH,F)_2$, Chondrodite $(Mg,Fe,Ti)_5(SiO_4)_2(F,OH,O)_2$, Reinhardbraunsite $Ca_5[(OH,F)_2|(SiO_4)_2]$, Ribbeite $Mn_5(SiO_4)_2(OH)_2$, Kumtyubeite $Ca_5(SiO_4)_2F_2$

- Humite series

  - Humite $(Mg,Fe)_7[(F,OH)_2|(SiO_4)_3]$, Leucophoenicite $(Mn,Ca,Mg,Zn)(SiO_4)_3(OH)_2$, Manganhumite $(Mn^{2+},Mg)_7[(OH)_2|(SiO_4)_3]$, Chegemite $Ca_7(SiO_4)_3(OH)_2$

- Clinohumite series

  - Clinohumite $(Mg,Fe^{2+})_9[(F,OH)_2|(SiO_4)_4]$, Jerrygibbsite $(Mn,Zn)_9(SiO_4)_4(OH)_2$, Sonolite $Mn_9(SiO_4)_4(F,OH)_2$, Hydroxylclinohumite $Mg_9[(F,OH)_2|(SiO_4)_4$

- Norbergite $Mg_3(SiO_4)(F,OH)_2$

- Chloritoid group

  - Chloritoid $(Fe^{2+},Mg,Mn)_2Al_4Si_2O_{10}(OH)_4$, Magnesiochloritoid $MgAl_2SiO_5(OH)_2$, Ottrelite $(Mn,Fe^{2+},Mg)_2Al_4Si_2O_{10}(OH)_4$, Carboirite-VIII $Fe^{2+}(Al,Ge)_2O[(Ge,Si)O_4](OH)_2$

Subclass '9.B': Sorosilicates

- Epidote supergroup, $\{A_2\}\{M_3\}[O|OH|SiO_4|Si_2O_7]$

  - Epidote group

    - Clinozoisite $\{Ca_2\}\{Al_3\}[O|OH|SiO_4|Si_2O_7]$, Clinozoisite-(Sr) $CaSrAl_3(Si_2O_7)(SiO_4)O(OH)$, Epidote $\{Ca_2\}\{Al_2Fe^{3+}\}[O|OH|SiO_4|Si_2O_7]$, Epidote-(Pb) $CaPbFe^{3+}Al_2(Si_2O_7)(SiO_4)O(OH)$, Mukhinite $\{Ca_2\}\{Al_2V^{3+}\}[O|OH|SiO_4|Si_2O_7]$, Piemontite $\{Ca_2\}\{Al_2Mn^{3+}\}[O|OH|SiO_4|Si_2O_7]$, Piemontite-(Sr) $\{CaSr\}\{Al_2Mn^{3+}\}[O|OH|SiO_4|Si_2O_7]$, Manganipiemontite-(Sr) $\{CaSr\}\{Mn^{3+}AlMn^{3+}\}[O|OH|SiO_4|Si_2O_7]$

  - Allanite group

    - Allanite-(Ce) $\{CaCe\}\{Al_2Fe^{2+}\}[O|OH|SiO_4|Si_2O_7]$, Allanite-(La) $\{CaLa\}\{Al_2Fe^{2+}\}[O|OH|SiO_4|Si_2O_7]$, Allanite-(Y) $\{CaY\}\{Al_2Fe^{2+}\}[O|OH|SiO_4|Si_2O_7]$, Dissakisite-(Ce) $\{CaCe\}\{Al_2Mg\}[O|OH|SiO_4|Si_2O_7]$, Dissakisite-(La) $\{CaLa\}\{Al_2Mg\}[O|OH|SiO_4|Si_2O_7]$, Ferriallanite-(Ce) $\{CaCe\}\{Fe^{3+}AlFe^{2+}\}[O|OH|SiO_4|Si_2O_7]$, Manganiandrosite-(Ce) $(Mn^{2+},Ca)(Ce,REE)Mn^{3+}AlMn^{2+}(Si_2O_7)(SiO_4)O(OH)$, Manganiandrosite-(La) $\{LaMn^{2+}\}\{Mn^{3+}AlMn^{2+}\}[O|OH|SiO_4|Si_2O_7]$, Vanadoandrosite-(Ce) $\{Mn^{2+}Ce\}\{V^{3+}AlMn^{2+}\}[O|OH|SiO_4|Si_2O_7]$

  - Dollaseite group

    - Dollaseite-(Ce) $\{CaCe\}\{MgAlMg\}[F|OH|SiO_4|Si_2O_7]$, Khristovite-(Ce) $\{CaCe\}\{MgAlMn^{2+}\}[F|OH|SiO_4|Si_2O_7]$, Mills et al. (2009)

Subclass '9.C': Cyclosilicates

- Tourmaline group

- Alkali-Deficient Tourmaline subgroup - Foitite subgroup

  - Foitite [ ][(Fe^{2+})$_2$(Al,Fe^{3+})][Al$_6$][(OH)$_3$|OH|(BO$_3$)$_3$|Si$_6$O$_{18}$], Magnesiofoitite [ ][Mg$_2$(Al,Fe^{3+})][Al$_6$][(OH)$_3$|OH|(BO$_3$)$_3$|Si$_6$O$_{18}$], Rossmanite [ ][LiAl$_2$][Al$_6$][(OH)$_3$|OH|(BO$_3$)$_3$|Si$_6$O$_{18}$], Oxy-Rossmanite [ ][LiAl$_2$]Al$_6$(OH)$_3$O(BO$_3$)$_3$[Si$_6$O$_{18}$]

- Calcic Tourmaline subgroup - Liddicoatite subgroup

  - Liddicoatite [Ca][Li$_2$Al][Al$_6$][(OH)$_3$|F|(BO$_3$)$_3$|Si$_6$O$_{18}$], Uvite CaMg$_3$(Al$_5$Mg)(Si$_6$O$_{18}$)(BO$_3$)$_3$(OH)$_4$, Feruvite [Ca][(Fe^{2+},Mg)$_3$][MgAl$_5$][(OH)$_3$|F|(BO$_3$)$_3$|Si$_6$O$_{18}$], Hydroxy-uvite (IMA2000-030 was not approved, but suspended) CaMg$_3$(Al$_5$Mg)(Si$_6$O$_{18}$)(BO$_3$)$_3$(OH)$_3$(OH)

- Ferric Tourmaline subgroup - Buergerite subgroup

  - Buergerite [Na][(Fe^{3+})$_3$][Al$_6$][O$_3$|F|(BO$_3$)$_3$|Si$_6$O$_{18}$], Povondraite [Na][(Fe^{3+})$_3$][(Fe^{3+})$_4$Mg$_2$][(OH)$_3$|O|(BO$_3$)$_3$|Si$_6$O$_{18}$]

- Lithian Tourmaline subgroup - Elbaite subgroup

  - Olenite NaAl$_9$B$_3$Si$_6$O$_{27}$O$_3$OH, Oxy-Dravite Na(MgAl$_2$)(MgAl$_5$)Si$_6$O$_{18}$(BO$_3$)$_3$(OH)$_3$O, Elbaite Na(Al$_{1.5}$Li$_{1.5}$)Al$_6$(OH)$_3$(OH)(BO$_3$)$_3$Si$_6$O$_{18}$

- Sodic Tourmaline subgroup - Schorl subgroup

  - Dravite [Na][Mg$_3$][Al$_6$][(OH)$_3$|OH|(BO$_3$)$_3$|Si$_6$O$_{18}$], Fluor-Dravite NaMg$_3$Al$_6$(OH)$_3$F(BO$_3$)$_3$(Si$_6$O$_{18}$), Schorl [Na][(Fe^{2+})$_3$][Al$_6$][(OH)$_3$|OH|(BO$_3$)$_3$|Si$_6$O$_{18}$], Schorl-(F) [Na][(Fe^{2+})$_3$][Al$_6$][(OH)$_3$|F,OH|(BO$_3$)$_3$|Si$_6$O$_{18}$], Chromdravite [Na][Mg$_3$][(Cr^{3+},Fe^{3+})$_6$][(OH)$_3$|OH|(BO$_3$)$_3$|Si$_6$O$_{18}$], Vanadiumdravite [Na][Mg$_3$][(V^{3+})$_6$][(OH)$_3$|OH|(BO$_3$)$_3$|Si$_6$O$_{18}$]

- Eudialyte group

- Carbokentbrooksite (Na,[ ])$_{12}$(Na,Ce)$_3$Ca$_6$Mn$_3$Zr$_3$Nb(Si$_{25}$O$_{73}$)(OH)$_3$(CO$_3$)•H$_2$O

- Eudialyte Na$_4$(Ca,Ce)$_2$(Fe^{2+},Mn,Y)ZrSi$_8$O$_{22}$(OH,Cl)$_2$ (?)

- Feklichevite Na$_{11}$Ca$_9$(Fe^{3+},Fe^{2+})$_2$Zr$_3$Nb[Si$_{25}$O$_{73}$](OH,H$_2$O,Cl,O)$_5$

- Ferrokentbrooksite Na$_{15}$Ca$_6$(Fe,Mn)$_3$Zr$_3$NbSi$_{25}$O$_{73}$(O,OH,H$_2$O)$_3$(Cl,F,OH)$_2$

- Georgbarsanovite Na$_{12}$(Mn,Sr,REE)$_3$Ca$_6$(Fe^{2+})$_3$Zr$_3$NbSi$_{25}$O$_{76}$Cl$_2$•H$_2$O

- Golyshevite (Na,Ca)$_{10}$Ca$_9$(Fe^{3+},Fe^{2+})$_2$Zr$_3$NbSi$_{25}$O$_{72}$(CO$_3$)(OH)$_3$•H$_2$O

- Ikranite (Na,H$_3$O)$_{15}$(Ca,Mn,REE)$_6$(Fe^{3+})$_2$Zr$_3$([ ],Zr)([ ],Si)Si$_{24}$O$_{66}$(O,OH)$_6$Cl•2-3H$_2$O

- Johnsenite-(Ce) Na$_{12}$(Ce,REE,Sr)$_3$Ca$_6$Mn$_3$Zr$_3$W(Si$_{25}$O$_{73}$)(CO$_3$)(OH,Cl)$_2$

- Kentbrooksite (Na,REE)$_{15}$(Ca,REE)$_6$Mn^{2+}Zr$_3$NbSi$_{25}$O$_{74}$F$_2$•2H$_2$O

- Khomyakovite Na$_{12}$Sr$_3$Ca$_6$Fe$_3$Zr$_3$W(Si$_{25}$O$_{73}$)(O,OH,H$_2$O)$_3$(OH,Cl)$_2$

- Manganokhomyakovite Na$_{12}$Sr$_3$Ca$_6$Mn$_3$Zr$_3$W(Si$_{25}$O$_{73}$)(O,OH,H$_2$O)$_3$(OH,Cl)$_2$

- Mogovidite Na$_9$(Ca,Na)$_6$Ca$_6$Fe$_2$Zr$_3$[ ]Si$_{25}$O$_{72}$(CO$_3$)(OH)$_4$

- Oneillite $Na_{15}Ca_3Mn_3(Fe^{2+})_3Zr_3Nb(Si_{25}O_{73})(O,OH,H_2O)_3(OH,Cl)_2$
- Raslakite $Na_{15}Ca_3Fe_3(Na,Zr)_3Zr_3(Si,Nb)(Si_{25}O_{73})(OH,H_2O)_3(Cl,OH)$
- Rastsvetaevite $Na_{27}K_8Ca_{12}Fe_3Zr_6Si_{52}O_{144}(O,OH,H_2O)_6Cl_2$
- Taseqite $Na_{12}Sr_3Ca_6Fe_3Zr_3NbSi_{25}O_{73}(O,OH,H_2O)_3Cl_2$
- Zirsilite-(Ce) $(Na,[\ ])_{12}(Ce,Na)_3Ca_6Mn_3Zr_3Nb(Si_{25}O_{73})(OH)_3(CO_3)\cdot H_2O$
- Alluaivite $Na_{19}(Ca,Mn^{2+})_6(Ti,Nb)_3(Si_3O_9)_2(Si_{10}O_{28})_2Cl\cdot 2H_2O$
- Andrianovite $Na_{12}(K,Sr,Ce)_3Ca_6Mn_3Zr_3Nb(Si_{25}O_{73})(O,H_2O,OH)_5$
- Aqualite $(H_3O)_8(Na,K,Sr)_5Ca_6Zr_3Si_{26}O_{66}(OH)_9Cl$
- Dualite $Na_{30}(Ca,Na,Ce,Sr)_{12}(Na,Mn,Fe,Ti)_6Zr_3Ti_3MnSi_{51}O_{144}(OH,H_2O,Cl)_9$
- Labyrinthite $(Na,K,Sr)_{35}Ca_{12}Fe_3Zr_6TiSi_{51}O_{144}(O,OH,H_2O)_9Cl_3$, Mills et al. (2009)

Subclass '9.D': Inosilicates

Single Chain Inosilicates

- Astrophyllite group
  - Astrophyllite $K_2Na(Fe^{2+},Mn)_7Ti_2Si_8O_{26}(OH)_4$
  - Magnesioastrophyllite $K_2Na[Na(Fe^{2+},Fe^{3+},Mn)Mg_2]Ti_2Si_8O_{26}(OH)_4F$
  - Hydroastrophyllite $(H_3O,K)_2Ca(Fe^{3+},Mn)_{5-6}Ti_2Si_8O_{26}(OH)_4F$
  - Niobophyllite $K_2Na(Fe^{2+},Mn)_7(Nb,Ti)_2Si_8O_{26}(OH)_4(F,O)$
  - Zircophyllite $K_2(Na,Ca)(Mn,Fe^{2+})_7(Zr,Nb)_2Si_8O_{26}(OH)_4F$
  - Kupletskite $K_2Na(Mn,Fe^{2+})_7(Ti,Nb)_2Si_8O_{26}(OH)_4F$
  - Kupletskite-(Cs) $(Cs,K)_2Na(Mn,Fe^{2+},Li)_7(Ti,Nb)_2Si_8O_{26}(OH)_4F$
  - Niobokupletskite $K_2Na(Mn,Zn,Fe)_7(Nb,Zr,Ti)_2Si_8O_{26}(OH)_4(O,F)$, Mills et al. (2009)
- Sapphirine supergroup
  - Sapphirine group
    - Khmaralite $(Mg,Al,Fe)_{16}(Al,Si,Be)_{12}O_{40}$, Sapphirine $(Mg,Al)_8(Al,Si)_6O_{20}$
  - Aenigmatite group
    - Aenigmatite $(Na,Ca)_4(Fe^{2+},Ti,Mg)_{12}Si_{12}O_{40}$, Krinovite $NaMg_2CrSi_3O_{10}$, Wilkinsonite $Na_2(Fe^{2+})_4(Fe^{3+})_2Si_6O_{20}$
  - Rhoenite group
    - Dorrite $Ca_2Mg_2(Fe^{3+})_4(Al,Fe^{3+})_4Si_2O_{20}$, Hogtuvaite $(Ca,Na)2(Fe^{2+},Fe^{3+},Ti,Mg,Mn)_6(Si,Be,Al)_6O_{20}$, Makarochkinite $Ca_2(Fe^{2+})_4Fe^{3+}TiSi_4BeAlO_{20}$, Rhonite $Ca_2(Mg,Fe^{2+},$

Fe^{3+},Ti)$_6$(Si,Al)$_6$O$_{20}$, Serendibite Ca$_2$(Mg,Al)$_6$(Si,Al,B)$_6$O$_{20}$, Welshite Ca$_4$Mg$_9$Sb$_3$O$_4$[Si$_6$Be$_3$AlFe$_2$O$_{36}$]

- Surinamite (Mg,Fe^{2+})$_3$Al$_4$BeSi$_3$O$_{16}$, Mills et al. (2009)

Pyroxene Supergroup

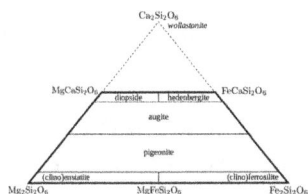

Pyroxene Quadrilateral

- Orthopyroxene group

  - Donpeacorite (Mn^{2+},Mg)Mg[SiO$_3$]$_2$, Enstatite MgSiO$_3$, Ferrosilite, FeSiO$_3$

- Clinopyroxene group

  - Aegirine NaFe^{3+}Si$_2$O$_6$, Augite (Ca,Na)(Mg,Fe^{2+},Al,Fe^{3+},Ti)[(Si,Al)$_2$O$_6$], Clinoenstatite MgSiO$_3$, Clinoferrosilite Fe^{2+}SiO$_3$, Diopside CaMg[Si$_2$O$_6$], Esseneite CaFe^{3+}[AlSiO$_6$], Grossmanite CaTi^{3+}AlSiO$_6$, Hedenbergite CaFe^{2+}[Si$_2$O$_6$], Jadeite Na(Al,Fe^{3+})[Si$_2$O$_6$], Jervisite (Na,Ca,Fe^{2+})(Sc,Mg,Fe^{2+})[Si$_2$O$_6$], Johannsenite CaMn^{2+}[Si$_2$O$_6$], Kanoite Mn^{2+}(Mg,Mn^{2+})[Si$_2$O$_6$], Kosmochlor NaCr[Si$_2$O$_6$], Kushiroite CaAl[Si$_2$O$_6$], Namansilite NaMn3+[Si$_2$O$_6$], Natalyite Na(V^{3+},Cr)[Si$_2$O$_6$], Petedunnite Ca(Zn,Mn^{2+},Mg,Fe^{2+})[Si$_2$O$_6$], Pigeonite (Mg,Fe^{2+},Ca)(Mg,Fe^{2+})Si$_2$O$_6$, Spodumene LiAlSi$_2$O$_6$

Multiple Chain Inosilicates

Note: the amphibole subcommittee (CNMNC/ IMA) published many reports (IMA 1978 s.p., IMA 1997 s.p., IMA 2003 s.p., IMA 2012 s.p.), renaming and redefining many minerals. Working draft: rruff.info, mindat.org and mineralienatlas.de are not up to date yet.

Amphibole Supergroup

- w(OH, F, Cl)-dominant amphibole: calcic subgroup

  - Cannilloite root name: fluoro-cannilloite CaCa$_2$Mg$_4$Al(Si$_5$Al$_3$)O$_{22}$(OH)$_2$ (1993-033, IMA 1997 s.p. Rd, IMA 2012 s.p. Rd Rn from cannilloite)

  - Edenite root name: edenite Na[Ca$_2$][Mg$_5$][(OH)$_2$|AlSi$_7$O$_{22}$] (1839, IMA 2012 s.p. Rd), ferro-edenite [Na][Ca$_2$][(Fe^{2+})$_5$][(OH)$_2$|AlSi$_7$O$_{22}$] (1946, IMA 1997 s.p. Rd, IMA 2012 s.p. Rd), fluoro-edenite Na[Ca$_2$][Mg$_5$][(F,OH)$_2$|AlSi$_7$O$_{22}$] (IMA 1994-059, IMA 2012 s.p. Rd)

  - Hastingsite root name: hastingsite [Na][Ca$_2$][(Fe^{2+})$_4$Fe^{3+}][(OH)$_2$|Al$_2$Si$_6$O$_{22}$] (1896, IMA 2012 s.p. Rd), magnesio-fluoro-hastingsite (Na,K)Ca$_2$(Mg,Fe^{3+},Ti)$_5$(Si,Al)$_8$O$_{22}$F$_2$ (IMA 2005-002, IMA 2012 s.p. Rd Rn from fluoro-magnesiohastingsite), magnesio-hastingsite Na[Ca$_2$][Mg$_4$Fe^{3+}][(OH)$_2$|Al$_2$Si$_6$O$_{22}$] (1928, IMA 1997 s.p. Rd, IMA 2012 s.p. Rd

Rn from magnesiohastingsite), potassic-fluoro-hastingsite $KCa2((Fe^{2+})_2,Mg_2,Fe^{3+})S_5(Si_6Al_2)_8O_{22}F_2$ (IMA 2005-006, IMA 2012 s.p. Rd Rn from fluoro-potassichastingsite), potassic-chloro-hastingsite $KCa_2(Mg_4Al)(Si_6Al_2)O_{22}Cl_2$ (IMA 2005-007, IMA 2012 s.p. Rd Rn from chloro-potassicpargasite, syn. dashkesanite), potassic-magnesio-hastingsite $(K,Na)Ca_2(Mg,Fe^{2+},Fe^{3+},Al)_5(Si,Al)_8O_{22}(OH,Cl)_2$ (IMA 2004-027b, IMA 2012 s.p. Rd Rn from potassic-magnesiohastingsite)

- Joesmithite $PbCa_2(Mg,Fe^{2+},Fe^{3+})_5Si_6Be_2O_{22}(OH)_2$ (1968, IMA 2012 s.p. Rd)

- Magnesio-hornblende root name: ferro-hornblende $[Ca_2][(Fe^{2+})_4Al][(OH)_2|AlSi_7O_{22}]$ (1930, IMA 1978 s.p., IMA 1997 s.p. Rn from ferro-hornblende, IMA 2012 s.p. Rd Rn from ferrohornblende), magnesio-hornblende $[Ca_2][Mg_4Al][(OH)_2|AlSi_7O_{22}]$ (1965, IMA 1997 s.p. Rd, IMA 2012 s.p. Rd Rn from magnesiohornblende)

- Pargasite root name: chromio-pargasite (IMA 2011-023, IMA 2012 s.p. Rd Rn from ehimeite), ferro-pargasite $[Na][Ca_2][(Fe^{2+})_4Al][(OH)_2|Al_2Si_6O_{22}]$ (1961, IMA 1997 s.p. Rd, IMA 2012 s.p. Rd Rn from ferropargasite), fluoro-pargasite $NaCa_2(Mg_3Fe^{2+}Al)_5(Si_6Al_2O_{22})F_2$ (IMA 2003-050, IMA 2012 s.p. Rd Rn from fluoropargasite), pargasite $[Na][Ca_2][Mg_4Al][(OH)_2|Al_2Si_6O_{22}]$ (1815, IMA 1997 s.p. Rd, IMA 2012 s.p. Rd), potassic-chloro-pargasite $KCa_2((Fe^{2+})_3MgFe^{3+})(Si_6Al_2)S_8O_{22}Cl_2$ (IMA 2001-036, IMA 2012 s.p. Rd Rn from chloro-potassichastingsite), potassic-ferro-pargasite $KCa_2((Fe^{2+})_4Al)Si_6Al_2O_{22}(OH)_2$ (IMA 2007-053, IMA 2012 s.p. Rd Rn from potassic-ferropargasite), potassic-fluoro-pargasite (IMA 2009-091, IMA 2012 s.p. Rd Rn from fluoro-potassic-pargasite), potassic-pargasite $[K][Ca_2][Mg_4Al][(OH)_2|Al_2Si_6O_{22}]$ (IMA 1994-046, IMA 2012 s.p. Rd Rn from potassicpargasite)

- Sadanagaite root name: potassic-ferro-ferri-sadanagaite $[K][Ca_2][(Fe^{2+})_3(Fe^{3+})_2][(OH)_2|Al_3Si_5O_{22}]$ (IMA 1997-035, IMA 2012 s.p. Rd Rn from potassic-ferrisadanagaite), potassic-ferro-sadanagaite $KCa_2Fe^{2+}_3(Al,Fe^{3+})_2(Si_5Al_3)O_{22}(OH)_2$ (IMA 1980-027, 2004 Rd, IMA 1997 s.p. Rd Rn from sadanagaite, IMA 2012 s.p. Rd Rn from potassicsadanagaite), potassic-sadanagaite $(K,Na)Ca_2(Mg,Fe^{2+},Al,Ti)_5[(Si,Al)_8O_{22}]$ (IMA 1982-102, 2004 Rd, IMA 2003 s.p. Rn from magnesio-sadanagaite, IMA 2012 s.p. Rd Rn from potassic-magnesiosadanagaite), sadanagaite $[Na][Ca_2][Mg_3Al_2][(OH)_2|Al_3Si_5O_{22}]$ (1984, IMA 1997 s.p. Rd Rn from sadanagaite, IMA 2002-051, 2004 Rn from potassic-magnesiosadanagaite, IMA 2012 s.p. Rd Rn from magnesiosadanagaite)

- Tremolite-actinolite root name: actinolite $Ca_2(Mg,Fe^{2+})_5(Si_8O_{22})(OH)_2$ (1794, IMA 2012 s.p. Rd), ferro-actinolite $[Ca_2][(Fe^{2+},Mg)_5][(OH)_2|Si_8O_{22}]$ (1946, IMA 1997 s.p. Rd, IMA 2012 s.p. Rd), tremolite $[Ca_2][Mg_5][(OH)_2|Si_8O_{22}]$ (1789, IMA 1997 s.p. Rd, IMA 2012 s.p. Rd)

- Tschermakite root name: tschermakite $[Ca_2][Mg_3Fe^{3+}Al][(OH)_2|Al_2Si_6O_{22}]$ (1945, IMA 1997 s.p. Rd, IMA 2012 s.p. Rd)

- w(OH, F, Cl)-dominant amphibole: lithium subgroup

  - Clino-holmquistite root name: clino-ferro-ferri-holmquistite $Li_2((Fe^{3+})_2(Fe^{2+})_3)Si_8O_{22}(OH)_2$ (IMA 1997 s.p., 2001-066, IMA 2012 s.p. Rd Rn from ferri-clinoferro-holmquistite)

- Holmquistite root name: ferro-holmquistite $(Li_2(Fe^{2+})_3Al_2)Si_8O_{22}(OH)_2$ (IMA 2004-030, IMA 2012 s.p. Rd), holmquistite $(Li_2Mg_3Al_2)Si_8O_{22}(OH)_2$ (1913, IMA 1997 s.p. Rd, IMA 2012 s.p. Rd)

- Pedrizite root name: ferri-pedrizite $NaLi_2((Fe^{3+})_2Mg_2Li)Si_8O_{22}(OH)_2$ (IMA 2001-032, IMA 2003 s.p. discredited, IMA 2012 s.p. Rd revalidated), ferro-ferri-pedrizite $NaLi_2((Fe^{3+})_2(Fe^{2+})_3)Si_8O_{22}(OH)_2$ (IMA 2003 s.p., IMA 2012 s.p. Rd Rn from sodic-ferro-ferripedrizite), ferro-fluoro-pedrizite $NaLi_2(Fe^{2+})_2Al_2Li)Si_8O_{22}F_2$ (IMA 2008-070, IMA 2012 s.p. Rd Rn from fluoro-sodic-ferropedrizite), fluoro-pedrizite $NaLi_2(Mg_2Al_2-Li)Si_8O_{22}F_2$ (IMA 2004-002, IMA 2012 s.p. Rd Rn from fluoro-sodic-pedrizite)

- w(OH, F, Cl)-dominant amphibole: Mg-Fe-Mn subgroup

  - Anthophyllite root name: anthophyllite $Mg_7Si_8O_{22}(OH)_2$ (1801, IMA 2012 s.p. Rd), ferro-anthophyllite $(Fe^{2+})_7Si_8O_{22}(OH)_2$ (1821, IMA 1997 s.p. Rd, IMA 2012 s.p. Rd), proto-anthophyllite $(Mg,Fe)_7Si_8O_{22}(OH)_2$ (IMA 2001-065, IMA 2012 s.p. Rd), proto-ferro-anthophyllite $(Fe^{2+},Mn^{2+})_2(Fe^{2+},Mg)_5(Si_4O_{11})_2(OH)_2$ (IMA 1986-006, IMA 2012 s.p. Rd), protomangano-ferro-anthophyllite $(Mn^{2+},Fe^{2+})_2(Fe^{2+},Mg)_5(Si_4O_{11})_2(OH)_2$ (IMA 1986-007, IMA 2012 s.p. Rd)

  - Gedrite root name: ferro-gedrite $(Fe^{2+})_5Al_2Si_6Al_2O_{22}(OH)_2$ (1939, IMA 1978 s.p. Rn, IMA 1997 s.p. Rd, IMA 2012 s.p. Rd Rn from ferrogedrite), gedrite $Mg_5Al_2Si_6Al_2O_{22}(OH)_2$ (1836, IMA 2012 s.p. Rd)

  - Minerals: cummingtonite $Mg_7Si_8O_{22}(OH)_2$ (1824, IMA 1997 s.p. Rd, IMA 2012 s.p. Rd), grunerite $(Fe^{2+})_7Si_8O_{22}(OH)_2$ (1853, IMA 1997 s.p. Rd, IMA 2012 s.p. Rd)

- w(OH, F, Cl)-dominant amphibole: sodic subgroup

  - Arfvedsonite root name: arfvedsonite $NaNa_2((Fe^{2+})_4Fe^{3+})Si_8O_{22}(OH)_2$ (1823, IMA 2012 s.p. Rd), magnesio-arfvedsonite $NaNa_2(Mg_4Fe^{2+})Si_8O_{22}(OH)_2$ (1957, IMA 2012 s.p. Rd), magnesio-fluoro-arfvedsonite $NaNa_2(Mg,Fe^{2+})_4Fe^{3+}[Si_8O_{22}](F,OH)_2$ (IMA 1998-056, IMA 2012 s.p. Rd Rn from fluoro-magnesio-arfvedsonite), potassic-arfvedsonite $KNa_2(Fe^{2+})_4Fe^{3+}Si_8O_{22}(OH)_2$ (IMA 2003-043, IMA 2012 s.p. Rd), potassic-magnesio-fluoro-arfvedsonite $KNa_2(Mg_4Fe^{3+})Si_8O_{22}F_2$ (IMA 1985-023, 2006 Rn from potassium fluor-magnesio-arfvedsonite, 2010 Rd, IMA 2012 s.p. Rd Rn from fluoro-potassic-magnesio-arfvedsonite)

  - Eckermannite root name: eckermannite $NaNa_2(Mg_4Al)Si_8O_{22}(OH)_2$ (1942, IMA 2012 s.p. Rd), mangano-ferri-eckermannite $NaNa_2(Mn^{2+})_4(Fe^{3+},Al)Si_8O_{22}(OH)_2$ (1968-028, IMA 1997 s.p., IMA 2012 s.p. Rd Rn from kôzulite)

  - Glaucophane root name: ferro-glaucophane $Na_2((Fe^{2+})_3Al_2)Si_8O_{22}(OH)_2$ (1957, IMA 1997 s.p. Rd, IMA 2012 s.p. Rd Rn from ferroglaucophane), glaucophane $Na_2(Mg_3Al_2)Si_8O_{22}(OH)_2$ (1963, IMA 1997 s.p. Rd, IMA 2012 s.p. Rd)

  - Leakeite root name: ferri-fluoro-leakeite $NaNa_2(Mg_2Fe^{3+}_2Li)Si_8O_{22}F_2$ (IMA 2009-085, IMA 2012 s.p. Rd Rn from fluoroleakeite), ferri-leakeite $Na(NaLi)(Mg_2(Fe^{3+})_2Li)Si_8O_{22}(OH)_2$ (IMA 2001-069, IMA 2012 s.p. Rd Rn from ferriwhittakerite), ferro-ferri-fluoro-leakeite $NaNa_2((Fe^{2+})_2(Fe^{3+})_2Li)Si_8O_{22}F_2$ (IMA 1993-026, IMA 2012 s.p. Rd Rn from fluoro-ferroleakeite), fluoro-leakeite $NaNa_2(Mg_2Al_2Li)Si_8O_{22}F_2$ (IMA 2009-012, IMA 2012 s.p. Rd Rn from fluoro-aluminoleakeite), potassic-ferri-leakeite $KNa_2Mg_2(Fe^{3+})_2LiSi_8O_{22}(OH)_2$ (IMA 2001-049, IMA 2012 s.p. Rd Rn from potassicleakeite), potassic-leakeite $KNa_2(Mg_2Al_2Li)Si_8O_{22}(OH)_2$ (2002, IMA 2012 s.p. Rd), potassic-mangani-leakeite $KNa_2(Mg_2Mn^{3+}_2Li)Si_8O_{22}(OH)_2$ (IMA 1992-032, IMA 2012 s.p. Rd Rn from kornite)

- Nybøite root name: ferro-ferri-nybøite $NaNa_2(Fe^{2+})_3(Fe^{3+})_2(Si_7Al)O_{22}(OH)_2$ (IMA 1997 s.p., IMA 2012 s.p. Rd Rn from ferric-ferronyboite), fluoro-nybøite $NaNa_2(Al_2Mg_3)(Si_7$-$Al)O_{22}(F,OH)_2$ (IMA 2002-010, IMA 2012 s.p. Rd Rn from fluoronybøite), nybøite $NaNa_2(Mg_3Al_2)Si_7AlO_{22}(OH)_2$ (1981, IMA 1997 s.p. Rd, IMA 2012 s.p. Rd)

- Riebeckite root name: fluoro-riebeckite $Na_2(Fe^{2+}_3Fe^{3+}_2)Si_8O_{22}F_2$ (1966, IMA 2012 s.p. Rd), magnesio-riebeckite $Na_2[(Mg,Fe^{2+})_3(Fe^{3+})_2]Si_8O_{22}(OH)_2$ (1957, IMA 1997 s.p. Rd, IMA 2012 s.p. Rd), riebeckite $Na_2((Fe^{2+})_3(Fe^{3+})_2)Si_8O_{22}(OH)_2$ (1888, IMA 1997 s.p. Rd, IMA 2012 s.p. Rd)

- w(OH, F, Cl)-dominant amphibole: sodic-calcic subgroup

  - Barroisite root name: barroisite $[CaNa][Mg_3AlFe^{3+}][(OH)_2|AlSi_7O_{22}]$ (1922, IMA 1997 s.p. Rd, IMA 2012 s.p. Rd)

  - Katophorite root name: ferri-fluoro-katophorite $Na(NaCa)(Mg_4Fe^{3+})(Si_7Al)O_{22}F_2$ (IMA 2012 s.p.), ferri-katophorite $Na_2Ca(Fe^{2+},Mg)_4Fe^{3+}(Si_7Al)O_{22}(OH)_2$ (IMA 1978 s.p., IMA 2012 s.p. Rd Rn from ferrikatophorite), ferro-katophorite $Na(NaCa)(Fe^{2+}_4Al)(Si_7Al)O_{22}(OH)_2$ (1894, IMA 2012 s.p. Rd Rn), katophorite $Na(NaCa)(Mg_4Al)(Si_7Al)O_{22}(OH)_2$ (1894, IMA 1997 s.p. Rd, IMA 2012 s.p. Rd)

  - Richterite root name: ferro-richterite $Na[CaNa][(Fe^{2+})_5][(OH)_2|Si_8O_{22}]$ (1946, IMA 1997 s.p. Rd, IMA 2012 s.p. Rd Rn from ferrorichterite), fluoro-richterite $Na(CaNa)Mg_5[Si_8O_{22}]F_2$ (IMA 1992-020, IMA 2012 s.p. Rd Rn from fluororichterite), potassic-fluoro-richterite $K[CaNa][Mg_5][(F,OH)_2|Si_8O_{22}]$ (IMA 1986-046, IMA 2004 s.p. Rn from potassium-fluorrichterite, IMA 2012 s.p. Rd Rn from fluoro-potassicrichterite), richterite $Na[CaNa][Mg_5][(OH)_2|Si_8O_{22}]$ (1865, IMA 1997 s.p. Rd, IMA 2012 s.p. Rd)

  - Taramite root name: ferro-taramite $Na[CaNa][(Fe^{2+})_3Al_2][(OH)_2|Al_2Si_6O_{22}]$ (IMA 2006-023, IMA 2012 s.p. Rd Rn from aluminotaramite), fluoro-taramite Na_2Ca-$Mg_3Al_2(Si_6Al_2)O_{22}F_2$ (IMA 2006-025, IMA 2012 s.p. Rd Rn from fluoro-alumino-magnesiotaramite), potassic-ferro-ferri-taramite $K(CaNa)(Fe^{2+}_3Fe^{3+}_2)(Si_6Al_2)O_{22}(OH)_2$ (IMA 1964-003, IMA 1978 s.p. Rn from mboziite, IMA 1997 s.p. Rn from ferri-taramite, IMA 2012 s.p. Rd Rn from ferritaramite, Erratum 2013 Rd Rn from ferro-ferri-taramite), potassic-ferro-taramite $K(CaNa)(Fe^{2+}_3Al_2)(Si_6Al_2)O_{22}(OH)_2$ (IMA 2007-015, IMA 2012 s.p. Rd Rn from potassic-aluminotaramite), taramite $Na_2CaMg_3Al_2(Si_6Al_2)O_{22}(OH)_2$ (IMA 2006-024, IMA 2012 s.p. Rd Rn from alumino-magnesiotaramite)

  - Winchite root name: ferri-winchite $(Na)(Na,Ca)_2(Mg,Fe^{2+})_4Fe^{3+}[Si_8O_{22}](OH)_2$ (IMA 2004-034, IMA 2012 s.p. Rd Rn from ferriwinchite), ferro-winchite $[CaNa][(Fe^{2+})_4(Al,Fe^{3+})][(OH)_2|Si_8O_{22}]$ (IMA 1978 s.p., IMA 1997 s.p. Rd, IMA 2012 s.p. Rd

Rn from ferrowinchite), winchite [CaNa][Mg$_4$(Al,Fe^{3+})][(OH)$_2$|Si$_8$O$_{22}$] (1906, IMA 1997 s.p. Rd, IMA 2012 s.p. Rd)

- w(OH, F, Cl)-dominant amphibole: Na-Mg-Fe-Mn subgroup

  - Ghoseite root name: ferri-ghoseite Na(NaMn^{2+})(Mg$_4$Fe^{3+})Si$_8$O$_{22}$(OH)$_2$ (IMA 2003-066, IMA 2012 s.p. Rd Rn from parvowinchite)

- w(O)-dominant amphibole group

  - Kaersutite root name: ferri-kaersutite NaCa$_2$(Mg$_3$Fe^{3+}Ti)(Si$_6$Al$_2$)O$_{22}$O$_2$ (IMA 2011-035, IMA 2012 s.p. Rd Rn from ferrikaersutite), kaersutite NaCa$_2$(Mg$_4$Ti)(Si$_6$Al$_2$)O$_{23}$(OH) (IMA 1997 s.p., IMA 2012 s.p. Rd)

  - Minerals: mangani-dellaventuraite NaNa$_2$(Mg$_2$,Mn^{3+},Li,Ti)Si$_8$O$_{22}$O$_2$ (IMA 2003-061, IMA 2012 s.p. Rd Rn from dellaventuraite), mangano-mangani-ungarettiite NaNa$_2$((Mn^{2+})$_2$(Mn^{3+})$_3$)Si$_8$O$_{22}$O$_2$ (IMA 1994-004, IMA 2012 s.p. Rd Rn from ungarettiite)

Subclass '9.E': Phyllosilicates

- Category:Clay minerals group

- Category:Medicinal clay

- Category:Mica group

- Category:Serpentine group/ Kaolinite-Serpentine group

- Category:Smectite group/ Montmorillonite group

- Pyrophyllite-Talc group

- Chlorite group: Clinochlore, Nimite, Pennantite, Baileychlore, Cookeite, Donbassite, Gonyerite, Odinite, Sudoite, Orthochamosite

Subclass '9.F': Tectosilicates

Tectosilicates without zeolitic H$_2$O

- Category:Quartz varieties

- Feldspar family

- Feldspathoid family

  - Sodalite group

  - Helvine group

  - Category:Cancrinite group

Tectosilicates with Zeolitic H$_2$O

- Category:Zeolite group

- Alflarsenite $NaCa_2Be_3Si_4O_{13}(OH)•2H_2O$

- Zeolites with T_5O_{10} Units – The Fibrous Zeolites

  - Natrolite subgroup

    - Gonnardite $(Na,Ca)_2(Si,Al)_5O_{10}·3H_2O$, Mesolite $Na_2Ca_2Si_9Al_6O_{30}·8H_2O$, Natrolite $Na_2Al_2Si_3O_{10}·2H_2O$, Paranatrolite $Na_2Al_2Si_3O_{10}·3H_2O$, Scolecite $CaAl_2Si_3O_{10}·3H_2O$

  - Tetranatrolite? $Na_2[Al_2Si_3O_{10}]•2H_2O$, Thomsonite-Sr $(Sr,Ca)_2Na[Al_5Si_5O_{20}]•7H_2O$, Thomsonite-Ca $NaCa_2Al_5Si_5O_{20}•6H_2O$, Kalborsite $K6Al_4Si_6BO_{20}(OH)_4Cl$, Edingtonite $BaAl_2Si_3O_{10}•4H_2O$

- Chains of single connected 4-membered rings

  - Ammonioleucite $(NH_4,K)AlSi_2O_6$, Leucite $KAlSi_2O_6$, Analcime $NaAlSi_2O_6•H_2O$, Hsianghualite $Ca_3Li_2Be_3(SiO_4)_3F_2$, Lithosite $K_6Al_4Si_8O_{25}•H_2O$, Pollucite $(Cs,Na)_2Al_2Si_4O_{12}•H_2O$, Wairakite $CaAl_2Si_4O_{12}•2H_2O$, Laumontite $CaAl_2Si_4O_{12}•4H_2O$, Yugawaralite $CaAl_2Si_6O_{16}•4H_2O$, Roggianite $Ca_2[Be(OH)_2Al_2Si_4O_{13}]•2.5H_2O$, Goosecreekite $CaAl_2Si_6O_{16}•5H_2O$, Montesommaite $(K,Na)_9Al_9Si_{23}O_{64}•10H_2O$, Partheite $Ca_2Al_4Si_4O_{15}(OH)_2•4H_2O$

- Chains of doubly connected 4-membered rings

  - Amicite $K_2Na_2Al_4Si_4O_{16}•5H_2O$, Garronite $Na_2Ca_5Al_{12}Si_{20}O_{64}•27H_2O$, Gobbinsite $(Na_2,Ca)_2K_2Al_6Si_{10}O_{32}•12H_2O$, Gismondine $Ca_2Al_4Si_4O_{16}•9H_2O$, Harmotome $(Ba,Na,K)_{(1-2)}(Si,Al)_8O_{16}•6H_2O$, Phillipsite-Na $(Na,K,Ca)_{(1-2)}(Si,Al)_8O_{16}•6H_2O$, Phillipsite-Ca $(Ca,K,Na)_{(1-2)}(Si,Al)_8O_{16}•6H_2O$, Phillipsite-K $(K,Na,Ca)_{(1-2)}(Si,Al)_8O_{16}•6H_2O$, Merlinoite $(K,Ca,Na,Ba)_7Si_{23}Al_9O_{64}•23H_2O$, Mazzite-Mg $K_2CaMg_2(Al,Si)_{36}O_{72}•28H_2O$, Mazzite-Na $Na_8Al_8Si_{28}O_{72}•30H_2O$, Perlialite $K_8Tl_4Al_{12}Si_{24}O_{72}•20H_2O$, Boggsite $NaCa_2(Al_5Si_{19}O_{48})•17H_2O$, Paulingite-Ca $(Ca,K,Na,Ba)_5[Al_{10}Si_{35}O_{84}]•34H_2O$, Paulingite-K $(K_2,Ca,Na_2,Ba)_5Al_{10}Si_{35}O_{90}•45H_2O$, Paulingite-Na $(Na_2,K_2,Ca,Ba)_5Al_{10}Si_{35}O_{90}•45H_2O$

- Chains of 6-membered rings – tabular zeolites

  - Gmelinite-Ca $(Ca,Na_2)Al_2Si_4O_{12}•6H_2O$, Gmelinite-K $(K,Na,Ca)_6(Al_7Si_{17}O_{48})•22H_2O$, Gmelinite-Na $(Na_2,Ca)Al_2Si_4O_{12}•6H_2O$, Chabazite-K $(K_2,Ca,Na_2,Mg)[Al_2Si_4O_{12}]•6H_2O$, Chabazite-Ca $(Ca_{0.5},Na,K)_4[Al_4Si_8O_{24}]•12H_2O$, Chabazite-Na $(Na_2,K_2,Ca,Mg)[Al_2Si_4O_{12}]•6H_2O$, Chabazite-Sr $(Sr,Ca,K_2,Na_2)[Al_2Si_4O_{12}]•6H_2O$, Herschelite? $(Na,Ca,K)AlSi_2O_6•3H_2O$, Willhendersonite $KCaAl_3Si_3O_{12}•5H_2O$, Levyne-Ca $(Ca,Na_2,K_2)Al_2Si_4O_{12}•6H_2O$, Levyne-Na $(Na_2,Ca,K_2)Al_2Si_4O_{12}•6H_2O$, Bellbergite $(K,Ba,Sr)_2Sr_2Ca_2(Ca,Na)_4Al_{18}Si_{18}O_{72}•30H_2O$, Erionite-Ca $(Ca,K_2,Na_2)_2[Al_4Si_{14}O_{36}]•15H_2O$, Erionite-K $(K_2,Ca,Na_2)_2[Al_4Si_{14}O_{36}]•15H_2O$, Erionite-Na $(Na_2,K_2,Ca)_2[Al_4Si_{14}O_{36}]•15H_2O$, Wenkite $Ba_4Ca_6(Si,Al)_{20}O_{39}(OH)_2(SO_4)_3•nH_2O$ (?), Offretite $(K_2,Ca,Mg)_{2.5}Al_5Si_{13}O_{36}•15H_2O$, Faujasite-Ca $(Ca,Na_2,Mg)_{3.5}[Al_7Si_{17}O_{48}]•32H_2O$, Faujasite-Mg $(Mg,Na_2,Ca)_{3.5}[Al_7Si_{17}O_{48}]•32H_2O$, Faujasite-Na $(Na_2,Ca,Mg)_{3.5}[Al_7Si_{17}O_{48}]•32H_2O$, Maricopaite $Pb_7Ca_2(Si,Al)_{48}O_{100}•32H_2O$, Mordenite $(Ca,Na_2,K_2)Al_2Si_{10}O_{24}•7H_2O$, Dachiardite-Ca $(Ca,Na_2,K_2)_4Al_{10}Si_{38}O_{96}•25H_2O$, Dachiardite-Na $(Na_2,Ca,K_2)_4Al_4Si_{20}O_{48}•13H_2O$, Epistilbite $CaAl_2Si_6O_{16}•5H_2O$, Ferrierite-K $(K,Na)_2Mg(Si,Al)_{18}O_{36}•9H_2O$, Ferrierite-Mg $(Mg,Na,K)_2Mg(Si,Al)_{18}O_{36}•9H_2O$, Ferrierite-Na $(Na,K)_2Mg(Si,Al)_{18}O_{36}•9H_2O$, Bikitaite $Li_2[Al_2Si_4O_{12}]•2H_2O$

- Chains of $T_{10}O_{20}$ Tetrahedra

- Clinoptilolite-Na $(Na,K,Ca)_{(2-3)}Al_3(Al,Si)_2Si_{13}O_{36} \cdot 12H_2O$, Clinoptilolite-K $(Na,K,Ca)_{(2-3)}$ $Al_3(Al,Si)_2Si_{13}O_{36} \cdot 12H_2O$, Clinoptilolite-Ca $(Ca,Na,K)_{(2-3)}Al_3(Al,Si)_2Si_{13}O_{36} \cdot 12H_2O$, Heulandite-Ba $(Ba,Ca,K,Na,Sr)_5Al_9Si_{27}O_{72} \cdot 22H_2O$, Heulandite-Ca $(Ca,Na)_{(2-3)}Al_3(Al,Si)_2Si_{13}O_{36} \cdot 12H_2O$, Heulandite-K $(K,Na,Ca)_{(2-3)}Al_3(Al,Si)_2Si_{13}O_{36} \cdot 12H_2O$, Heulandite-Na $(Na,Ca)_{(2-3)}Al_3(Al,Si)_2Si_{13}O_{36} \cdot 12H_2O$, Heulandite-Sr $(Sr,Na,Ca)_{(2-3)}Al_3(Al,Si)_2Si_{13}O_{36} \cdot 12H_2O$, Stilbite-Ca $NaCa_4[Al_8Si_{28}O_{72}] \cdot nH_2O$ (n=28-32), Stilbite-Na $Na_3Ca_3[Al_8Si_{28}O_{72}] \cdot nH_2O$ (n=28-32), Barrerite $(Na,K,Ca)_2Al_2Si_7O_{18} \cdot 6H_2O$, Stellerite $CaAl_2Si_7O_{18} \cdot 7H_2O$, Brewsterite-Ba $(Ba,Sr)Al_2Si_6O_{16} \cdot 5H_2O$, Brewsterite-Sr $(Sr,Ba)Al_4Si_{12}O_{32} \cdot 10H_2O$

- Other Rare Zeolites

  - Terranovaite $(Na,Ca)_8(Si_{68}Al_{12})O_{160} \cdot 29H_2O$, Gottardiite $Na_3Mg_3Ca_5Al_{19}Si_{117}O_{272} \cdot 93H_2O$, Lovdarite $K_2Na_6Be_4Si_{14}O_{36} \cdot 9H_2O$, Gaultite $Na_4Zn_2Si_7O_{18} \cdot 5H_2O$, Chiavennite $CaMnBe_2Si_5O_{13}(OH)_2 \cdot 2H_2O$, Tschernichite $(Ca,Na)(Si_6A_{12})O_{16} \cdot (4-8)H_2O$, Mutinaite $Na_3Ca_4Si_{85}Al_{11}O_{192} \cdot 60H_2O$, Tschortnerite $Ca_4(Ca,Sr,K,Ba)_3Cu_3(OH)_8[Si_{12}Al_{12}O_{48}] \cdot xH_2O$, x>=20, Thornasite $Na_{12}Th_3[Si_8O_{19}]_4 \cdot 18H_2O$, Direnzoite $NaK_6MgCa_2(Al_{13}Si_{47}O_{120}) \cdot 36H_2O$

- Unclassified zeolites

  - Cowlesite $CaAl_2Si_3O_{10} \cdot (5-6)H_2O$, Mountainite $(Ca,Na_2,K_2)_2Si_4O_{10} \cdot 3H_2O$

Native Element Minerals

Native copper

Native gold

Native silver

Native sulfur

Diamond

Native element minerals are those elements that occur in nature in uncombined form with a distinct mineral structure. The elemental class includes metals and intermetallic elements, naturally occurring alloys, semi-metals and non-metals. The Nickel–Strunz classification system also includes the naturally occurring phosphides, silicides, nitrides and carbides.

Elements

The following elements occur as native element minerals or alloys (list of minerals (complete)):

- Aluminium
- Antimony
- Arsenic
- Bismuth
- Carbon
- Cadmium
- Chromium

- Copper
- Gold
- Indium
- Iron
- Iridium
- Lead
- Mercury
- Nickel
- Osmium
- Palladium
- Platinum
- Rhenium
- Rhodium
- Selenium
- Silver
- Silicon
- Sulfur
- Tantalum
- Tellurium
- Tin
- Titanium
- Vanadium
- Zinc

Nickel–Strunz Classification -01- Native elements

This list uses the Classification of Nickel–Strunz (mindat.org, 10 ed, pending publication).

- Abbreviations:
    - "*" - discredited (IMA/CNMNC status).
    - "?" - questionable/doubtful (IMA/CNMNC status).
    - "REE" - Rare-earth element (Sc, Y, La, Ce, Pr, Nd, Pm, Sm, Eu, Gd, Tb, Dy, Ho, Er, Tm, Yb, Lu)

- "PGE" - Platinum-group element (Ru, Rh, Pd, Os, Ir, Pt)
- 03.C Aluminofluorides, 06 Borates, 08 Vanadates (04.H V$^{[5,6]}$ Vanadates), 09 Silicates:
  - Neso: insular
  - Soro: grouping
  - Cyclo: ring
  - Ino: chain
  - Phyllo: sheet
  - Tecto: three-dimensional framework
- Nickel–Strunz code scheme: NN.XY.##x
  - NN: Nickel–Strunz mineral class number
  - X: Nickel–Strunz mineral division letter
  - Y: Nickel–Strunz mineral family letter
  - ##x: Nickel–Strunz mineral/group number, x add-on letter

Class: Native Elements

- 01.A Metals and Intermetallic Alloys
  - 01.AA Copper-cupalite family: 05 Native copper, 05 Lead, 05 Native gold, 05 Native silver, 05 Nickel, 05 Aluminium; 10a Auricupride, 10b Tetra-auricupride; 15 Novodneprite, 15 Khatyrkite, 15 Anyuiite; 20 Cupalite, 25 Hunchunite
  - 01.AB Zinc-brass family (Cu-Zn alloys): 05 Cadmium, 05 Zinc, 05 Titanium*, 05 Rhenium*; 10a Brass*, 10a Zhanghengite, 10b Danbaite, 10b Tongxinite*
  - 01.AC Indium-tin family: 05 Indium, 10 Tin; 15 Yuanjiangite, 15 Sorosite
  - 01.AD Mercury-amalgam family: 00 Amalgam*, 05 Mercury; 10 Belendorffite, 10 Kolymite; 15a Paraschachnerite, 15a Schachnerite, 15b Luanheite, 15c Eugenite, 15d Moschellandsbergite; 20a Weishanite, 20b Goldamalgam*; 25 Potarite, 30 Leadamalgam
  - 01.AE Iron-chromium family: 05 Kamacite? (Iron var.), 05 Iron, 05 Chromium; 10 Antitaenite*, 10 Taenite, 10 Tetrataenite; 15 Chromferide, 15 Wairauite, 15 Ferchromide; 20 Awaruite, 25 Jedwabite
  - 01.AF Platinum-group elements: 05 Osmium, 05 Rutheniridosmine, 05 Ruthenium; 10 Palladium, 10 Iridium, 10 Rhodium, 10 Platinum
  - 01.AG PGE-metal alloys: 05 Garutiite, 05 Hexaferrum; 10 Atokite, 10 Zvyagintsevite, 10 Rustenburgite; 15 Taimyrite, 15 Tatyanaite; 20 Paolovite; 25 Plumbopalladinite, 25 Stannopalladinite; 30 Cabriite; 35 Chengdeite, 35 Isoferroplatinum; 40 Ferronickelplatinum, 40 Tetraferroplatinum, 40 Tulameenite; 45 Hongshiite*, 45 Skaergaardite; 50 Yixunite, 55 Damiaoite, 60 Niggliite, 65 Bortnikovite, 70 Nielsenite

- 01.B Metallic Carbides, Silicides, Nitrides and Phosphides

  - 01.BA Carbides: 05 Cohenite; 10 Isovite, 10 Haxonite; 15 Tongbaite; 20 Khamrabae-vite, 20 Niobocarbide, 20 Tantalcarbide; 25 Qusongite, 30 Yarlongite

  - 01.BB Silicides: Zangboite; 05 Mavlyanovite, 05 Suessite; 10 Perryite, 15 Fersilicite*, 20 Ferdisilicite*, 25 Luobusaite, 30 Gupeiite, 35 Hapkeite, 40 Xifengite

  - 01.BC Nitrides: 05 Roaldite, 10 Siderazot, 15 Carlsbergite, 15 Osbornite

  - 01.BD Phosphides: 05 Schreibersite, 05 Nickelphosphide; 10 Barringerite, 10 Monip-ite; 15 Allabogdanite, 15 Florenskyite, 15 Andreyivanovite; 20 Melliniite

- 01.C Metalloids and Nonmetals

  - 01.CA Arsenic group elements: 05 Bismuth, 05 Antimony, 05 Arsenic, 05 Stibarsen; 10 Arsenolamprite, 10 Pararsenolamprite; 15 Paradocrasite

  - 01.CB Carbon-silicon family: 05a Graphite, 05b Chaoite, 05c Fullerite; 10a Diamond, 10b Lonsdaleite, 15 Silicon

  - 01.CC Sulfur-selenium-iodine: 05 Sulfur, 05 Rosickyite; 10 Tellurium, 10 Selenium

- 01.D Nonmetallic Carbides and Nitrides

  - 01.DA Nonmetallic carbides: 05 Moissanite

  - 01.DB Nonmetallic nitrides: 05 Nierite, 10 Sinoite

- 01.X Unclassified Strunz Elements (Metals and intermetallic alloys; metalloids and non-metals; carbides, silicides, nitrides, phosphides)

  - 01.XX Unknown: 00 Hexamolybdenum, 00 Tantalum*, 00 Brownleeite

Oxide Minerals

The oxide mineral class includes those minerals in which the oxide anion (O^{2-}) is bonded to one or more metal ions. The hydroxide bearing minerals are typically included in the oxide class. The minerals with complex anion groups such as the silicates, sulfates, carbonates and phosphates are classed separately.

Oxide mineral exhibit at the Museum of Geology in South Dakota

Simple oxides:

- X_2O and XO
  - Cuprite Cu_2O
  - Ice H_2O
  - Periclase group
    - Periclase MgO
    - Manganosite MnO
  - Zincite group
    - Zincite ZnO
    - Bromellite BeO
    - Tenorite CuO
    - Litharge PbO
- X_2O_3
  - Hematite group
    - Corundum Al_2O_3
    - Hematite Fe_2O_3
    - Ilmenite $FeTiO_3$
- XO_2
  - Rutile group
    - Rutile TiO_2
    - Pyrolusite MnO_2
    - Cassiterite SnO_2
  - Baddeleyite ZrO_2
  - Uraninite UO_2
  - Thorianite ThO_2
- XY_2O_4
  - Spinel group
    - Spinel $MgAl_2O_4$
    - Gahnite $ZnAl_2O_4$
    - Magnetite Fe_3O_4 ($Fe^{2+}Fe^{3+}_2O_4$)
    - Franklinite $(Zn,Fe,Mn)(Fe,Mn)_2O_4$

- Chromite $FeCr_2O_4$
- Chrysoberyl $BeAl_2O_4$
- Columbite$(Fe,Mn)(Nb,Ta)_2O_6$

Hydroxide subgroup:

- Brucite $Mg(OH)_2$
- Manganite $MnO(OH)$
- Romanechite $BaMn^{2+}Mn^{4+}_8O_{16}(OH)_4$
- Goethite group
  - Diaspore $\alpha AlO(OH)$
  - Goethite $\alpha FeO(OH)$

Nickel–Strunz Classification -04- Oxides

IMA-CNMNC proposes a new hierarchical scheme (Mills et al., 2009). This list uses it to modify the Classification of Nickel–Strunz (mindat.org, 10 ed, pending publication).

- Abbreviations:
  - "*" - discredited (IMA/CNMNC status).
  - "?" - questionable/doubtful (IMA/CNMNC status).
  - "REE" - Rare-earth element (Sc, Y, La, Ce, Pr, Nd, Pm, Sm, Eu, Gd, Tb, Dy, Ho, Er, Tm, Yb, Lu)
  - "PGE" - Platinum-group element (Ru, Rh, Pd, Os, Ir, Pt)
  - 03.C Aluminofluorides, 06 Borates, 08 Vanadates (04.H $V^{[5,6]}$ Vanadates), 09 Silicates:
    - Neso: insular
    - Soro: grouping
    - Cyclo: ring
    - Ino: chain
    - Phyllo: sheet
    - Tekto: three-dimensional framework
- Nickel–Strunz code scheme: NN.XY.##x
  - NN: Nickel–Strunz mineral class number
  - X: Nickel–Strunz mineral division letter
  - Y: Nickel–Strunz mineral family letter

- ##x: Nickel–Strunz mineral/group number, x add-on letter

Class: Oxides

- 04.A Metal:Oxygen = 2.1 and 1:1

  - 04.AA Cation:Anion (M:O) = 2:1 (and 1.8:1): 05 Ice, 10 Cuprite, 15 Paramelaconite

  - 04.AB M:O = 1:1 (and up to 1:1.25); with small to medium-sized cations only: 05 Crednerite, 10 Tenorite; 15 Delafossite, 15 Mcconnellite; 20 Bromellite, 20 Zincite; 25 Lime, 25 Bunsenite, 25 Monteponite, 25 Manganosite, 25 Periclase, 25 Wustite

  - 04.AC M:O = 1:1 (and up to 1:1.25); with large cations (± smaller ones): 05 Swedenborgite; 10 Brownmillerite, 10 Srebrodolskite; 15 Montroydite, 20 Litharge, 20 Romarchite, 25 Massicot

- 04.B Metal:Oxygen = 3:4 and similar

  - 04.BA With small and medium-sized cations: 05 Chrysoberyl, 10 Manganostibite

  - 04.BB With only medium-sized cations: 05 Filipstadite, 05 Donathite?, 05 Gahnite, 05 Galaxite, 05 Hercynite, 05 Spinel, 05 Cochromite, 05 Chromite, 05 Magnesiochromite, 05 Manganochromite, 05 Nichromite, 05 Zincochromite, 05 Magnetite, 05 Cuprospinel, 05 Franklinite, 05 Jacobsite, 05 Magnesioferrite, 05 Trevorite, 05 Brunogeierite, 05 Coulsonite, 05 Magnesiocoulsonite, 05 Qandilite, 05 Ulvospinel, 05 Vuorelainenite; 10 Hydrohetaerolite, 10 Hausmannite, 10 Iwakiite, 10 Hetaerolite; 15 Maghemite, 20 Tegengrenite, 25 Xieite

  - 04.BC With medium-sized and large cations: 05 Marokite, 10 Dmitryivanovite

  - 04.BD With only large cations: 05 Minium

04.C Metal:Oxygen = 2:3, 3:5, and Similar

  - 04.CB With medium-sized cations: 05 Tistarite, 05 Auroantimonate*, 05 Brizziite-VII, 05 Brizziite-III, 05 Corundum, 05 Eskolaite, 05 Hematite, 05 Karelianite, 05 Geikielite, 05 Ecandrewsite, 05 Ilmenite, 05 Pyrophanite, 05 Melanostibite, 05 Romanite*; 10 Bixbyite, 10 Avicennite; 15 Armalcolite, 15 Mongshanite*, 15 Pseudobrookite; 20 Magnesiohogbomite-6N6S, 20 Magnesiohogbomite-2N3S, 20 Magnesiohogbomite-2N2S, 20 Zincohogbomite-2N2S, 20 Ferrohogbomite-2N2S; 25 Pseudorutile, 25 Ilmenorutile; 30 Oxyvanite, 30 Berdesinskiite; 35 Olkhonskite, 35 Schreyerite; 40 Kamiokite, 40 Nolanite, 40 Rinmanite; 45 Stibioclaudetite, 45 Claudetite; 50 Arsenolite, 50 Senarmontite; 55 Valentinite, 60 Bismite, 65 Sphaerobismoite, 70 Sillenite, 75 Kyzylkumite

  - 04.CC With large and medium-sized cations: 05 Chrombismite, 10 Freudenbergite, 15 Grossite, 20 Mayenite, 25 Yafsoanite; 30 Barioperovskite, 30 Lakargiite, 30 Natroniobite, 30 Latrappite, 30 Lueshite, 30 Perovskite; 35 Macedonite, 35 Isolueshite, 35 Loparite-(Ce), 35 Tausonite; 40 Crichtonite, 40 Dessauite, 40 Davidite-(Ce), 40 Davidite-(La), 40 Mathiasite, 40 Lindsleyite, 40 Landauite, 40 Loveringite, 40 Loveringite, 40 Cleusonite, 40 Gramaccioliite-(Y); 45 Hawthorneite, 45 Magnetoplumbite, 45 Haggertyite, 45 Batiferrite, 45 Hibonite, 45 Nezilovite, 45 Yimengite, 45 Diaoyudaoite,

45 Lindqvistite, 45 Plumboferrite; 50 Jeppeite, 55 Zenzenite, 60 Mengxianminite*

04.D Metal:Oxygen = 1:2 and similar

- 04.DA With small cations

- (moved to -09- Subclass: tektosilicates)

- 04.DB With medium-sized cations; chains of edge-sharing octahedra: 05 Tripuhyite, 05 Tugarinovite, 05 Varlamoffite*, 05 Argutite, 05 Cassiterite, 05 Rutile, 05 Pyrolusite, 05 Plattnerite, 05 Squawcreekite?; 10 Bystromite, 10 Ordonezite, 10 Tapiolite-(Fe), 10 Tapiolite-(Mn), 10 Tapiolite*, 15a Paramontroseite, 15a Ramsdellite, 15b Akhtenskite, 15c Nsutite; 20 Scrutinyite; 25 Ixiolite, 25 Ishikawaite, 25 Srilankite, 25 Samarskite-(Y), 25 Samarskite-(Yb), 25 Yttrocolumbite-(Y); 30 Heftetjernite, 30 Wolframoixiolite*, 30 Krasnoselskite*, 30 Ferberite, 30 Hubnerite, 30 Sanmartinite, 30 Wolframite*; 35 Tantalite-(Mg), 35 Tantalite-(Fe), 35 Tantalite-(Mn), 35 Columbite-(Mg), 35 Columbite-(Fe), 35 Columbite-(Mn), 35 Qitianlingite; 40 Ferrowodginite, 40 Lithiotantite, 40 Lithiowodginite, 40 Tantalowodginite*, 40 Titanowodginite, 40 Wodginite, 40 Ferrotitanowodginite; 45 Tivanite, 50 Carmichaelite, 55 Alumotantite, 60 Biehlite

- 04.DC With medium-sized cations; sheets of edge-sharing octahedra: 05 Bahianite, 10 Simpsonite

- 04.DD With medium-sized cations; frameworks of edge-sharing octahedra: 05 Anatase, 10 Brookite

- 04.DE With medium-sized cations; with various polyhedra: 05 Downeyite, 10 Koragoite; 15 Koechlinite, 15 Russellite, 15 Tungstibite; 20 Tellurite, 25 Paratellurite; 30 Cervantite, 30 Bismutotantalite, 30 Bismutocolumbite, 30 Clinocervantite, 30 Stibiocolumbite, 30 Stibiotantalite; 35 IMA2007-058, 35 Baddeleyite

- 04.DF With large (± medium-sized) cations; dimers and trimers of edge-sharing octahedra: 05 Nioboaeschynite-(Y), 05 Aeschynite-(Ce), 05 Aeschynite-(Nd), 05 Aeschynite-(Y), 05 Nioboaeschynite-(Ce), 05 Nioboaeschynite-(Nd), 05 Tantalaeschynite-(Y), 05 Rynersonite, 05 Vigezzite, 10 Changbaiite, 15 Murataite

- 04.DG With large (± medium-sized) cations; chains of edge-sharing octahedra: 05 Euxenite-(Y), 05 Loranskite-(Y), 05 Polycrase-(Y), 05 Uranopolycrase, 05 Fersmite, 05 Kobeite-(Y), 05 Tanteuxenite-(Y), 05 Yttrocrasite-(Y); 10 Fergusonite-beta-(Nd), 10 Fergusonite-beta-(Y), 10 Fergusonite-beta-(Ce), 10 Yttrotantalite-(Y); 15 Foordite, 15 Thoreaulite; 20 Raspite

- 04.DH With large (± medium-sized) cations; sheets of edge-sharing octahedra:

    - IMA/CNMNC revised the Pyrochlore supergroup 2010 (04.DH.15 and 04.DH.20)

    - 05 Brannerite, 05 Orthobrannerite, 05 Thorutite; 10 Kassite, 10 Lucasite-(Ce)

    - Pyrochlore group: Fluorcalciopyrochlore, Fluorkenopyrochlore, Fluornatropyrochlore, Fluorstrontiopyrochlore, Hydropyrochlore, Hydroxycalciopyrochlore,

Kenoplumbopyrochlore, Oxycalciopyrochlore, Oxynatropyrochlore, Oxy-plumbopyrochlore, Oxyyttropyrochlore-(Y)

- Microlite group: Fluorcalciomicrolite, Fluornatromicrolite, Hydrokenomicrolite, Hydromicrolite, Hydroxykenomicrolite, Kenoplumbomicrolite, Oxycalciomicrolite, Oxystannomicrolite, Oxystibiomicrolite

- Romeite group: Cuproromeite, Fluorcalcioromeite, Fluornatroromeite, Hydroxycalcioromeite, Oxycalcioromeite, Oxyplumboromeite, Stibiconite

- Betafite group: Calciobetafite, Oxyuranobetafite

- Elsmoreite group: Hydrokenoelsmoreite

- 25 Rosiaite; 30 Zirconolite-3O, 30 Zirconolite-3T, 30 Zirconolite-2M, 30 Zirconolite; 35 Liandratite, 35 Petscheckite; 40 Ingersonite, 45 Pittongite

- Discredited minerals 04.DH.15: Bariomicrolite (of Hogarth 1977), Bariopyrochlore (of Hogarth 1977), Betafite (of Hogarth 1977), Bismutomicrolite (of Hogarth 1977), Ceriopyrochlore (of Hogarth 1977), Jixianite, Natrobistantite, Plumbomicrolite (of Hogarth 1977), Plumbobetafite (of Hogarth 1977), Stannomicrolite (of Hogarth 1977), Stibiobetafite (of Černý et al.), Yttrobetafite (of Hogarth 1977), Yttropyrochlore (of Hogarth 1977), Bismutopyrochlore (of Chukanov et al.) and Bismutostibiconite 04.DH.20

- 04.DJ With large (± medium-sized) cations; polyhedral frameworks: 05 Calciotantite, 05 Irtyshite, 05 Natrotantite

- 04.DK With large (± medium-sized) cations; tunnel structures: 05 Ankangite, 05 Coronadite, 05 Hollandite, 05 Manjiroite, 05 Mannardite, 05 Redledgeite, 05 Priderite, 05 Henrymeyerite, 05 Akaganeite, 10 Cryptomelane, 10 Romanechite, 10 Strontiomelane, 10 Todorokite

- 04.DL With large (± medium-sized) cations; fluorite-type structures: 05 Cerianite-(Ce), 05 Zirkelite, 05 Thorianite, 05 Uraninite; 10 Calzirtite, 10 Hiarneite, 10 Tazheranite

- 04.DM With large (± medium-sized) cations; unclassified: 05 Sosedkoite, 05 Rankamaite; 15 Cesplumtantite, 20 Eyselite, 25 Kuranakhite

- 04.E Metal:Oxygen = < 1:2

  - 04.E: IMA2008-040

  - 04.EA Oxides with metal : oxygen < 1:2 (M_2O_5, MO_3): 05 Tantite, 10 Krasnogorite*, 10 Molybdite

- 04.X Unclassified Strunz Oxides

  - 04.XX Unknown: 00 Allendeite, 00 Ashanite?, 00 Hongquiite*, 00 Psilomelane?, 00 Uhligite?, 00 Clinobirnessite*, 00 Kleberite*, 00 Chubutite*, 00 Struverite?, 00 IMA2000-016, 00 IMA2000-026

Class: Hydroxides

- 04.F Hydroxides (without V or U)

  - 04.FA Hydroxides with OH, without H_2O; corner-sharing tetrahedra: 05a Behoite, 05b Clinobehoite; 10 Sweetite, 10 Wulfingite, 10 Ashoverite

  - 04.FB Hydroxides with OH, without H_2O; insular octahedra: 05 Shakhovite; 10 Cualstibite, 10 Zincalstibite

  - 04.FC Hydroxides with OH, without H_2O; corner-sharing octahedra: 05 Dzhalindite, 05 Sohngeite, 05 Bernalite; 10 Burtite, 10 Mushistonite, 10 Natanite, 10 Vismirnovite, 10 Schoenfliesite, 10 Wickmanite; 15 Jeanbandyite, 15 Mopungite, 15 Stottite; 15 Tetrawickmanite; 20 Ferronigerite-6N6S, 20 Ferronigerite-2N1S, 20 Magnesionigerite-6N6S, 20 Magnesionigerite-2N1S; 25 Magnesiotaaffeite-6N3S, 25 Magnesiotaaffeite-2N2S, 25 Ferrotaaffeite-6N3S

  - 04.FD Hydroxides with OH, without H_2O; chains of edge-sharing octahedra: 05 Spertiniite; 10 Bracewellite, 10 Diaspore, 10 Guyanaite, 10 Groutite, 10 Goethite, 10 Montroseite, 10 Tsumgallite; 15 Manganitev; 20 Cerotungstite-(Ce), 20 Yttrotungstite-(Y), 20 Yttrotungstite-(Ce); 25 Frankhawthorneite; 30 Khinite, 30 Parakhinite

  - 04.FE Hydroxides with OH, without H_2O; sheets of edge-sharing octahedra: 05 Amakinite, 05 Brucite, 05 Portlandite, 05 Pyrochroite, 05 Theophrastite, 05 Fougerite; 10 Bayerite, 10 Doyleite, 10 Gibbsite, 10 Nordstrandite; 15 Boehmite, 15 Lepidocrocite; 20 Grimaldiite, 20 Heterogenite-2H, 20 Heterogenite-3R; 25 Feitknechtite, 25 Lithiophorite; 30 Quenselite, 35 Ferrihydrite; 40 Feroxyhyte, 40 Vernadite; 45 Quetzalcoatlite

  - 04.FF Hydroxides with OH, without H_2O; various polyhedra: 05 Hydroromarchite

  - 04.FG Hydroxides with OH, without H_2O; unclassified: 05 Janggunite, 10 Cesarolite, 15 Kimrobinsonite

  - 04.FH Hydroxides with $H_2O \pm (OH)$; insular octahedra: 05 Bottinoite, 05 Brandholzite

  - 04.FJ Hydroxides with $H_2O \pm (OH)$; corner-sharing octahedra: 05 Sidwillite, 05 Meymacite; 10 Tungstite; 15 Ilsemannite, 15 Hydrotungstite; 20 Parabariomicrolite

  - 04.FK Hydroxides with $H_2O \pm (OH)$; chains of edge-sharing octahedra: 05 Bamfordite

  - 04.FL Hydroxides with $H_2O \pm (OH)$; sheets of edge-sharing octahedra: 05 Meixnerite, 05 Jamborite, 05 Iowaite, 05 Woodallite, 05 Akdalaite, 05 Muskoxite; 10 Hydrocalumite, 15 Kuzelite; 20 Aurorite, 20 Chalcophanite, 20 Ernienickelite, 20 Jianshuiite; 25 Woodruffite, 30 Asbolane; 40 Takanelite, 40 Rancieite; 45 Birnessite, 55 Cianciulliite, 60 Jensenite, 65 Leisingite, 75 Cafetite, 80 Mourite, 85 Deloryite

  - 04.FM Hydroxides with $H_2O \pm (OH)$; Unclassified: 15 Franconite, 15 Hochelagaite, 15 Ternovite; 25 Belyankinite, 25 Gerasimovskite, 25 Manganbelyankinite; 30 Silhydrite, 35 Cuzticite, 40 Cyanophyllite

  - 04.FN: 05 Menezesite

- 04.G Uranyl Hydroxides

  - 04.GA Without additional cations: IMA2008-022; 05 Metaschoepite, 05 Paraschoepite, 05 Schoepite; 10 Ianthinite; 15 Metastudtite, 15 Studtite

  - 04.GB With additional cations (K, Ca, Ba, Pb, etc.); with mainly $UO_2(O,OH)_5$ pentagonal polyhedra: 05 Compreignacite, 05 Agrinierite, 05 Rameauite; 10 Billietite, 10 Becquerelite, 10 Protasite; 15 Richetite; 20 Calciouranoite, 20 Bauranoite, 20 Metacalciouranoite; 25 Fourmarierite, 30 Wolsendorfite, 35 Masuyite; 40 Metavandendriesscheite, 40 Vandendriesscheite; 45 Vandenbrandeite, 50 Sayrite, 55 Curite, 60 Iriginite, 65 Uranosphaerite, 70 Holfertite

  - 04.GC With additional cations; with $UO_2(O,OH)_6$ hexagonal polyhedra: 05 Clarkeite, 10 Umohoite, 15 Spriggite

- 04.H V Vanadates

  - (moved to -08- Class: vanadates)

- 04.I Ice group

- 04.X Unclassified Strunz Oxides (Hydroxides)

  - 04.XX Unknown: 00 Ungursaite*, 00 Scheteligite?

Halide Minerals

The halide mineral class include those minerals with a dominant halide anion (F^-, Cl^-, Br^- and I^-). Complex halide minerals may also have polyatomic anions in addition to or that include halides.

Halite

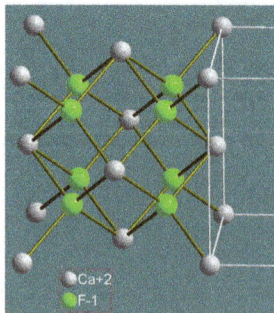

Fluorite structure

Examples include the following:

- Halite NaCl

- Sylvite KCl

- Chlorargyrite AgCl and bromargyrite AgBr

- Fluorite CaF_2

- Atacamite $Cu_2Cl(OH)_3$

- Bischofite $(MgCl_2 \cdot 6H_2O)$

- Carnallite $KMgCl_3 \cdot 6H_2O$

- Cryolite Na_3AlF_6

- Cryptohalite $(\alpha)(NH_4)_2SiF_6$

- Bararite $(\beta)(NH_4)_2SiF_6$

Commercially Significant Halide Minerals

Two commercially important halide minerals are halite and fluorite. The former is a major source of sodium chloride, in parallel with sodium chloride extracted from sea water or brine wells. Fluorite is a major source of hydrogen fluoride, complementing the supply obtained as a byproduct of the production of fertilizer. Carnallite and bischofite are important sources of magnesium. Natural cryolite was historically required for the production of aluminium, however, currently most cryolite used is produced synthetically.

Many of the halide minerals occur in marine evaporite deposits. The Atacama Desert also has large quantities of halide minerals as well as chlorates, iodates, oxyhalides and the like as well as nitrates, borates and other water-soluble minerals—not only underground but it crusts on the surface due to the low rainfall—the Atacama is the world's driest desert as well as one of the oldest (25 million years)

Nickel–Strunz Classification -03- Halides

IMA-CNMNC proposes a new hierarchical scheme (Mills et al., 2009). This list uses the Classification of Nickel–Strunz (mindat.org, 10 ed, pending publication).

- Abbreviations:

  - "*" - discredited (IMA/CNMNC status).

  - "?" - questionable/doubtful (IMA/CNMNC status).

  - "REE" - Rare-earth element (Sc, Y, La, Ce, Pr, Nd, Pm, Sm, Eu, Gd, Tb, Dy, Ho, Er, Tm, Yb, Lu)

  - "PGE" - Platinum-group element (Ru, Rh, Pd, Os, Ir, Pt)

  - 03.C Aluminofluorides, 06 Borates, 08 Vanadates (04.H V[5,6] Vanadates), 09 Silicates:

- Neso: insular

- Soro: grouping

- Cyclo: ring

- Ino: chain

- Phyllo: sheet

- Tekto: three-dimensional framework

- Nickel–Strunz code scheme: NN.XY.##x

  - NN: Nickel–Strunz mineral class number

  - X: Nickel–Strunz mineral division letter

  - Y: Nickel–Strunz mineral family letter

  - ##x: Nickel–Strunz mineral/group number, x add-on letter

Class: Halides

Halide specimens at Museum of Geology, South Dakota

- 03.A Simple halides, without H_2O

  - 03.AA M:X = 1:1, 2:3, 3:5, etc.: Panichiite; 05 Nantokite, 05 Marshite, 05 Miersite; 10 Iodargyrite, 10 Tocornalite; 15 Bromargyrite, 15 Embolite*, 15 Chlorargyrite; 20 Carobbiite, 20 Griceite, 20 Halite, 20 Sylvite, 20 Villiaumite; 25 Salammoniac, 25 Lafossaite; 30 Calomel, 30 Kuzminite, 30 Moschelite; 35 Neighborite; 40 Chlorocalcite, 45 Kolarite, 50 Radhakrishnaite; 55 Hephaistosite, 55 Challacolloite

  - 03.AB M:X = 1:2: 05 Tolbachite, 10 Coccinite, 15 Sellaite; 20 Chloromagnesite*, 20 Lawrencite, 20 Scacchite; 25 Frankdicksonite, 25 Fluorite; 30 Tveitite-(Y); 35 Gagarinite-(Y); 35 Zajacite-(Ce)

  - 03.AC M:X = 1:3: 05 Zharchikhite, 10 Molysite; 15 Fluocerite-(Ce), 15 Fluocerite-(La), 20 Gananite

- 03.B Simple Halides, with H_2O

  - 03.BA M:X = 1:1 and 2:3: 05 Hydrohalite, 10 Carnallite

- 03.BB M:X = 1:2: 05 Eriochalcite, 10 Rokuhnite, 15 Bischofite, 20 Nickelbischofite, 25 Sinjarite, 30 Antarcticite, 35 Tachyhydrite

- 03.BC M:X = 1:3: 05 Chloraluminite

- 03.BD Simple Halides with H_2O and additional OH: 05 Cadwaladerite, 10 Lesukite, 15 Korshunovskite, 20 Nepskoeite, 25 Koenenite

- 03.C Complex Halides

  - 03.C: Steropesite, IMA2008-032, IMA2008-039

  - 03.CA Borofluorides: 05 Ferruccite; 10 Avogadrite, 10 Barberiite

  - 03.CB Neso-aluminofluorides: 05 Cryolithionite; 15 Cryolite, 15 Elpasolite, 15 Simmonsite; 20 Colquiriite, 25 Weberite, 30 Karasugite, 35 Usovite; 40 Pachnolite, 40 Thomsenolite; 45 Carlhintzeite, 50 Yaroslavite

  - 03.CC Soro-aluminofluorides: 05 Gearksutite; 10 Acuminite, 10 Tikhonenkovite; 15 Artroeite; 20 Calcjarlite, 20 Jarlite, 20 Jorgensenite

  - 03.CD Ino-aluminofluorides: 05 Rosenbergite, 10 Prosopite

  - 03.CE Phyllo-aluminofluorides: 05 Chiolite

  - 03.CF Tekto-aluminofluorides: 05 Ralstonite, 10 Boldyrevite?, 15 Bogvadite

  - 03.CG Aluminofluorides with CO_3, SO_4, PO_4: 05 Stenonite; 10 Chukhrovite-(Nd), 10 Chukhrovite-(Ce), 10 Chukhrovite-(Y), 10 Meniaylovite; 15 Creedite, 20 Boggildite, 25 Thermessaite

  - 03.CH: 05 Malladrite, 10 Bararite; 15 Cryptohalite, 15 Hieratite; 20 Demartinite, 25 Knasibfite

  - 03.CJ With MX_6 complexes; M = Fe, Mn, Cu: 05 Chlormanganokalite, 05 Rinneite; 10 Erythrosiderite, 10 Kremersite; 15 Mitscherlichite, 20 Douglasite, 30 Zirklerite

- 03.D Oxyhalides, Hydroxyhalides and Related Double Halides

  - 03.DA With Cu, etc., without Pb: 05 Melanothallite; 10a Atacamite, 10a Kempite, 10a Hibbingite, 10b Botallackite, 10b Clinoatacamite, 10b Belloite, 10c Gillardite, 10c Kapellasite, 10c Haydeeite, 10c Paratacamite, 10c Herbertsmithite; 15 Claringbullite, 20 Simonkolleite; 25 Buttgenbachite, 25 Connellite; 30 Abhurite, 35 Ponomarevite; 40 Calumetite, 40 Anthonyite; 45 Khaidarkanite, 50 Bobkingite, 55 Avdoninite, 60 Droninoite

  - 03.DB With Pb, Cu, etc.: 05 Diaboleite, 10 Pseudoboleite, 15 Boleite, 20 Cumengite, 25 Bideauxite, 30 Chloroxiphite, 35 Hematophanite; 40 Asisite, 40 Parkinsonite; 45 Murdochite, 50 Yedlinite

  - 03.DC With Pb (As, Sb, Bi), without Cu: 05 Laurionite, 05 Paralaurionite; 10 Fiedlerite, 15 Penfieldite, 20 Laurelite; 25 Zhangpeishanite, 25 Matlockite, 25 Rorisite, 25 Daubreeite, 25 Bismoclite, 25 Zavaritskite; 30 Nadorite, 30 Perite; 35 Aravaipaite, 37

Calcioaravaipaite, 40 Thorikosite, 45 Mereheadite, 50 Blixite, 55 Pinalite, 60 Symesite; 65 Ecdemite, 65 Heliophyllite; 70 Mendipite, 75 Damaraite, 80 Onoratoite, 85 Cotunnite, 90 Pseudocotunnite, 95 Barstowite

- 03.DD With Hg: 05 Eglestonite, 05 Kadyrelite; 10 Poyarkovite, 15 Hanawaltite, 20 Terlinguaite, 25 Pinchite; 30 Mosesite, 30 Gianellaite; 35 Kleinite, 40 Tedhadleyite, 45 Vasilyevite, 50 Aurivilliusite, 55 Terlinguacreekite, 60 Kelyanite, 65 Comancheite

- 03.DE With Rare-Earth Elements: 05 Haleniusite-(La)

- 03.X Unclassified Strunz Halogenides

- 03.XX Unknown: 00 Hydrophilite?, 00 Hydromolysite?, 00 Yttrocerite*, 00 Lorettoite?, 00 IMA2009-014, 00 IMA2009-015

Sulfate Minerals

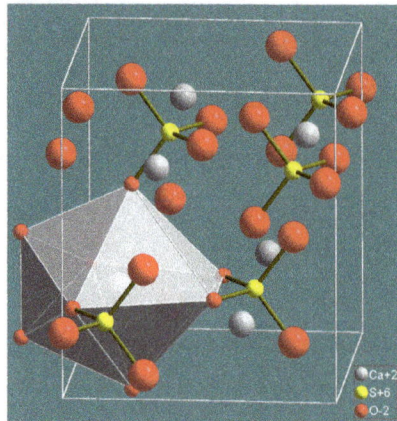

Anhydrite crystal structure

The sulfate minerals are a class of minerals which include the sulfate ion (SO_4^{2-}) within their structure. The sulfate minerals occur commonly in primary evaporite depositional environments, as gangue minerals in hydrothermal veins and as secondary minerals in the oxidizing zone of sulfide mineral deposits. The chromate and manganate minerals have a similar structure and are often included with the sulfates in mineral classification systems.

Barite with cerussite

Sulfate minerals include:

- Anhydrous sulfates

  - Barite $BaSO_4$

  - Celestite $SrSO_4$

  - Anglesite $PbSO_4$

  - Anhydrite $CaSO_4$

  - Hanksite $Na_{22}K(SO_4)_9(CO_3)_2Cl$

- Hydroxide and hydrous sulfates

  - Gypsum $CaSO_4 \cdot 2H_2O$

  - Chalcanthite $CuSO_4 \cdot 5H_2O$

  - Kieserite $MgSO_4 \cdot H_2O$

  - Starkeyite $MgSO_4 \cdot 4H_2O$

  - Hexahydrite $MgSO_4 \cdot 6H_2O$

  - Epsomite $MgSO_4 \cdot 7H_2O$

  - Meridianiite $MgSO_4 \cdot 11H_2O$

  - Melanterite $FeSO_4 \cdot 7H_2O$

  - Antlerite $Cu_3SO_4(OH)_4$

  - Brochantite $Cu_4SO_4(OH)_6$

  - Alunite $KAl_3(SO_4)_2(OH)_6$

  - Jarosite $KFe_3(SO_4)_2(OH)_6$

Nickel–Strunz Classification -07- Sulfates

Hanksite, one of the rare minerals that is a sulfate and carbonate

IMA-CNMNC proposes a new hierarchical scheme (Mills et al., 2009). This list uses it to modify the Classification of Nickel–Strunz (mindat.org, 10 ed, pending publication).

- Abbreviations:

    - "*" - discredited (IMA/CNMNC status).

    - "?" - questionable/doubtful (IMA/CNMNC status).

    - "REE" - Rare-earth element (Sc, Y, La, Ce, Pr, Nd, Pm, Sm, Eu, Gd, Tb, Dy, Ho, Er, Tm, Yb, Lu)

    - "PGE" - Platinum-group element (Ru, Rh, Pd, Os, Ir, Pt)

    - 03.C Aluminofluorides, 06 Borates, 08 Vanadates (04.H $V^{[5,6]}$ Vanadates), 09 Silicates:

        - Neso: insular

        - Soro: grouping

        - Cyclo: ring

        - Ino: chain

        - Phyllo: sheet

        - Tekto: three-dimensional framework

- Nickel–Strunz code scheme: NN.XY.##x

  - NN: Nickel–Strunz mineral class number

  - X: Nickel–Strunz mineral division letter

  - Y: Nickel–Strunz mineral family letter

  - ##x: Nickel–Strunz mineral/group number, x add-on letter

Class: Sulfates, Selenates, Tellurates

- 07.A Sulfates (selenates, etc.) without Additional Anions, without H_2O

    - 07.AB With medium-sized cations: 05 Millosevichite, 05 Mikasaite; 10 Chalcocyanite, 10 Zincosite*

    - 07.AC With medium-sized and large cations: IMA2008-029, 05 Vanthoffite; 10 Efremovite, 10 Manganolangbeinite, 10 Langbeinite; 15 Eldfellite, 15 Yavapaiite; 20 Godovikovite, 20 Sabieite; 25 Thenardite, 35 Aphthitalite

    - 07.AD With only large cations: 05 Arcanite, 05 Mascagnite; 10 Mercallite, 15 Misenite, 20 Letovicite, 25 Glauberite, 30 Anhydrite; 35 Anglesite, 35 Barite, 35 Celestine, 35 Radiobarite*, 35 Olsacherite; 40 Kalistrontite, 40 Palmierite

- 07.B Sulfates (selenates, etc.) with Additional Anions, without H_2O

    - 07.BB With medium-sized cations: 05 Caminite, 10 Hauckite, 15 Antlerite, 20 Dolerophanite, 25 Brochantite, 30 Vergasovaite, 35 Klebelsbergite, 40 Schuetteite, 45 Paraotwayite, 50 Xocomecatlite, 55 Pauflerite

- 07.BC With medium-sized and large cations: 05 Dansite; 10 Alunite, 10 Ammonioalunite, 10 Ammoniojarosite, 10 Beaverite, 10 Argentojarosite, 10 Huangite, 10 Dorallcharite, 10 Jarosite, 10 Hydroniumjarosite, 10 Minamiite, 10 Natrojarosite, 10 Natroalunite, 10 Osarizawaite, 10 Plumbojarosite, 10 Walthierite, 10 Schlossmacherite; 15 Yeelimite; 20 Atlasovite, 20 Nabokoite; 25 Chlorothionite; 30 Fedotovite, 30 Euchlorine; 35 Kamchatkite, 40 Piypite; 45 Klyuchevskite-Duplicate, 45 Klyuchevskite, 45 Alumoklyuchevskite; 50 Caledonite, 55 Wherryite, 60 Mammothite; 65 Munakataite, 65 Schmiederite, 65 Linarite; 70 Chenite, 75 Krivovichevite

- 07.BD With only large cations: 05 Sulphohalite; 10 Galeite, 10 Schairerite; 15 Kogarkoite; 20 Cesanite, 20 Caracolite; 25 Burkeite, 30 Hanksite, 35 Cannonite, 40 Lanarkite, 45 Grandreefite, 50 Itoite, 55 Chiluite, 60 Hectorfloresite, 65 Pseudograndreefite, 70 Sundiusite

- 07.C Sulfates (selenates, etc.) without Additional Anions, with H_2O

  - 07.CB With only medium-sized cations: 05 Gunningite, 05 Dwornikite, 05 Kieserite, 05 Szomolnokite, 05 Szmikite, 05 Poitevinite, 05 Cobaltkieserite; 07 Sanderite, 10 Bonattite, 15 Boyleite, 15 Aplowite, 15 Ilesite, 15 Rozenite, 15 Starkeyite, 15 IMA2002-034; 20 Chalcanthite, 20 Jokokuite, 20 Pentahydrite, 20 Siderotil; 25 Bianchite, 25 Ferrohexahydrite, 25 Chvaleticeite, 25 Hexahydrite, 25 Moorhouseite, 25 Nickelhexahydrite; 30 Retgersite; 35 Bieberite, 35 Boothite, 35 Mallardite, 35 Melanterite, 35 Zincmelanterite, 35 Alpersite; 40 Epsomite, 40 Goslarite, 40 Morenosite; 45 Alunogen, 45 Meta-alunogen; 50 Coquimbite, 50 Paracoquimbite; 55 Rhomboclase, 60 Kornelite, 65 Quenstedtite, 70 Lausenite; 75 Lishizhenite, 75 Romerite; 80 Ransomite; 85 Bilinite, 85 Apjohnite, 85 Dietrichite, 85 Halotrichite, 85 Pickeringite, 85 Redingtonite, 85 Wupatkiite; 90 Meridianiite, 95 Caichengyunite

  - 07.CC With medium-sized and large cations: 05 Krausite, 10 Tamarugite; 15 Mendozite, 15 Kalinite; 20 Lonecreekite, 20 Alum-(K), 20 Alum-(Na), 20 Lanmuchangite, 20 Tschermigite; 25 Pertlikite, 25 Monsmedite?, 25 Voltaite, 25 Zincovoltaite; 30 Krohnkite, 35 Ferrinatrite, 40 Goldichite, 45 Loweite; 50 Blodite, 50 Changoite, 50 Nickelblodite; 55 Mereiterite, 55 Leonite; 60 Boussingaultite, 60 Cyanochroite, 60 Mohrite, 60 Picromerite, 60 Nickelboussingaultite; 65 Polyhalite; 70 Leightonite, 75 Amarillite, 80 Konyaite, 85 Wattevilleite

  - 07.CD With only large cations: 05 Matteuccite, 10 Mirabilite, 15 Lecontite, 20 Hydroglauberite, 25 Eugsterite, 30 Gorgeyite; 35 Koktaite, 35 Syngenite; 40 Gypsum, 45 Bassanite, 50 Zircosulfate, 55 Schieffelinite, 60 Montanite, 65 Omongwaite

- 07.D Sulfates (selenates, etc.) with additional anions, with H_2O

  - 07.DB With only medium-sized cations; insular octahedra and finite groups: 05 Svyazhinite, 05 Aubertite, 05 Magnesioaubertite; 10 Rostite, 10 Khademite; 15 Jurbanite; 20 Minasragrite, 20 Anorthominasragrite, 20 Orthominasragrite; 25 Bobjonesite; 30 Amarantite, 30 Hohmannite, 30 Metahohmannite; 35 Aluminocopiapite, 35 Copiapite, 35 Calciocopiapite, 35 Cuprocopiapite, 35 Ferricopiapite, 35 Magnesiocopiapite, 35 Zincocopiapite

- 07.DC With only medium-sized cations; chains of corner-sharing octahedra: 05 Aluminite, 05 Meta-aluminite; 10 Butlerite, 10 Parabutlerite; 15 Fibroferrite, 20 Xitieshanite; 25 Botryogen, 25 Zincobotryogen; 30 Chaidamuite, 30 Guildite

- 07.DD With only medium-sized cations; sheets of edge-sharing octahedra: 05 Basaluminite?, 05 Felsobanyaite, 07.5 Kyrgyzstanite, 08.0 Zn-Schulenbergite; 10 Langite, 10 Posnjakite, 10 Wroewolfeite; 15 Spangolite, 20 Ktenasite, 25 Christelite; 30 Campigliaite, 30 Devilline, 30 Orthoserpierite, 30 Niedermayrite, 30 Serpierite; 35 Motukoreaite, 35 Mountkeithite, 35 Glaucocerinite, 35 Honessite, 35 Hydrowoodwardite, 35 Hydrohonessite, 35 Shigaite, 35 Natroglaucocerinite, 35 Wermlandite, 35 Nikischerite, 35 Zincaluminite, 35 Woodwardite, 35 Carrboydite, 35 Zincowoodwardite, 35 Zincowoodwardite-3R, 35 Zincowoodwardite-1T; 40 Lawsonbauerite, 40 Torreyite, 45 Mooreite, 50 Namuwite, 55 Bechererite, 60 Ramsbeckite, 65 Vonbezingite, 70 Redgillite; 75 Chalcoalumite, 75 Nickelalumite*; 80 Guarinoite, 80 Theresemagnanite, 80 Schulenbergite; 85 Montetrisaite

- 07.DE With only medium-sized cations; unclassified: 05 Mangazeite; 10 Carbonate-cyanotrichite, 10 Cyanotrichite; 15 Schwertmannite, 20 Tlalocite, 25 Utahite, 35 Coquandite, 40 Osakaite, 45 Wilcoxite, 50 Stanleyite, 55 Mcalpineite, 60 Hydrobasaluminite, 65 Zaherite, 70 Lautenthalite, 75 Camérolaite, 80 Brumadoite

- 07.DF With large and medium-sized cations: 05 Uklonskovite, 10 Kainite, 15 Natrochalcite; 20 Metasideronatrite, 20 Sideronatrite; 25 Despujolsite, 25 Fleischerite, 25 Schaurteite, 25 Mallestigite; 30 Slavikite, 35 Metavoltine; 40 Lannonite, 40 Vlodavetsite; 45 Peretaite, 50 Gordaite, 55 Clairite, 60 Arzrunite, 65 Elyite, 70 Yecoraite, 75 Riomarinaite, 80 Dukeite, 85 Xocolatlite

- 07.DG With large and medium-sized cations; with NO_3, CO_3, $B(OH)_4$, SiO_4 or IO_3: 05 Darapskite; 10 Clinoungemachite, 10 Ungemachite, 10 Humberstonite; 15 Bentorite, 15 Charlesite, 15 Ettringite, 15 Jouravskite, 15 Sturmanite, 15 Thaumasite, 15 Carraraite, 15 Buryatite; 20 Rapidcreekite, 25 Tatarskite, 30 Nakauriite, 35 Chessexite; 40 Carlosruizite, 40 Fuenzalidaite; 45 Chelyabinskite*

- 07.E Uranyl Sulfates

  - 07.EA Without cations: 05 Uranopilite, 05 Metauranopilite, 10 Jachymovite

  - 07.EB With medium-sized cations: 05 Johannite, 10 Deliensite

  - 07.EC With medium-sized and large cations: 05 Cobaltzippeite, 05 Magnesiozippeite, 05 Nickelzippeite, 05 Natrozippeite, 05 Zinc-zippeite, 05 Zippeite; 10 Rabejacite, 15 Marecottite, 20 Pseudojohannite

- 07.J Thiosulfates

  - 07.JA Thiosulfates with Pb: 05 Sidpietersite

- 07.X Unclassified Strunz Sulfates (Selenates, Tellurates)

  - 07.XX Unknown: 00 Aiolosite, 00 Steverustite, 00 Grandviewite, 00 IMA2009-008, 00 Adranosite, 00 Blakeite

Class: Chromates

- 07.F Chromates

    - 07.FA Without additional anions: 05 Tarapacaite, 10 Chromatite, 15 Hashemite, 20 Crocoite

    - 07.FB With additional O, V, S, Cl: 05 Phoenicochroite, 10 Santanaite, 15 Wattersite, 20 Deanesmithite, 25 Edoylerite

    - 07.FC With PO_4, AsO_4, SiO_4: 05 Vauquelinite; 10 Fornacite, 10 Molybdofornacite; 15 Hemihedrite, 15 Iranite; 20 Embreyite, 20 Cassedanneite;

    - 07.FD Dichromates: 05 Lopezite

Class: Molybdates, Wolframates And Niobates

- 07.G Molybdates, wolframates and niobates

    - 07.GA Without additional anions or H_2O: 05 Fergusonite-(Ce), 05 Fergusonite-(Nd) [N], 05 Fergusonite-(Y), 05 Powellite, 05 Wulfenite, 05 Stolzite, 05 Scheelite; 10 Formanite-(Y), 10 Iwashiroite-(Y); 15 Paraniite-(Y)

    - 07.GB With additional anions and/or H_2O: 05 Lindgrenite, 10 Szenicsite, 15 Cuprotungstite, 20 Phyllotungstite, 25 Rankachite, 30 Ferrimolybdite, 35 Anthoinite, 35 Mpororoite, 40 Obradovicite-KCu, 45 Mendozavilite-NaFe, 45 Paramendozavilite, 50 Tancaite-(Ce)

- 07.H Uranium and uranyl molybdates and wolframates

    - 07.HA With U^{4+}: 05 Sedovite, 10 Cousinite, 15 Moluranite

    - 07.HB With U^{6+}: 15 Calcurmolite, 20 Tengchongite, 25 Uranotungstite

Phosphate Minerals

Phosphate minerals are those minerals that contain the tetrahedrally coordinated phosphate (PO_4^{3-}) anion along with the freely substituting arsenate (AsO_4^{3-}) and vanadate (VO_4^{3-}). Chlorine (Cl^-), fluorine (F^-), and hydroxide (OH^-) anions also fit into the crystal structure.

The phosphate class of minerals is a large and diverse group, however, only a few species are relatively common.

Examples

Phosphate minerals include:

- triphylite $Li(Fe,Mn)PO_4$

- monazite $(Ce,La,Y,Th)PO_4$

- hinsdalite $PbAl_3(PO_4)(SO_4)(OH)_6$

- pyromorphite $Pb_5(PO_4)_3Cl$
- vanadinite $Pb_5(VO_4)_3Cl$
- erythrite $Co_3(AsO_4)_2 \cdot 8H_2O$
- amblygonite $LiAlPO_4F$
- lazulite $(Mg,Fe)Al_2(PO_4)_2(OH)_2$
- wavellite $Al_3(PO_4)_2(OH)_3 \cdot 5H_2O$
- turquoise $CuAl_6(PO_4)_4(OH)_8 \cdot 5H_2O$
- autunite $Ca(UO_2)_2(PO_4)_2 \cdot 10\text{-}12H_2O$
- carnotite $K_2(UO_2)_2(VO_4)_2 \cdot 3H_2O$
- phosphophyllite $Zn_2(Fe,Mn)(PO_4)_2 \bullet 4H_2O$
- struvite $(NH_4)MgPO_4 \cdot 6H_2O$
- Xenotime-Y $Y(PO_4)$
- Apatite group $Ca_5(PO_4)_3(F,Cl,OH)$
  - hydroxylapatite $Ca_5(PO_4)_3OH$
  - fluorapatite $Ca_5(PO_4)_3F$
  - chlorapatite $Ca_5(PO_4)_3Cl$
  - bromapatite
- Mitridatite group:
  - Arseniosiderite-mitridatite   series $(Ca_2(Fe^{3+})_3[(O)_2|(AsO_4)_3] \cdot 3H_2O$ --
    $Ca_2(Fe^{3+})_3[(O)_2|(PO_4)_3] \cdot 3H_2O)$
  - Arseniosiderite-robertsite   series $(Ca_2(Fe^{3+})_3[(O)_2|(AsO_4)_3] \cdot 3H_2O$ --
    $Ca_3(Mn^{3+})_4[(OH)_3|(PO_4)_2]_2 \cdot 3H_2O)$

Applications

Phosphate rock is a general term that refers to rock with high concentration of phosphate minerals, most commonly of the apatite group. It is the major resource mined to produce phosphate fertilizers for the agriculture sector. Phosphate is also used in animal feed supplements, food preservatives, anti-corrosion agents, cosmetics, fungicides, ceramics, water treatment and metallurgy.

The largest use of minerals mined for their phosphate content is the production of fertilizer.

Phosphate minerals are often used for control of rust and prevention of corrosion on ferrous materials applied with electrochemical conversion coatings.

Nickel–Strunz Classification -08- Phosphates

IMA-CNMNC proposes a new hierarchical scheme (Mills et al., 2009). This list uses it to modify the Classification of Nickel–Strunz (mindat.org, 10 ed, pending publication).

- Abbreviations:
  - "*" - discredited (IMA/CNMNC status).
  - "?" - questionable/doubtful (IMA/CNMNC status).
  - "REE" - Rare-earth element (Sc, Y, La, Ce, Pr, Nd, Pm, Sm, Eu, Gd, Tb, Dy, Ho, Er, Tm, Yb, Lu)
  - "PGE" - Platinum-group element (Ru, Rh, Pd, Os, Ir, Pt)
  - 03.C Aluminofluorides, 06 Borates, 08 Vanadates (04.H $V^{[5,6]}$ Vanadates), 09 Silicates:
    - Neso: insular
    - Soro: grouping
    - Cyclo: ring
    - Ino: chain
    - Phyllo: sheet
    - Tekto: three-dimensional framework
- Nickel–Strunz code scheme: NN.XY.##x
  - NN: Nickel–Strunz mineral class number
  - X: Nickel–Strunz mineral division letter
  - Y: Nickel–Strunz mineral family letter
  - ##x: Nickel–Strunz mineral/group number, x add-on letter

Class: Phosphates

- 08.A Phosphates, etc. without additional anions, without H_2O
  - 08.AA With small cations (some also with larger ones): 05 Berlinite, 05 Rodolicoite; 10 Beryllonite, 15 Hurlbutite, 20 Lithiophosphate, 25 Nalipoite, 30 Olympite
  - 08.AB With medium-sized cations: 05 Farringtonite; 10 Ferrisicklerite, 10 Heterosite, 10 Natrophilite, 10 Lithiophilite, 10 Purpurite, 10 Sicklerite, 10 Simferite, 10 Triphylite; 15 Chopinite, 15 Sarcopside; 20 Beusite, 20 Graftonite
  - 08.AC With medium-sized and large cations: 10 IMA2008-054, 10 Alluaudite, 10 Hagendorfite, 10 Ferroalluaudite, 10 Maghagendorfite, 10 Varulite, 10 Ferrohagendorfite*; 15 Bobfergusonite, 15 Ferrowyllieite, 15 Qingheiite, 15 Rosemaryite, 15 Wyllieite, 15 Ferrorosemaryite; 20 Maricite, 30 Brianite, 35 Vitusite-(Ce); 40 Olgite?, 40 Bario-olgite; 45 Ferromerrillite, 45 Bobdownsite, 45 Merrillite-(Ca)*, 45 Merrillite, 45

Merrillite-(Y)*, 45 Whitlockite, 45 Tuite, 45 Strontiowhitlockite; 50 Stornesite-(Y), 50 Xenophyllite, 50 Fillowite, 50 Chladniite, 50 Johnsomervilleite, 50 Galileiite; 55 Harrisonite, 60 Kosnarite, 65 Panethite, 70 Stanfieldite, 90 IMA2008-064

- 08.AD With only large cations: 05 Nahpoite, 10 Monetite, 15 Archerite, 15 Biphosphammite; 20 Phosphammite, 25 Buchwaldite; 35 Pretulite, 35 Xenotime-(Y), 35 Xenotime-(Yb); 45 Ximengite, 50 Monazite-(Ce), 50 Monazite-(La), 50 Monazite-(Nd), 50 Monazite-(Sm), 50 Brabantite?

- 08.B Phosphates, etc. with Additional Anions, without H_2O

  - 08.BA With small and medium-sized cations: 05 Vayrynenite; 10 Hydroxylherderite, 10 Herderite; 15 Babefphite

  - 08.BB With only medium-sized cations, (OH, etc.):RO_4 £1:1: 05 Amblygonite, 05 Natromontebrasite?, 05 Montebrasite?, 05 Tavorite; 10 Zwieselite, 10 Triplite, 10 Magniotriplite?, 10 Hydroxylwagnerite; 15 Joosteite, 15 Stanekite, 15 Triploidite, 15 Wolfeite, 15 Wagnerite; 20 Satterlyite, 20 Holtedahlite; 25 Althausite; 30 Libethenite, 30 Zincolibethenite; 35 Tarbuttite; 40 Barbosalite, 40 Hentschelite, 40 Scorzalite, 40 Lazulite; 45 Trolleite, 55 Phosphoellenbergerite; 90 Zinclipscombite, 90 Lipscombite, 90 Richellite

  - 08.BC With only medium-sized cations, (OH, etc.):RO_4 > 1:1 and < 2:1: 10 Plimerite, 10 Frondelite, 10 Rockbridgeite

  - 08.BD With only medium-sized cations, (OH, etc.):RO_4 = 2:1: 05 Pseudomalachite, 05 Reichenbachite, 10 Gatehouseite, 25 Ludjibaite

  - 08.BE With only medium-sized cations, (OH, etc.):RO_4 > 2:1: 05 Augelite, 10 Grattarolaite, 15 Cornetite, 30 Raadeite, 85 Waterhouseite

  - 08.BF With medium-sized and large cations, (OH, etc.):RO_4 < 0.5:1: 05 Arrojadite, 05 Arrojadite-(BaFe), 05 Arrojadite-(KFe), 05 Arrojadite-(NaFe), 05 Arrojadite-(SrFe), 05 Arrojadite-(KNa), 05 Arrojadite-(PbFe), 05 Arrojadite-(BaNa), 05 Fluorarrojadite-(BaNa), 05 Fluorarrojadite-(KNa), 05 Fluorarrojadite-(BaFe), 05 Ferri-arrojadite-(BaNa), 05 Dickinsonite, 05 Dickinsonite-(KNa), 05 Dickinsonite-(KMnNa), 05 Dickinsonite-(KNaNa), 05 Dickinsonite-(NaNa); 10 Samuelsonite, 15 Griphite, 20 Nabiasite

  - 08.BG With medium-sized and large cations, (OH, etc.):RO_4 = 0.5:1: 05 Bearthite, 05 Goedkenite, 05 Tsumebite; 10 Melonjosephite, 15 Tancoite

  - 08.BH With medium-sized and large cations, (OH,etc.):RO_4 = 1:1: 05 Thadeuite; 10 Lacroixite, 10 Isokite, 10 Panasqueiraite; 15 Drugmanite; 20 Bjarebyite, 20 Kulanite, 20 Penikisite, 20 Perloffite, 20 Johntomaite; 25 Bertossaite, 25 Palermoite; 55 Jagowerite, 60 Attakolite

  - 08.BK With medium-sized and large cations, (OH, etc.): 05 Brazilianite, 15 Curetonite, 25 Lulzacite

  - 08.BL With medium-sized and large cations, (OH, etc.):RO_4 = 3:1: 05 Corkite, 05 Hinsdalite, 05 Orpheite, 05 Woodhouseite, 05 Svanbergite; 10 Kintoreite, 10 Benauite, 10 Crandallite, 10 Goyazite, 10 Springcreekite, 10 Gorceixite; 10 Lusungite?, 10

Plumbogummite, 10 Ferrazite?; 13 Eylettersite, 13 Florencite-(Ce), 13 Florencite-(La), 13 Florencite-(Nd), 13 Waylandite, 13 Zairite; 15 Viitaniemiite, 20 Kuksite, 25 Pattersonite

- 08.BM With medium-sized and large cations, (OH, etc.):RO_4 = 4:1: 10 Paulkellerite, 15 Brendelite

- 08.BN With only large cations, (OH, etc.):RO_4 = 0.33:1: 05 IMA2008-068, 05 Phosphohedyphane, 05 IMA2008-009, 05 Alforsite, 05 Apatite*, 05 Apatite-(CaOH), 05 Apatite-(CaCl), 05 Apatite-(CaF), 05 Apatite-(SrOH), 05 Apatite-(CaOH)-M, Carbonate-fluorapatite?, 05 Carbonate-hydroxylapatite?, 05 Belovite-(Ce), 05 Belovite-(La), 05 Fluorcaphite, 05 Pyromorphite, 05 Hydroxylpyromorphite, 05 Deloneite-(Ce), 05 Kuannersuite-(Ce), 10 Arctite

- 08.BO With only large cations, (OH, etc.):RO_4 1:1: 05 Nacaphite, 10 Petitjeanite, 15 Smrkovecite, 25 Heneuite, 30 Nefedovite, 40 Artsmithite

- 08.C Phosphates without Additional Anions, with H_2O

  - 08.CA With small and large/medium cations: 05 Fransoletite, 05 Parafransoletite; 10 Ehrleite, 15 Faheyite; 20 Gainesite, 20 Mccrillisite, 20 Selwynite; 25 Pahasapaite, 30 Hopeite, 40 Phosphophyllite; 45 Parascholzite, 45 Scholzite; 65 Gengenbachite, 70 Parahopeite

  - 08.CB With only medium-sized cations, RO_4:H_2O = 1:1: 05 Serrabrancaite, 10 Hureaulite

  - 08.CC With only medium-sized cations, RO_4:H_2O = 1:1.5: 05 Garyansellite, 05 Kryzhanovskite, 05 Landesite, 05 Phosphoferrite, 05 Reddingite

  - 08.CD With only medium-sized cations, RO_4:H_2O = 1:2: 05 Kolbeckite, 05 Metavariscite, 05 Phosphosiderite; 10 Strengite, 10 Variscite; 20 Ludlamite

  - 08.CE With only medium-sized cations, RO_4:H_2O £1:2.5: 10 Newberyite, 20 Phosphorrosslerite; 25 Metaswitzerite, 25 Switzerite; 35 Bobierrite; 40 Arupite, 40 Baricite, 40 Vivianite, 40 Pakhomovskyite; 50 Cattiite, 55 Koninckite; 75 IMA2008-046, 75 Malhmoodite; 80 Santabarbaraite, 85 Metavivianite

  - 08.CF With large and medium-sized cations, RO_4:H_2O > 1:1: 05 Tassieite, 05 Wicksite, 05 Bederite; 10 Haigerachite

  - 08.CG With large and medium-sized cations, RO_4:H_2O = 1:1: 05 Collinsite, 05 Cassidyite, 05 Fairfieldite, 05 Messelite, 05 Hillite, (05 Uranophane-beta but Uranophane 09.AK.15); 20 Phosphogartrellite

  - 08.CH With large and medium-sized cations, RO_4:H_2O < 1:1: 10 Anapaite, 20 Dittmarite, 20 Niahite, 25 Francoanellite, 25 Taranakite, 30 Schertelite, 35 Hannayite, 40 Hazenite, 40 Struvite, 40 Struvite-(K), 45 Rimkorolgite, 50 Bakhchisaraitsevite, 55 IMA2008-048

  - 08.CJ With only large cations: 05 Stercorite, 10 Mundrabillaite, 10 Swaknoite, 15 Nastrophite, 15 Nabaphite, 45 Brockite, 45 Grayite, 45 Rhabdophane-(Ce), 45 Rhabdophane-(La), 45 Rhabdophane-(Nd), 45 Tristramite, 50 Brushite, 50 Churchite-(Dy)*, 50 Churchite-(Nd), 50 Churchite-(Y), 50 Ardealite, 60 Dorfmanite, 70 Catalanoite, 80 Ningyoite

- 08.D Phosphates

    - 08.DA With small (and occasionally larger) cations: 05 Moraesite, 10 Footemineite, 10 Ruifrancoite, 10 Guimaraesite, 10 Roscherite, 10 Zanazziite, 10 Atencioite, 10 Greifensteinite; 15 Uralolite, 20 Weinebeneite, 25 Tiptopite, 30 Veszelyite, 35 Kipushite, 40 Spencerite, 45 Glucine

    - 08.DB With only medium-sized cations, (OH, etc.):RO_4 < 1:1: 05 Diadochite, 10 Vashegyite, 15 Schoonerite, 20 Sinkankasite, 25 Mitryaevaite, 30 Sanjuanite, 50 Giniite, 55 Sasaite, 60 Mcauslanite, 65 Goldquarryite, 70 Birchite

    - 08.DC With only medium-sized cations, (OH, etc.):RO_4 = 1:1 and < 2:1: 05 Nissonite; 15 Kunatite, 15 Earlshannonite, 15 Whitmoreite; 17 Kleemanite, 20 Bermanite, 20?? Oxiberaunite*, 22 Kovdorskite; 25 Ferrostrunzite, 25 Ferristrunzite, 25 Metavauxite, 25 Strunzite; 27 Beraunite; 30 Gordonite, 30 Laueite, 30 Sigloite, 30 Paravauxite, 30 Ushkovite, 30 Ferrolaueite, 30 Mangangordonite, 30 Pseudolaueite, 30 Stewartite, 30 Kastningite, 35 Vauxite, 37 Vantasselite, 40 Cacoxenite; 45 Gormanite, 45 Souzalite; 47 Kingite; 50 Wavellite, 50 Allanpringite, 52 Kribergite, 60 Nevadaite

    - 08.DD With only medium-sized cations, (OH, etc.):RO_4 = 2:1: 15 Aheylite, 15 Chalcosiderite, 15 Faustite, 15 Planerite, 15 Turquoise; 20 Ernstite, 20 Childrenite, 20 Eosphorite

    - 08.DE With only medium-sized cations, (OH, etc.):RO_4 = 3:1: 05 Senegalite, 10 Fluellite, 20 Zapatalite, (35 Alumoakermanite, Mindat.org: 09.BB.10), 35 Aldermanite

    - 08.DF With only medium-sized cations, (OH,etc.):RO_4 > 3:1: 05 Hotsonite-VII, 05 Hotsonite-VI; 10 Bolivarite, 10 Evansite, 10 Rosieresite, 25 Sieleckiite, 40 Gladiusite

    - 08.DG With large and medium-sized cations, (OH, etc.):RO_4 < 0.5:1: 05 Sampleite

    - 08.DH With large and medium-sized cations, (OH, etc.):RO_4 < 1:1: 05 Minyulite; 10 Leucophosphite, 10 Spheniscidite, 10 Tinsleyite; 15 Kaluginite*, 15 Keckite, 15 Jahnsite-(CaMnFe), 15 Jahnsite-(CaMnMg), 15 Jahnsite-(CaMnMn), 15 Jahnsite-(MnMnMn)*, 15 Jahnsite-(CaFeFe), 15 Jahnsite-(NaFeMg), 15 Jahnsite-(CaMgMg), 15 Jahnsite-(NaMnMg), 15 Rittmannite, 15 Whiteite-(MnFeMg), 15 Whiteite-(CaFeMg), 15 Whiteite-(CaMnMg); 20 Manganosegelerite, 20 Overite, 20 Segelerite, 20 Wilhelmvierlingite, 20 Juonniite; 25 Calcioferrite, 25 Kingsmountite, 25 Montgomeryite, 25 Zodacite; 30 Lunokite, 30 Pararobertsite, 30 Robertsite, 30 Mitridatite; 35 Matveevite?, 35 Mantienneite, 35 Paulkerrite, 35 Benyacarite, 40 Xanthoxenite, 55 Englishite

    - 08.DJ With large and medium-sized cations, (OH, etc.):RO_4 = 1:l: 05 Johnwalkite, 05 Olmsteadite, 10 Gatumbaite, 20 Meurigite-Na, 20 Meurigite-K, 20 Phosphofibrite, 25 Jungite, 30 Wycheproofite, 35 Ercitite, 40 Mrazekite

    - 08.DK With large and medium-sized cations, (OH, etc.):RO_4 > 1:1 and < 2:1: 15 Matioliite, 15 IMA2008-056, 15 Dufrenite, 15 Burangaite, 15 Natrodufrenite; 20 Kidwellite, 25 Bleasdaleite, 30 Matulaite, 35 Krasnovite

    - 08.DL With large and medium-sized cations, (OH, etc.):RO_4 = 2:1: 05 Foggite; 10 Cyrilovite, 10 Millisite, 10 Wardite; 15 Petersite-(Y), 15 Calciopetersite; 25 Angastonite

- 08.DM With large and medium-sized cations, (OH, etc.):RO_4 > 2:1: 05 Morinite, 15 Melkovite, 25 Gutsevichite?, 35 Delvauxite

- 08.DN With only large cations: 05 Natrophosphate, 10 Isoclasite, 15 Lermontovite, 20 Vyacheslavite

- 08.DO With CO_3, SO_4, SiO_4: 05 Girvasite, 10 Voggite, 15 Peisleyite, 20 Perhamite, 25 Saryarkite-(Y), 30 Micheelsenite, 40 Parwanite, 45 Skorpionite

- 08.E Uranyl Phosphates

  - 08.EA UO_2:RO_4 = 1:2: 05 Phosphowalpurgite, 10 Parsonsite, 15 Ulrichite, 20 Lakebogaite

  - 08.EB UO_2:RO_4 = 1:1: 05 Autunite, 05 Uranocircite, 05 Torbernite, 05 Xiangjiangite, 05 Saleeite; 10 Bassetite, 10 Meta-autunite, 10 Metauranocircite, 10 Metatorbernite, 10 Lehnerite, 10 Przhevalskite; 15 Chernikovite, 15 Meta-ankoleite, 15 Uramphite; 20 Threadgoldite, 25 Uranospathite, 30 Vochtenite, 35 Coconinoite, 40 Ranunculite, 45 Triangulite, 50 Furongite, 55 Sabugalite

  - 08.EC UO_2:RO_4 = 3:2: 05 Francoisite-(Ce), 05 Francoisite-(Nd), 05 Phuralumite, 05 Upalite; 10 Kivuite?, 10 Yingjiangite, 10 Renardite, 10 Dewindtite, 10 Phosphuranylite; 15 Dumontite; 20 Metavanmeersscheite, 20 Vanmeersscheite; 25 Althupite, 30 Mundite, 35 Phurcalite, 40 Bergenite

  - 08.ED Unclassified: 05 Moreauite, 10 Sreinite, 15 Kamitugaite

- 08.F Polyphosphates

  - 08.FA Polyphosphates, without OH and H_2O; dimers of corner-sharing RO_4 tetrahedra: 20 Pyrocoproite*, 20 Pyrophosphite*

  - 08.FC Polyphosphates, with H_2O only: 10 Canaphite, 20 Arnhemite*, 25 Wooldridgeite, 30 Kanonerovite

- 08.X Unclassified Strunz Phosphates

  - 08.XX Unknown: 00 Sodium-autunite, 00 Pseudo-autunite*, 00 Cheralite-(Ce)?, 00 Laubmannite?, 00 Spodiosite?, 00 Sodium meta-autunite, 00 Kerstenite?, 00 Lewisite, 00 Coeruleolactite, 00 Viseite, 00 IMA2009-005

Organic Compounds (Minerals)

Some organic compounds are valid minerals, recognized by the CNMNC (IMA).

Nickel–Strunz classification –10- Organic compounds

- Abbreviations:

  - "*" – discredited (IMA/CNMNC status).

  - "?" – questionable/doubtful (IMA/CNMNC status).

- Nickel–Strunz code scheme: NN.XY.##x

  - NN: Nickel–Strunz mineral class number

  - X: Nickel–Strunz mineral division letter

  - Y: Nickel–Strunz mineral family letter

  - ##x: Nickel–Strunz mineral/group number, x add-on letter

Class: Organic Compounds

- 10.A Salts of organic acids

  - 10.AA Formates, Acetates, etc.: 05 formicaite, 10 dashkovaite, 20 acetamide, 25 calclacite, 30 paceite, 35 hoganite

  - 10.AB Oxalates: 05 humboldtine, 05 lindbergite; 10 glushinskite, 15 moolooite, 20 stepanovite, 25 minguzzite, 30 wheatleyite, 35 zhemchuzhnikovite, 40 weddellite, 45 whewellite, 50 caoxite, 55 oxammite, 60 natroxalate, 65 coskrenite-(Ce), 70 levinsonite-(Y), 75 zugshunstite-(Ce), 80 novgorodovaite

  - 10.AC Benzene Salts: 05 mellite, 10 earlandite, 15 pigotite?

  - 10.AD Cyanates: 05 julienite*, 10 kafehydrocyanite*

- 10.B Hydrocarbons

  - 10.BA Hydrocarbons: 05 fichtelite, 10 hartite, 15 dinite*, 20 idrialite, 25 kratochvilite, 30 karpatite, 35 phylloretine?, 40 ravatite, 45 simonellite, 50 evenkite

- 10.C Miscellaneous organic minerals

  - 10.C amber*

  - 10.CA Miscellaneous organic materials: 05 refikite, 10 flagstaffite, 15 hoelite, 20 abelsonite, 25 kladnoite; 30 tinnunculite*, 30 guanine; 35 urea, 40 uricite

Classification of Non-Silicate Minerals

This list gives an overview of the classification of non-silicate minerals and includes mostly IMA recognized minerals and its groupings. This list complements the alphabetical list on List of minerals (complete) and List of minerals. Rocks, ores, mineral mixtures, not IMA approved minerals, not named minerals are mostly excluded. Mostly major groups only, or groupings used by *New Dana Classification* and *Mindat*.

Classification of Minerals

Introduction

The grouping of the New Dana Classification and of the mindat.org is similar only, and so this classification is an overview only. Consistency is missing too on the group name endings (group, sub-group, series) between New Dana Classification and mindat.org. Category, class and supergroup name endings are used as layout tools in the list as well.

- Abbreviations:
  - "*" - Mineral not IMA Approved.
  - "?" - IMA Discredited Mineral Name.
  - "REE" - Rare earth element (Sc, Y, La, Ce, Pr, Nd, Pm, Sm, Eu, Gd, Tb, Dy, Ho, Er, Tm, Yb, Lu)
  - "PGE" - Platinum group element (Ru, Rh, Pd, Os, Ir, Pt)

Category 01

- Elements: Metals and Alloys, Carbides, Silicides, Nitrides, Phosphides
  - Category:Diamond
  - Category:Gold

Class: Native Elements

- Category:Carbide minerals
  - Osbornite group carbides and nitrides
    - Osbornite TiN, Khamrabaevite (Ti,V,Fe)C, Niobocarbide (Nb,Ta)C, Tantalcarbide TaC
- Category:Phosphide minerals
  - Barringerite group phosphides
    - Barringerite $(Fe,Ni)_2P$, Schreibersite $(Fe,Ni)_3P$, Nickelphosphide $(Ni,Fe)_3P$, Allabogdanite $(Fe,Ni)_2P$, Melliniite $(Ni,Fe)_4P$, Monipite MoNiP
- Copper group/ Gold group
  - Gold Au, Silver Ag, Copper Cu, Lead Pb, Aluminium Al, Maldonite Au_2Bi
- Silver Amalgam Alloys
  - Amalgam* Ag_2Hg_3, Moschellandsbergite Ag_2Hg_3, Schachnerite $Ag_{1.1}Hg_{0.9}$, Paraschachnerite Ag_3Hg_2, Luanheite Ag_3Hg, Eugenite Ag_9Hg_2, Weishanite $(Au,Ag)_3Hg_2$
- Iron-Nickel group
  - Iron Fe, Kamacite? alpha-(Fe,Ni), Taenite gamma-(Fe,Ni), Tetrataenite FeNi, Awaruite Ni_2Fe to Ni_3Fe, Nickel Ni, Wairauite CoFe
- Suessite group silicides
  - Suessite $(Fe,Ni)_3Si$, Gupeiite Fe_3Si, Xifengite Fe_5Si_3, Hapkeite Fe_2Si, Luobusaite $Fe_{0.}$

$_{83}Si_2$, Mavlyanovite Mn_5Si_3, Brownleeite MnSi

- Platinum group (Space group Fm3m)

  - Platinum Pt, Iridium (Ir,Os,Ru,Pt), Rhodium (Rh,Pt), Palladium Pd,Pt

- Osmium group (Space group P63/mmc)

  - Osmium (Os,Ir), Ruthenium (Ru,Ir,Os), Rutheniridosmine (Ir,Os,Ru), Hexaferrum (Fe,Os,Ru,Ir), Hexamolybdenum (Mo,Ru,Fe,Ir,Os), IMA2008-055 (Ni,Fe,Ir)

- Tetraferroplatinum group (Space group P4/mmm)

  - Tetraferroplatinum PtFe, Tulameenite Pt_2FeCu, Ferronickelplatinum Pt_2FeNi, Potarite PdHg

- Isoferroplatinum group (Space group Pm3m)

  - Isoferroplatinum $(Pt,Pd)_3(Fe,Cu)$, Rustenburgite $(Pt,Pd)_3Sn$, Atokite $(Pd,Pt)_3Sn$, Zvyagintsevite Pd_3Pb, Chengdeite Ir_3Fe, Yixunite Pt_3In

- Arsenic group

  - Arsenic As, Antimony Sb, Stibarsen SbAs, Bismuth Bi, Stistaite SnSb

- Carbon polymorph group (IMA-CNMNC discourages a grouping of diamond and graphite, Mills et al. (2009))

  - Graphite C, Chaoite C, Fullerite C60, (Diamond C, Lonsdaleite C)

Category 02

- Sulfides, Sulfosalts, Sulfarsenates, Sulfantimonates, Selenides, Tellurides

Class: Sulfide Minerals - Including Selenides and Tellurides

- Chalcocite-Digenite group ($[Cu]_{2-x}$ S] formulae)

  - Chalcocite Cu_2S, Djurleite $Cu_{31}S_{16}$, Digenite Cu_9S_5, Roxbyite $Cu_{1.78}S$, Anilite Cu_7S_4, Geerite Cu_8S_5, Spionkopite $Cu_{1.4}S$

- Joseite group (Trigonal: R-3m)

  - Joseite $Bi_4(S,Te)_3$, Joseite-B $Bi_4(S,Te)_3$, Ikunolite $Bi_4(S,Se)_3$, Laitakarite $Bi_4(Se,S)_3$, Pilsenite Bi_4Te_3, Poubaite $PbBi_2Se_2(Te,S)_2$, Rucklidgeite (Bi,Pb)3Te$_4$, Babkinite $Pb_2Bi_2(S,Se)_3$

- Pentlandite group (Isometric: Fm3m)

  - Pentlandite $(Fe,Ni)_9S_8$, Argentopentlandite $Ag(Fe,Ni)_8S_8$, Cobaltpentlandite Co_9S_8, Shadlunite $(Pb,Cd)(Fe,Cu)_8S_8$, Manganoshadlunite $(Mn,Pb)(Cu,Fe)_8S_8$, Geffroyite $(Ag,Cu,Fe)_9(Se,S)_8$

- Galena group (Isometric: Fm3m, IMA-CNMNC discourages the use of this grouping, Mills et al. (2009))

- Galena PbS, Clausthalite PbSe, Altaite PbTe, Alabandite MnS, Oldhamite (Calcium sulfide) $(Ca,Mg,Fe)S$, Niningerite $(Mg,Fe^{2+},Mn)S$, Borovskite Pd_3SbTe_4, Crerarite $(Pt,Pb)Bi_3(S,Se)_{4-x}$ (x~0.7), Keilite $(Fe,Mn,Mg,Ca,Cr)S$

- Sphalerite group (Isometric: F4-3m)

  - Sphalerite $(Zn,Fe)S$, Stilleite ZnSe, Metacinnabar HgS, Tiemannite HgSe, Coloradoite HgTe, Hawleyite CdS, Rudashevskyite $(Fe,Zn)S$

- Wurtzite group (Hexagonal: P63mc)

  - Wurtzite $(Zn,Fe)S$, Greenockite CdS, Cadmoselite CdSe, Rambergite MnS

- Nickeline group (Hexagonal: P63/mmc)

  - Nickeline NiAs, Breithauptite NiSb, Sederholmite NiSe, Hexatestibiopanickelite $(Ni,Pd)(Te,Sb)$, Sudburyite $(Pd,Ni)Sb$, Kotulskite $Pd(Te,Bi)$, Sobolevskite PdBi, Stumpflite $Pt(Sb,Bi)$, Langisite $(Co,Ni)As$, Freboldite CoSe, Achavalite FeSe, Sorosite $Cu(Sn,Sb)$, Vavrinite Ni_2SbTe_2

- Chalcopyrite group (Tetragonal: I-42d)

  - Chalcopyrite $CuFeS_2$, Eskebornite $CuFeSe_2$, Gallite $CuGaS_2$, Roquesite $CuInS_2$, Lenaite $AgFeS_2$, Laforetite $AgInS_2$

- Stannite group (Tetragonal: I-42m) A_2BCS type

  - Stannite Cu_2FeSnS_4, Cernyite Cu_2CdSnS_4, Briartite $Cu_2(Zn,Fe)GeS_4$, Kuramite Cu_3SnS_4, Sakuraiite $(Cu,Zn,Fe,In,Sn)_4S_4$, Hocartite Ag_2FeSnS_4, Pirquitasite Ag_2ZnSnS_4, Velikite Cu_2HgSnS_4, Kesterite $Cu_2(Zn,Fe)SnS_4$, Ferrokesterite $Cu_2(Fe,Zn)SnS_4$, Barquillite $Cu2CdGeS_4$

- Thiospinel group, AB_2X_4 (Isometric: Fd3m)

  - Bornhardtite $Co^{2+}(Co^{3+})_2Se_4$, Cadmoindite $CdIn_2S_4$, Carrollite $Cu(Co,Ni)_2S_4$, Cuproiridsite $CuIr_2S_4$, Cuprorhodsite $CuRh_2S_4$, Daubréelite $Fe^{2+}Cr_2S_4$, Ferrorhodsite $(Fe,Cu)(Rh,Ir,Pt)_2S_4$, Fletcherite (mineral) $Cu(Ni,Co)_2S_4$, Florensovite $Cu(Cr_{1.5}Sb_{0.5})S_4$, Greigite $Fe^{2+}(Fe^{3+})_2S_4$, Indite $Fe^{2+}In_2S_4$, Kalininite $ZnCr_2S_4$, Linnaeite $Co^{2+}(Co^{3+})_2S_4$, Malanite $Cu(Pt,Ir)_2S_4$, Polydymite $NiNi_2S_4$, Siegenite $(Ni,Co)_3S_4$, Violarite $Fe^{2+}(Ni^{3+})_2S_4$, Trustedtite Ni_3Se_4, Tyrrellite $(Cu,Co,Ni)_3Se_4$

- Tetradymite group (Trigonal: R-3m)

  - Tetradymite Bi_2Te_2S, Tellurobismuthite Bi_2Te_3, Tellurantimony Sb_2Te_3, Paraguanajuatite $Bi_2(Se,S)_3$, Kawazulite $Bi_2(Te,Se,S)_3$, Skippenite $Bi_2Se_2(Te,S)$, Vihorlatite $Bi_{24}Se_{17}Te_4$

- Pyrite group (Isometric: Pa3)

  - Pyrite FeS_2, Vaesite NiS_2, Cattierite CoS_2, Penroseite $(Ni,Co,Cu)Se_2$, Trogtalite $CoSe_2$, Villamaninite $(Cu,Ni,Co,Fe)S_2$, Fukuchilite Cu_3FeS_8, Krutaite $CuSe_2$, Hauerite MnS_2, Laurite RuS_2, Aurostibite $AuSb_2$, Krutovite $NiAs_2$, Sperrylite $PtAs_2$, Geversite $Pt(Sb,Bi)_2$, Insizwaite $Pt(Bi,Sb)_2$, Erlichmanite OsS_2, Dzharkenite $FeSe_2$, Gaotaiite Ir_3Te_8, Mayingite IrBiTe

- Marcasite group (Orthorhombic: Pnnm)

  - Marcasite FeS_2, Ferroselite $FeSe_2$, Frohbergite $FeTe_2$, Hastite? $CoSe_2$, Mattagamite $CoTe_2$, Kullerudite $NiSe_2$, Omeiite $(Os,Ru)As_2$, Anduoite $(Ru,Os)As_2$, Lollingite $FeAs_2$, Seinajokite $(Fe,Ni)(Sb,As)_2$, Safflorite $(Co,Fe)As_2$, Rammelsbergite $NiAs_2$, Nisbite $NiSb_2$

- Cobaltite group (Cubic or pseudocubic crystals)

  - Cobaltite $CoAsS$, Gersdorffite $NiAsS$, Ullmannite $NiSbS$, Willyamite $(Co,Ni)SbS$, Tolovkite $IrSbS$, Platarsite $(Pt,Rh,Ru)AsS$, Irarsite $(Ir,Ru,Rh,Pt)AsS$, Hollingworthite $(Rh,Pt,Pd)AsS$, Jolliffeite $(Ni,Co)AsSe$, Padmaite $PdBiSe$, Michenerite $(Pd,Pt)BiTe$, Maslovite $PtBiTe$, Testibiopalladite $PdTe(Sb,Te)$, Changchengite $IrBiS$, Milotaite $PdSbSe$, Kalungaite $PdAsSe$

- Arsenopyrite group (Monoclinic: P21/c (Pseudo-orthorhombic))

  - Arsenopyrite $FeAsS$, Gudmundite $FeSbS$, Osarsite $(Os,Ru)AsS$, Ruarsite $RuAsS$, Iridarsenite $(Ir,Ru)As_2$, Clinosafflorite $(Co,Fe,Ni)As_2$

- Molybdenite group

  - Drysdallite $Mo(Se,S)_2$, Molybdenite MoS_2, Tungstenite WS_2

- Skutterudite group

  - Ferroskutterudite $(Fe,Co)As_3$; Nickelskutterudite $NiAs_{2-3}$; Skutterudite $(Co,Fe,Ni)As_{2-3}$; Kieftite $CoSb_3$

Class: Sulfosalt Minerals

- Colusite group

  - Colusite $Cu_{12-13}V(As,Sb,Sn,Ge)_3S_{16}$, Germanocolusite $Cu_{13}V(Ge,As)_3S_{16}$, Nekrasovite $Cu^+_{26}V_2(Sn,As,Sb)_6S_{32}$, Stibiocolusite $Cu_{13}V(Sb,As,Sn)_3S_{16}$

- Cylindrite group

  - Cylindrite $Pb_3Sn_4FeSb_2S_{14}$, Franckeite $(Pb,Sn)_6Fe^{2+}Sn_2Sb_2S_{14}$, Incaite $Pb_4Sn_4FeSb_2S_{15}$, Potosiite $Pb_6Sn_2FeSb_2S_{14}$, Abramovite $Pb_2SnInBiS_7$, Coiraite $(Pb,Sn)_{12.5}As_3Sn_5FeS_{28}$

- Hauchecornite group (Tetragonal: P4/nnn or I4/mmm)

  - Hauchecornite $Ni_9Bi(Sb,Bi)S_8$, Bismutohauchecornite $Ni_9Bi_2S_8$, Tellurohauchecornite Ni_9BiTeS_8, Arsenohauchecornite $Ni_18Bi3AsS16$, Tucekite $Ni_9Sb_2S_8$

- Tetrahedrite group (Isometric: I-43m)

  - Tetrahedrite $(Cu,Fe)_{12}Sb_4S_{13}$, Tennantite $(Cu,Fe)_{12}As_4S_{13}$, Freibergite $(Ag,Cu,Fe)_{12}(Sb,As)_4S_{13}$, Hakite $(Cu,Hg)_3(Sb,As)(Se,S)_3$, Giraudite $(Cu,Zn,Ag)_{12}(As,Sb)_4(Se,S)_{13}$, Goldfieldite $Cu_{12}(Te,Sb,As)_4S_{13}$, Argentotennantite $(Ag,Cu)_{10}(Zn,Fe)_2(As,Sb)_4S_{13}$

- Proustite group

  - Proustite Ag_3AsS_3, Pyrargyrite Ag_3SbS_3

- Aikinite group (Orthorhombic containing Pb, Cu, Bi, and S)

  - Aikinite $PbCuBiS_3$, Krupkaite $PbCuBi_3S_6$, Gladite $PbCuBi_5S_9$, Hammarite $Pb_2Cu_2Bi_4S_9$ (?), Friedrichite $Pb_5Cu_5Bi_7S_{18}$, Pekoite $PbCuBi_{11}(S,Se)_{18}$, Lindstromite $Pb_3Cu_3Bi_7S_{15}$, Salzburgite $Cu_{1.6}Pb_{1.6}Bi_{6.4}S_{12}$

- Lillianite group (Orthorhombic, $AmBnS_6$ where A=Pb, Ag, Mn and B=Sb, Bi)

  - Lillianite $Pb_3Bi_2S_6$, Bursaite? $Pb_5Bi_4S_{11}$, Gustavite $PbAgBi_3S_6$ (?), Andorite $PbAgSb_3S_6$, Uchucchacuaite $AgPb_3MnSb_5S_{12}$, Ramdohrite $Ag_3Pb_6Sb_{11}S_{24}$, Roshchinite $Ag_{19}Pb_{10}Sb_{51}S_{96}$ or $Pb(Ag,Cu)_2(Sb,As)_5S_{10}$, Fizelyite $Pb_{14}Ag_5Sb_{21}S_{48}$

- Matildite group

  - Matildite $AgBiS_2$, Bohdanowiczite $AgBiSe_2$, Volynskite $AgBiTe_2$, Zlatogorite $CuNiSb_2$

- Sartorite group

  - Sartorite $PbAs_2S_4$, Guettardite $Pb(Sb,As)_2S_4$, Twinnite $Pb(Sb,As)_2S_4$, Marumoite $Pb_{32}As_{40}S_{92}$

- Pavonite group (Monoclinic: C/2c bismuth sulfosalts)

  - Pavonite $(Ag,Cu)(Bi,Pb)_3S_5$, Makovickyite $Ag_{1.5}Bi_{5.5}S_9$, Benjaminite $(Ag,Cu)_3(Bi,Pb)_7S_{12}$, Mummeite $Cu_{0.58}Ag_{3.11}Pb_{1.10}Bi_{6.65}S_{13}$, Borodaevite $Ag_5(Bi,Sb)_9S_{16}$, Cupropavonite $AgPbCu_2Bi_5S_{10}$, Cupromakovickyite $Cu_4AgPb_2Bi_9S_{18}$, Kudriavite $(Cd,Pb)Bi_2S_4$, IMA2008-058 $Ag_5Bi_{13}S_{22}$, IMA2005-036 $Cu_8Pb_4Ag_3Bi_{19}S_{38}$

Category 03

- Halogenides, Oxyhalides, Hydroxyhalides

  - Atacamite group

    - Polymorphs of $Cu_2[(OH)_3|Cl]$: Atacamite, Botallackite, Clinoatacamite, Paratacamite

    - Gillardite $Cu_3Ni(OH)_6Cl_2$, Haydeeite $Cu_3Mg(OH)_6Cl_2$, Herbertsmithite $Cu_3Zn[(OH)_3|Cl]_2$, Kapellasite $Cu_3Zn[(OH)_3|Cl]_2$

  - Fluorite group

    - Fluorite CaF_2, Fluorocronite PbF_2, Frankdicksonite BaF_2, Tveitite-(Y) $Ca_{1-x}Y_xF_{2+x}$ (x~0.3), IMA2009-014 SrF_2

  - Halite group (IMA-CNMNC discourages the use of this grouping, Mills et al. (2009))

    - Halite NaCl, Sylvite KCl, Villiaumite NaF, Carobbiite KF, Griceite LiF

  - Chlorargyrite group

    - Bromargyrite AgBr, Chlorargyrite AgCl, Marshite CuI, Miersite (Ag,Cu)I, Nantokite CuCl

  - Lawrencite group

- Chloromagnesite $MgCl_2$, Lawrencite $(Fe^{2+},Ni)Cl_2$, Scacchite $MnCl_2$, Tolbachite $CuCl_2$

- Matlockite group

  - Bismoclite $(BiO)Cl$, Daubréeite $(BiO)(OH,Cl)$, Laurionite $PbCl(OH)$, Paralaurionite $PbCl(OH)$, Rorisite $CaFCl$, Zavaritskite $(BiO)F$, Matlockite $PbFCl$

- Challacolloite group

  - Challacolloite KPb_2Cl_5, Hephaistosite $TlPb_2Cl_5$, Steropesite Tl_3BiCl_6, Panichiite $(NH_4)_2SnCl_6$

- Chukhrovite group

  - Chukhrovite-(Y) $Ca_3(Y,Ce)Al_2(SO_4)F_{13}\bullet10H_2O$, Chukhrovite-(Ce) $Ca_3(Ce,Y)Al_2(SO_4)F_{13}\bullet10H_2O$, Meniaylovite $Ca_4AlSi(SO_4)F_{13}\bullet12H_2O$, Chukhrovite-(Nd) $Ca_3(Nd,Y)Al_2(SO_4)F_{13}\bullet12H_2O$

Category 04

- Oxides and Hydroxides, Vanadates, Arsenites, Antimonites, Bismuthites, Sulfites, Iodates

  - Category:Vanadate minerals

  - Periclase group (Isometric: Fm3m, IMA-CNMNC discourages the use of this grouping, Mills et al. (2009))

    - Periclase MgO, Bunsenite NiO, Manganosite MnO, Monteponite CdO, Lime CaO, Wustite FeO, Hongquiite* TiO

  - Hematite group/ Corundum group (Rhombohedral: R-3c)

    - Corundum Al_2O_3 (Sapphire, Ruby), Eskolaite Cr_2O_3, Hematite Fe_2O_3, Karelianite V_2O_3, Tistarite Ti_2O_3

  - Perovskite group

    - Perovskite $CaTiO_3$, Latrappite $(Ca,Na)(Nb,Ti,Fe)O_3$, Loparite-(Ce) $(Ce,Na,Ca)_2(Ti,Nb)_2O_6$, Lueshite $NaNbO_3$, Tausonite $SrTiO_3$, Isolueshite $(Na,La,Ca)(Nb,Ti)O_3$, Barioperovskite $BaTiO_3$, Lakargiite $CaZrO_3$

  - Ilmenite group

    - Ilmenite $Fe^{2+}TiO_3$, Geikielite $MgTiO_3$, Pyrophanite $MnTiO_3$, Ecandrewsite $(Zn,Fe^{2+},Mn^{2+})TiO_3$, Melanostibite $Mn(Sb^{5+},Fe^{3+})O_3$, Brizziite-III $NaSb^{5+}O_3$, Akimotoite $(Mg,Fe)SiO_3$

  - Rutile group (Tetragonal: P4/mnm)

    - Rutile TiO_2, Ilmenorutile $(Ti,Nb,Fe^{3+})O_2$, Struverite? $(Ti,Ta,Fe^{3+})O_2$, Pyrolusite MnO_2, Cassiterite SnO_2, Plattnerite PbO_2, Argutite GeO_2, Squawcreekite? $(Fe^{3+},Sb^{5+},W^{6+})O_4\bullet H_2O$, Stishovite SiO_2

- Multiple Oxides with O_{19} groups/ Magnetoplumbite group

  - Hibonite $(Ca,Ce)(Al,Ti,Mg)_{12}O_{19}$, Yimengite $K(Cr,Ti,Fe,Mg)_{12}O_{19}$, Hawthorneite $Ba[Ti_3Cr_4Fe_4Mg]O_{19}$, Magnetoplumbite $Pb(Fe^{3+},Mn^{3+})_{12}O_{19}$, Haggertyite $Ba[(Fe^{2+})_6Ti_5Mg]O_{19}$, Nezilovite $PbZn_2(Mn^{4+},Ti^{4+})_2(Fe3+)_8O_{19}$, Batiferrite $Ba(Ti_2(Fe^{3+})_8(Fe^{2+})_2)O_{19}$, Barioferrite $Ba(Fe^{3+})_{12}O_{19}$, Plumboferrite $Pb_2(Fe^{3+})_{(11-x)}(Mn^{2+})_xO_{(19-2x)}$ x = 1/3, IMA2009-027 $(Fe,Mg)Al_{12}O_{19}$

- Cryptomelane group (Hard black, fine-grained)

  - Hollandite $Ba(Mn^{4+},Mn^{2+})_8O_{16}$, Cryptomelane $K(Mn^{4+},Mn^{2+})_8O_{16}$, Manjiroite $(Na,K)(Mn^{4+},Mn^{2+})_8O_{16} \cdot nH_2O$, Coronadite $Pb(Mn^{4+},Mn^{2+})_8O_{16}$, Strontiomelane $Sr(Mn^{4+})_6Mn3+_2O_{16}$, Henrymeyerite $BaFe^{2+}Ti_7O_{16}$

- Aeschynite group

  - Aeschynite-(Ce) $(Ce,Ca,Fe)(Ti,Nb)_2(O,OH)_6$, Nioboaeschynite-(Ce) $(Ce,Ca)(Nb,Ti)_2(O,OH)_6$, Aeschynite-(Y) $(Y,Ca,Fe)(Ti,Nb)_2(O,OH)_6$, Tantalaeschynite-(Y) $(Y,Ce,Ca)(Ta,Ti,Nb)_2O_6$, Aeschynite-(Nd) $(Nd,Ce)(Ti,Nb)_2(O,OH)_6$, Nioboaeschynite-(Nd) $(Nd,Ce)(Nb,Ti)_2(O,OH)_6$, Nioboaeschynite-(Y) $[(Y,REE),Ca,Th,Fe](Nb,Ti,Ta)_2(O,OH)_6$

- Crichtonite group $(ABC_{18}T_2O_{38})$

  - Landauite $NaMnZn_2(Ti,Fe^{3+})_6Ti_{12}O_{38}$, Loveringite $(Ca,Ce)(Ti,Fe^{3+},Cr,Mg)_{21}O_{38}$, Crichtonite $(Sr,La,Ce,Y)(Ti,Fe^{3+},Mn)_{21}O_{38}$, Senaite $Pb(Ti,Fe,Mn)_{21}O_{38}$, Davidite-(La) $(La,Ce,Ca)(Y,U)(Ti,Fe^{3+})_{20}O_{38}$, Davidite-(Ce) $(Ce,La)(Y,U)(Ti,Fe^{3+})_{20}O_{38}$, Mathiasite $(K,Ca,Sr)(Ti,Cr,Fe,Mg)_{21}O_{38}$, Lindsleyite $(Ba,Sr)(Ti,Cr,Fe,Mg)_{21}O_{38}$, Dessauite $(Sr,Pb)(Y,U)(Ti,Fe^{3+})_{20}O_{38}$, Cleusonite $Pb(U^{4+},U^{6+})(Ti,Fe^{2+},Fe^{3+})_{20}(O,OH)_{38}$, Gramaccioliite-(Y) $(Pb,Sr)(Y,Mn)Fe_2(Ti,Fe)_{18}O_{38}$

Spinel Group

- AB_2O_4

  - Aluminum subgroup

    - Spinel $MgAl_2O_4$, Galaxite $(Mn,Mg)(Al,Fe^{3+})_2O_4$, Hercynite $Fe^{2+}Al_2O_4$, Gahnite $ZnAl_2O_4$

  - Iron subgroup

    - Magnesioferrite $MgFe^{3+}_2O_4$, Jacobsite $(Mn^{2+},Fe^{2+},Mg)(Fe^{3+},Mn^{3+})_2O_4$, Magnetite $Fe^{2+}(Fe^{3+})_2O_4$, Franklinite $(Zn,Mn^{2+},Fe^{2+})(Fe^{3+},Mn^{3+})_2O_4$, Trevorite $Ni(Fe^{3+})_2O_4$, Cuprospinel $(Cu,Mg)(Fe^{3+})_2O_4$, Brunogeierite $(Ge^{2+},Fe^{2+})(Fe^{3+})_2O_4$

  - Chromium subgroup

    - Magnesiochromite $MgCr_2O_4$, Manganochromite $(Mn,Fe^{2+})(Cr,V)_2O_4$, Chromite $Fe^{2+}Cr_2O_4$, Nichromite $(Ni,Co,Fe^{2+})(Cr,Fe^{3+},Al)_2O_4$, Cochromite $(Co,Ni,Fe^{2+})(Cr,Al)_2O_4$, Zincochromite $ZnCr_2O_4$

  - Vanadium subgroup

- Vuorelainenite $(Mn^{2+},Fe^{2+})(V^{3+},Cr^{3+})_2O_4$, Coulsonite $Fe^{2+}(V^{3+})_2O_4$, Magnesiocoulsonite $Mg(V^{3+})_2O_4$

- Titanium subgroup

  - Qandilite $(Mg,Fe^{2+})_2(Ti,Fe^{3+},Al)O_4$, Ulvospinel $Ti(Fe^{2+})_2O_4$

- Taaffeite group

  - Magnesiotaaffeite-2N2S $Mg_3Al_8BeO_{16}$, Magnesiotaaffeite-6N3S $(Mg,Fe^{2+},Zn)_2Al_6BeO_{12}$, Ferrotaaffeite-6N3S $(Fe^{2+},Zn,Mg)_2Al_6BeO_{12}$

- Kusachiite $CuBi_2O_4$, Iwakiite $Mn^{2+}(Fe^{3+},Mn^{3+})_2O_4$, Hausmannite $Mn^{2+}(Mn^{3+})_2O_4$, Hetaerolite $Zn(Mn^{3+})_2O_4$, Hydrohetaerolite $Zn_2(Mn^{3+})_4O_8{\cdot}H_2O$, Minium $(Pb^{2+})_2Pb^{4+}O_4$, Chrysoberyl $BeAl_2O_4$, Marokite $Ca(Mn^{3+})_2O_4$, Filipstadite $(Mn^{2+},Mg)_4Sb^{5+}Fe^{3+}O_8$, Tegengrenite $(Mg,Mn^{2+})_2(Sb^{5+})_{0.5}(Mn^{3+},Si,Ti)_{0.5}O_4$, Yafsoanite $Ca_3Te_2Zn_3O_{12}$, Xieite $FeCr_2O_4$

Nickel-Strunz 04.DH Mineral Family

IMA/CMNMC revised the Pyrochlore supergroup 2010.

- Pyrochlore supergroup

  - Pyrochlore group (D atom is Nb)

    - Fluorcalciopyrochlore $(Ca,[\ ])_2Nb_2(O,OH)_6F$, Fluorkenopyrochlore $([\ ],Na,Ce,Ca)_2(Nb,Ti)_2O_6F$, Fluornatropyrochlore $(Na,REE,Ca)_2Nb_2(O,OH)_6F$, Fluorstrontiopyrochlore $(Sr,[\ ])_2Nb_2(O,OH)_6F$, Hydropyrochlore $(H_2O,[\ ])_2Nb_2(O,OH)_6(H_2O)$, Hydroxycalciopyrochlore $(Ca,[\ ])_2Nb_2(O,OH)_6(OH)$, Kenoplumbopyrochlore $(Pb,[\ ])Nb_2O_6([\ ],O)$, Oxycalciopyrochlore $Ca_2Nb_2O_6O$, Oxynatropyrochlore $(Na,Ca,U)_2Nb_2O_6(O,OH)$, Oxyplumbopyrochlore $Pb_2Nb_2O_6O$, Oxyyttropyrochlore-(Y) $(Y,[\ ])_2Nb_2O_6O$

  - Microlite group (D atom is Ta)

    - Fluorcalciomicrolite $(Ca,Na)_2Ta_2O_6F$, Fluornatromicrolite $(Na,Ca,Bi)_2Ta_2O_6F$, Hydrokenomicrolite $([\ ],H_2O)_2Ta_2(O,OH)_6H_2O$, Hydromicrolite $(H_2O,[\ ])_2Ta_2(O,OH)_6H_2O$, Hydroxykenomicrolite $([\ ],Na,Sb^{3+})_2Ta_2O_6(OH)$, Kenoplumbomicrolite $(Pb,[\ ])_2Ta_2O_6([\ ],O,OH)$, Oxycalciomicrolite $Ca_2Ta_2O_6O$, Oxystannomicrolite $Sn_2Ta_2O_6O$, Oxystibiomicrolite $(Sb^{3+},Ca)_2Ta_2O_6O$

  - Romeite group (D atom is Sb)

    - Cuproromeite $Cu_2Sb_2(O,OH)_7$, Fluorcalcioromeite $(Ca,Sb^{3+})_2(Sb^{5+},Ti)_2O_6F$, Fluornatroromeite $(Na,Ca)_2Sb_2(O,OH)_6F$, Hydroxycalcioromeite $(Ca,Sb^{3+})_2(Sb^{5+},Ti)_2O_6(OH)$, Oxycalcioromeite $Ca_2Sb_2O_6O$, Oxyplumboromeite $Pb_2Sb_2O_6O$, Stibiconite $Sb^{3+}Sb^{+6}_2O_6(OH)$

  - Betafite group (D atom is Ti): Calciobetafite $Ca_2(Ti,Nb)_2O_6O$, Oxyuranobetafite $(U,Ca,[\ ])_2(Ti,Nb)_2O_6O$

- Elsmoreite group (D atom is W): Hydrokenoelsmoreite $[\ ]_2W_2O_6(H_2O)$
- Cesstibtantite group
  - Cesstibtantite $(Cs,Na)SbTa_4O_{12}$, Natrobistantite $(Na,Cs)Bi(Ta,Nb,Sb)_4O_{12}$
- Brannerite-Thorutite series, Orthobrannerite-Thorutite series:
  - Brannerite $(U^{4+},REE,Th,Ca)(Ti,Fe^{3+},Nb)_2(O,OH)_6$, Orthobrannerite $U^{4+}U^{6+}Ti_4O_{12}(OH)_2$, Thorutite $(Th,U,Ca)Ti_2(O,OH)_6$

Class: Hydroxides and Oxides Containing Hydroxyl

- Diaspore group (Orthorhombic, Pnma or Pnmd)
  - Diaspore $AlO(OH)$, Goethite $Fe^{3+}O(OH)$, Groutite $Mn^{3+}O(OH)$, Montroseite $(V^{3+},Fe^{3+},V^{4+})O(OH)$, Bracewellite $Cr^{3+}O(OH)$, Tsumgallite $GaO(OH)$
- Brucite group (Rhombohedral: P-3m1)
  - Brucite $Mg(OH)_2$, Amakinite $(Fe^{2+},Mg)(OH)_2$, Pyrochroite $Mn(OH)_2$, Portlandite $Ca(OH)_2$, Theophrastite $Ni(OH)_2$
- Wickmanite group
  - (Cubic or Trigonal, 2^+ cations containing Sn)
    - Wickmanite $Mn^{2+}Sn^{4+}(OH)_6$, Schoenfliesite $MgSn^{4+}(OH)_6$, Natanite $Fe^{2+}Sn^{4+}(OH)_6$, Vismirnovite $ZnSn^{4+}(OH)_6$, Burtite $CaSn(OH)_6$, Mushistonite $(Cu,Zn,Fe)Sn^{4+}(OH)_6$
  - (Tetragonal: P42/n)
    - Stottite $Fe^{2+}Ge(OH)_6$, Tetrawickmanite $Mn^{2+}Sn^{4+}(OH)_6$, Jeanbandyite $(Fe^{3+},Mn^{2+})Sn^{4+}(OH)_6$, Mopungite $NaSb(OH)_6$

Category 05

- Carbonates and Nitrates
  - Calcite group (Trigonal: R-3c)
    - Calcite $CaCO_3$, Magnesite $MgCO_3$, Siderite $Fe^{2+}CO_3$, Rhodochrosite $MnCO_3$, Spherocobaltite $CoCO_3$, Smithsonite $ZnCO_3$, Otavite $CdCO_3$, Gaspeite $(Ni,Mg,Fe^{2+})CO_3$
  - Aragonite group (Orthorhombic: Pmcn)
    - Aragonite $CaCO_3$, Witherite $BaCO_3$, Strontianite $SrCO_3$, Cerussite $PbCO_3$
  - Dolomite group
    - Ankerite $Ca(Fe^{2+},Mg,Mn^{2+})(CO_3)_2$, Dolomite $CaMg(CO_3)_2$, Kutnohorite $Ca(Mn,Mg,Fe)(CO_3)_2$, Minrecordite $CaZn(CO_3)_2$

- Burbankite group

  - Hexagonal

    - Burbankite $(Na,Ca)_3(Sr,Ba,Ce)_3(CO_3)_5$, Khanneshite $(NaCa)_3(Ba,Sr,Ce,Ca)_3(CO_3)_5$, Calcioburbankite $Na_3(Ca,REE,Sr)_3(CO_3)_5$, Sanromanite $Na_2CaPb_3(CO_3)_5$

  - Monoclinic

    - Rémondite-(Ce) $Na_3(Ce,La,Ca,Na,Sr)_3(CO_3)_5$, Petersenite-(Ce) $(Na,Ca)_4(Ce,La,Nd)_2(CO_3)_5$, Rémondite-(La) $Na_3(La,Ce,Ca)_3(CO_3)_5$

- Rosasite group

  - Rosasite $(Cu,Zn)_2(CO_3)(OH)_2$, Glaukosphaerite $(Cu,Ni)_2(CO_3)(OH)_2$, Kolwezite $(Cu,Co)_2(CO_3)(OH)_2$, Zincrosasite $(Zn,Cu)_2(CO_3)(OH)_2$, Mcguinnessite $(Mg,Cu)_2(CO_3)(OH)_2$

- Malachite group

  - Malachite $Cu_2(CO_3)(OH)_2$, Nullaginite $Ni_2(CO_3)(OH)_2$, Pokrovskite $Mg_2(CO_3)(OH)_2 \cdot 0.5H_2O$, Chukanovite $Fe_2(CO_3)(OH)_2$

- Ancylite group

  - Ancylite-(Ce) $SrCe(CO_3)_2(OH) \cdot H_2O$, Calcioancylite-(Ce) $CaCe(CO_3)_2(OH) \cdot H_2O$, Calcioancylite-(Nd) $CaNd(CO_3)_2(OH) \cdot H_2O$, Gysinite-(Nd) $Pb(Nd,La)(CO_3)_2(OH) \cdot H_2O$, Ancylite-(La) $Sr(La,Ce)(CO_3)_2(OH) \cdot H_2O$, Kozoite-(Nd) $(Nd,La,Sm,Pr)(CO_3)(OH)$, Kozoite-(La) $La(CO_3)(OH)$

- Sjogrenite-Hydrotalcite group

  - Sjogrenite subgroup: Hexagonal

    - Manasseite $Mg_6Al_2[(OH)_{16}|CO_3] \cdot 4H_2O$, Barbertonite $Mg_6(Cr,Al)_2[(OH)_{16}|CO_3] \cdot 4H_2O$, Sjogrenite $Mg_6(Fe^{3+})_2[(OH)_{16}|CO_3] \cdot 4H_2O$, Zaccagnaite $Zn_4Al_2(OH)_{12}(CO_3) \cdot 3H_2O$, Fougerite $(Fe^{2+},Mg)_6(Fe^{3+})_2(OH)_{18} \cdot 4H_2O$

  - Hydrotalcite subgroup: Rhombohedral I, $Mg_6(R^{3+})_2(OH)_{16}CO_3 \cdot 4H_2O$, where R^{3+} = Al, Cr, or Fe

    - Hydrotalcite $Mg_6Al_2[(OH)_{16}CO_3] \cdot 4H_2O$, Stichtite $Mg_6Cr_2[(OH)_{16}|CO_3] \cdot 4H_2O$, Pyroaurite $Mg6Fe^{3+}_2[(OH)_{16}|CO_3] \cdot 4H_2O$, Desautelsite $Mg_6(Mn^{3+})_2[(OH)_{16}|CO_3] \cdot 4H_2O$, Droninoite $Ni_3Fe^{3+}Cl(OH)_8 \cdot 2H_2O$, Hydrowoodwardite $Cu_{1-x}Al_x[(OH)_2|(SO_4)_{x/2}] \cdot nH_2O$, Iowaite $Mg_4Fe(OH)_8O-Cl \cdot 4H_2O$

  - Hydrotalcite subgroup: Rhombohedral II

    - Reevesite $Ni_6(Fe^{3+})_2(CO_3)(OH)16 \cdot 4H_2O$, Takovite $Ni_6Al_2(OH)_{16}(CO_3,OH) \cdot 4H_2O$, Comblainite $(Ni^{2+})_6(Co^{3+})_2(CO_3)(OH)_{16} \cdot 4H_2O$

- Tundrite group

  - Tundrite-(Ce) $Na_2Ce_2TiO_2(SiO_4)(CO_3)_2$, Tundrite-(Nd) $Na_3(Nd,La)_4(Ti,Nb)_2(SiO_4)_2(CO_3)_3O_4(OH)\bullet2H_2O$

- Category:Nitrate minerals

Category 06

- Borates

  - Ludwigite group (Space group: Pbam)

    - Ludwigite $Mg_2Fe^{3+}BO_5$, Vonsenite $Fe^{2+}_2Fe^{3+}BO_5$, Azoproite $(Mg,Fe^{2+})_2(Fe^{3+},Ti,Mg)BO_5$, Bonaccordite $Ni_2Fe^{3+}BO_5$, Chestermanite $Mg_2(Fe^{3+},Mg,Al,Sb^{5+})BO_3O_2$, Fredrikssonite $Mg_2(Mn^{3+},Fe^{3+})O_2(BO_3)$

  - Boracite group (Tecto-heptaborates)

    - (Orthorhombic: Rca21)

      - Boracite $Mg_3B_7O_{13}Cl$, Ericaite $(Fe^{2+},Mg,Mn)_3B_7O_{13}Cl$, Chambersite $Mn_3B_7O_{13}Cl$

    - (Trigonal: R3c)

      - Congolite $(Fe^{2+},Mg,Mn)_3B_7O_{13}Cl$, Trembathite $(Mg,Fe^{2+})_3B_7O_{13}Cl$

  - Inderite group (Neso-triborates)

    - Inyoite $Ca_2B_6O_6(OH)_{10}\bullet8H_2O$, Inderborite $CaMg[B_3O_3(OH)_5]_2\bullet6H_2O$, Inderite $MgB_3O_3(OH)_5\bullet5H_2O$, Kurnakovite $Mg(H_4B_3O_7)(OH)\cdot5H_2O$, Meyerhofferite $Ca_2(H_3B_3O_7)_2\cdot4H_2O$, Solongoite $Ca_2(H_3B_3O_7)(OH)Cl$

  - Santite group (Neso-pentaborates)

    - Santite $KB_5O_6(OH)_4\bullet2(H_2O)$, Ramanite-(Rb) $Rb[B5O_6(OH)_4]\bullet2H_2O$, Ramanite-(Cs) $Cs[B_5O_6(OH)_4]\bullet2H_2O$

  - Hilgardite group (Tecto-pentaborates)

    - Hilgardite $Ca_2B_5O_9Cl\bullet H_2O$, Kurgantaite $CaSr[B_5O_9]Cl\bullet H_2O$, IMA2007-047 $Pb_2[B_5O_9]Cl\bullet0.5H_2O$

  - Pringleite group

    - Pringleite $Ca_9B_{26}O_{34}(OH)_{24}Cl_4\bullet13H_2O$, Ruitenbergite $Ca_9B_{26}O_{34}(OH)_{24}Cl_4\bullet13H_2O$, Brianroulstonite $Ca_3[B_5O_6(OH)_6](OH)Cl_2\bullet8H_2O$, Penobsquisite $Ca_2Fe_{2+}[B_9O_{13}(OH)_6]Cl\bullet4H_2O$, Walkerite $Ca_{16}(Mg,Li,[\ ])_2[B_{13}O_{17}(OH)_{12}]_4Cl_6\bullet28H_2O$

Category 07

- Sulfates, Selenates, Chromates, Molybdates, Wolframates, Niobates

  - Barite group

- Barite $BaSO_4$, Celestine $SrSO_4$, Anglesite $PbSO_4$

- Blodite group

  - Blodite $Na_2Mg(SO_4)_2 \cdot 4H_2O$, Nickelblodite $Na_2(Ni,Mg)(SO_4)_2 \cdot 4H_2O$, Leonite $K_2Mg(SO_4)_2 \cdot 4H_2O$, Mereiterite $K_2Fe^{2+}(SO_4)_2 \cdot 4H_2O$, Changoite $Na_2Zn(SO_4)_2 \cdot 4H_2O$

- Alum group, $XAl(SO_4)_2 \cdot 12H_2O$

  - Alum-(K) $KAl[SO_4]_2 \cdot 12H_2O$, Alum-(Na) $NaAl[SO_4]_2 \cdot 12H_2O$, Tschermigite (NH_4) $Al(SO_4)_2 \cdot 12H_2O$, Lonecreekite $(NH_4)(Fe^{3+},Al)(SO_4)_2 \cdot 12H_2O$, Lanmuchangite $TlAl(SO_4)_2 \cdot 12H_2O$

- Voltaite group

  - Voltaite $K_2(Fe^{2+})_5(Fe^{3+})_3Al(SO_4)_{12} \cdot 18H_2O$, Zincovoltaite $K_2Zn_5(Fe^{3+})_3Al(SO_4)_{12} \cdot 18H_2O$, Pertlikite $K_2(Fe^{2+},Mg)_2(Mg,Fe^{3+})_4(Fe^{3+})_2Al(SO_4)_{12} \cdot 18H_2O$

- Aluminite group

  - Aluminite $Al_2(SO_4)(OH)_4 \cdot 7(H_2O)$, Mangazeite $Al_2(SO_4)(OH)_4 \cdot 3H_2O$

- Zippeite group

  - Zippeite $K_4(UO_2)_6(SO_4)_3(OH)_{10} \cdot 4H_2O$, Natrozippeite $Na_4(UO_2)_6(SO_4)_3(OH)_{10} \cdot 4H_2O$, Magnesiozippeite $Mg(H_2O)_{3.5}(UO_2)_2(SO_4)O_2$, Nickelzippeite $(Ni^{2+})_2(UO_2)_6(SO_4)_3(OH)_{10} \cdot 16H_2O$, Zinc-zippeite $(Zn^{2+})_2(UO_2)_6(SO_4)_3(OH)_{10} \cdot 16H_2O$, Cobaltzippeite $(Co^{2+})_2(UO_2)_6(SO_4)_3(OH)_{10} \cdot 16H_2O$, Marecottite $Mg_3(H_2O)_{18}[(UO_2)_4O_3(SO_4)_2]_2 \cdot 10H_2O$, Pseudojohannite $Cu_{6.5}[(UO_2)_4O_4(SO_4)_2]_2(OH)_5 \cdot 25H_2O$, IMA2009-008 $Y_2[(UO_2)_8O_6(SO_4)_4(OH)_2] \cdot 26H_2O$

- Copiapite group

  - Copiapite $Fe^{2+}(Fe^{3+})_4(SO_4)_6(OH)_2 \cdot 20H_2O$, Magnesiocopiapite $Mg(Fe^{3+})_4(SO_4)_6(OH)_2 \cdot 20H_2O$, Cuprocopiapite $Cu(Fe^{3+})_4(SO_4)_6(OH)_2 \cdot 20H_2O$, Ferricopiapite $(Fe^{3+})_{2/3}(Fe^{3+})_4(SO_4)_6(OH)_2 \cdot 20H_2O$, Calciocopiapite $Ca(Fe^{3+})_4(SO_4)_6(OH)_2 \cdot 19H_2O$, Zincocopiapite $Zn(Fe^{3+})_4(SO_4)_6(OH)_2 \cdot 18H_2O$, Aluminocopiapite $Al_{2/3}(Fe^{3+})_4(SO_4)_6O(OH)_2 \cdot 20H_2O$

- Pb, Zn tellurates

  - Cheremnykhite $Zn_3Pb_3Te^{4+}O_6(VO_4)_2$, Kuksite $Pb_3Zn_3Te^{6+}O_6(PO_4)_2$, Dugganite $Pb_3Zn_3Te(As,V,Si)_2(O,OH)_{14}$, Joelbruggerite $Pb_3Zn_3Sb^{5+}As_2O_{13}(OH)$

- "Halotrichite" supergroup

- Hydrated acid and sulfates where $A(B)_2(XO_4)_4 \cdot x(H_2O)$

  - Halotrichite group

- Pickeringite $MgAl_2(SO_4)_4 \cdot 22H_2O$, Halotrichite $Fe^{2+}Al_2(SO_4)_4 \cdot 22H_2O$, Apjohnite $MnAl_2(SO_4)_4 \cdot 22H_2O$, Dietrichite $(Zn,Fe^{2+},Mn)Al_2(SO_4)_4 \cdot 22H_2O$, Bilinite $Fe^{2+}(Fe^{3+})_2(SO_4)_4 \cdot 22H_2O$, Redingtonite $(Fe^{2+},Mg,Ni)(Cr,Al)_2(SO_4)_4 \cdot 22H_2O$, Wupatkiite

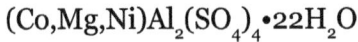

$(Co,Mg,Ni)Al_2(SO_4)_4 \cdot 22H_2O$

- Ransomite $Cu(Fe^{3+})_2(SO_4)_4 \cdot 6H_2O$, Romerite $Fe^{2+}(Fe^{3+})_2(SO_4)_4 \cdot 14H_2O$, Lishizhenite $Zn(Fe^{3+})_2(SO_4)_4 \cdot 14H_2O$

"Kieserite" supergroup

- Hydrated acid and sulfates where $AXO_4 \cdot x(H_2O)$

  - Kieserite group

    - Kieserite $MgSO_4 \cdot H_2O$, Szomolnokite $Fe^{2+}SO_4 \cdot H_2O$, Szmikite $MnSO_4 \cdot H_2O$, Poitevinite $(Cu,Fe^{2+},Zn)SO_4 \cdot H_2O$, Gunningite $(Zn,Mn)SO_4 \cdot H_2O$, Dwornikite $(Ni,Fe^{2+})SO_4 \cdot H_2O$, Cobaltkieserite $CoSO_4 \cdot H_2O$

  - Rozenite group (Monoclinic)

    - Rozenite $Fe^{2+}SO_4 \cdot 4H_2O$, Starkeyite $MgSO_4 \cdot 4H_2O$, Ilesite $(Mn,Zn,Fe^{2+})SO_4 \cdot 4H_2O$, Aplowite $(Co,Mn,Ni)SO_4 \cdot 4H_2O$, Boyleite $(Zn,Mg)SO_4 \cdot 4H_2O$, IMA2002-034 $CdSO_4 \cdot 4H_2O$

  - Chalchanthite group (Triclinic: P-1)

    - Chalcanthite $CuSO_4 \cdot 5H_2O$, Siderotil $Fe^{2+}SO_4 \cdot 5H_2O$, Pentahydrite $MgSO_4 \cdot 5H_2O$, Jokokuite $MnSO_4 \cdot 5H_2O$

  - Hexahydrite group (Space group: C2/c)

    - Hexahydrite $MgSO_4 \cdot 6H_2O$, Bianchite $(Zn,Fe^{2+})(SO_4) \cdot 6H_2O$, Ferrohexahydrite $Fe^{2+}SO_4 \cdot 6H_2O$, Nickelhexahydrite $(Ni,Mg,Fe^{2+})(SO_4) \cdot 6H_2O$, Moorhouseite $(Co,Ni,Mn)SO_4 \cdot 6H_2O$, Chvaleticeite $(Mn^{2+},Mg)SO_4 \cdot 6H_2O$

  - Melanterite group (Heptahydrates, Monoclinic: P21/c)

    - Melanterite $Fe^{2+}SO_4 \cdot 7H_2O$, Boothite $CuSO_4 \cdot 7H_2O$, Zincmelanterite $(Zn,Cu,Fe^{2+})SO_4 \cdot 7H_2O$, Bieberite $CoSO_4 \cdot 7H_2O$, Mallardite $Mn^{2+}SO_4 \cdot 7H_2O$, Alpersite $(Mg,Cu)SO_4 \cdot 7H_2O$

  - Epsomite group

    - Epsomite $MgSO_4 \cdot 7H_2O$, Goslarite $ZnSO_4 \cdot 7H_2O$, Morenosite $NiSO_4 \cdot 7H_2O$

  - Minasragrite group

    - (Monclinic and Triclinic)

      Minasragrite $VO(SO_4) \cdot 5H_2O$, Bobjonesite $VO(SO_4)H_2O_3$, Anorthominasragrite $V^{4+}O(SO_4)H_2O_5$

    - (Orthorhombic)

      Stanleyite $(V^{4+}O)SO_4 \cdot 6(H_2O)$, Orthominasragrite $VO(SO_4) \cdot 5(H_2O)$

  - Bassanite $2CaSO_4 \cdot H_2O$, Gypsum $CaSO_4 \cdot 2H_2O$, Sanderite $MgSO_4 \cdot 2H_2O$, Bonattite $CuSO_4 \cdot 3H_2O$, Retgersite $NiSO_4 \cdot 6H_2O$, Meridianiite $MgSO_4 \cdot 11H_2O$

Alunite Supergroup - Part I

- Category:Alunite group, $A^{1+}(B)_3(SO_4)_2(OH)_6$

  - Alunite $KAl_3[(OH)_3|SO_4]_2$, Ammonioalunite $(NH_4)Al_3[(OH)_3|SO_4]_2$, Ammonioja-rosite $(NH_4)(Fe^{3+})_3(SO_4)_2(OH)_6$, Argentojarosite $Ag(Fe^{3+})_3(SO_4)_2(OH)_6$, Beaverite-Cu $Pb(Fe^{3+},Cu)_3(SO_4)_2(OH)_6$ (Fe^{3+}:Cu ≈ 2:1), Beaverite-Zn $Pb((Fe^{3+})_2Zn)(SO_4)_2(OH)_6$, Dorallcharite $(Tl,K)(Fe^{3+})_3(SO_4)_2(OH)_6$, Huangite $Ca_{0.5}Al_3(SO_4)_2(OH)_6$, Hydroni-umjarosite $(H_3O)(Fe^{3+})_3(SO_4)_2(OH)_6$, Jarosite $K(Fe^{3+})_3[(OH)_3|SO_4]_2$, Natroalunite $(Na,K)Al_3[(OH)_3|SO_4]$, Natrojarosite $Na(Fe^{3+})_3(SO_4)_2(OH)_6$, Osarizawaite PbC-$uAl_2(SO_4)_2(OH)_6$, Plumbojarosite $Pb(Fe^{3+})_6(SO_4)_4(OH)_{12}$, Schlossmacherite (H_3O,Ca) $Al_3(AsO_4,SO_4)_2(OH)_6$, Walthierite $Ba_{0.5}Al_3(SO_4)_2(OH)_6$, Mills et al. (2009)

Category 08

- Phosphates, Arsenates, Polyvanadates

  - Category:Arsenate minerals

Class: Anydrous Phosphates

- Triphylite group

  - Triphylite $LiFe^{2+}PO_4$, Lithiophilite $LiMnPO_4$, Natrophilite $NaMnPO_4$

- Retzian series

  - Retzian-(Ce) $Mn_2Ce(AsO_4)(OH)_4$, Retzian-(Nd) $Mn_2(Nd,Ce,La)(AsO_4)(OH)_4$, Ret-zian-(La) $(Mn,Mg)_2(La,Ce,Nd)(AsO_4)(OH)_4$

"Alluaudite-Wyllieite" Supergroup

- Anhydrous phosphates, etc. $(A^+ B^{2+})_5 (XO_4)_3$

  - Berzeliite group

    - Berzeliite $(Ca,Na)_3(Mg,Mn)_2(AsO_4)_3$, Manganberzeliite $(Ca,Na)_3(Mn,Mg)_2(AsO_4)_3$, Palenzonaite $(Ca,Na)_3Mn^{2+}(V^{5+},As^{5+},Si)_3O_{12}$, Schaferite $NaCa_2Mg_2(VO_4)_3$

  - Alluaudite-Wyllieite group (Alluaudite subgroup I)

    - Caryinite $(Na,Pb)(Ca,Na)(Ca,Mn^{2+})(Mn^{2+},Mg)_2(AsO_4)_3$, Arseniopleite (Ca,Na) $(Na,Pb)Mn^{2+}(Mn^{2+},Mg,Fe^{2+})_2(AsO_4)_3$

  - Alluaudite-Wyllieite group (Alluaudite subgroup II/ Hagendorfite subgroup)

    - Ferrohagendorfite* $(Na,Ca)_2Fe^{2+}(Fe^{2+},Fe^{3+})_2(PO_4)_3$, Hagendorfite $NaCaMn(Fe^{2+},Fe^{3+},Mg)_2(PO_4)_3$, Varulite $NaCaMn(Mn,Fe^{2+},Fe^{3+})_2(-PO_4)_3$, Maghagendorfite $NaMgMn(Fe^{2+},Fe^{3+})_2(PO_4)_3$, Ferroalluaudite $NaCaFe^{2+}(Fe^{2+},Mn,Fe^{3+},Mg)_2(PO_4)_3$, Alluaudite $NaCaFe^{2+}(Mn,Fe^{2+},Fe^{3+},Mg)_2(-PO_4)_3$, Odanielite $Na(Zn,Mg)_3H_2(AsO_4)_3$, Johillerite $Na(Mg,Zn)_3Cu(AsO_4)_3$,

Nickenichite $Na_{0.8}Ca_{0.4}(Mg,Fe^{3+},Al)_3Cu_{0.4}(AsO_4)_3$, Yazganite $Na(Fe^{3+})_2(Mg,Mn)$ $(AsO_4)_3 \cdot H_2O$, IMA2008-054 $NaCaMn_2(PO_4)[PO_3(OH)]_2$, IMA2008-064 $Na_{16}(Mn^{2+})_{25}Al_8(PO_4)_{30}$

- Alluaudite-Wyllieite group (Wyllieite subgroup)

  - Ferrowyllieite $(Na,Ca,Mn)(Fe^{2+},Mn)(Fe^{2+},Fe^{3+},Mg)Al(PO_4)_3$, Wyllieite $(Na,-Ca,Mn^{2+})(Mn^{2+},Fe^{2+})(Fe^{2+}, Fe^{3+},Mg)Al(PO_4)_3$, Rosemaryite (Na,Ca,Mn^{2+}) $(Mn^{2+},Fe^{2+})(Fe^{3+},Fe^{2+},Mg)Al(PO_4)_3$, Qingheiite $Na_2(Mn^{2+},Mg,Fe^{2+})(Al,Fe^{3+})(PO_4)_3$, Bobfergusonite $Na_2(Mn^{2+})_5Fe^{3+}Al(PO_4)_6$, Bradaczekite $NaCu_4(AsO_4)_3$, Ferrorose-maryite [ ]$NaFe^{2+}Fe^{3+}Al(PO_4)_3$

- Fillowite group

  - Fillowite $Na_2Ca(Mn,Fe^{2+})_7(PO_4)_6$, Johnsomervilleite $Na_2Ca(Mg,Fe^{2+},Mn)_7(PO_4)_6$, Chladniite $Na_2Ca(Mg,Fe^{2+})_7(PO_4)_6$, Galileiite $Na(Fe^{2+})_4(PO_4)_3$, Xenophyllite $Na_4Fe_7(PO_4)_6$, Stornesite-(Y) $(Y, Ca)[ ]_2Na_6(Ca,Na)_8(Mg,Fe)_{43}(PO_4)_{36}$

- Nabiasite $BaMn_9[(V,As)O_4]_6(OH)_2$

"Whitlockite" Supergroup

- Anhydrous phosphates, etc. $(A^+ B^{2+})_3 (XO_4)_2$

  - Sarcopside group

    - Sarcopside $(Fe^{2+},Mn,Mg)_3(PO_4)_2$, Farringtonite $Mg_3(PO_4)_2$, Chopinite $(Mg,Fe)_3(PO_4)_2$

  - Whitlockite group

    - Whitlockite $Ca_9(Mg,Fe^{2+})(PO_4)_6(PO_3OH)$, Strontiowhitlockite $Sr_7(Mg,-Ca)_3(PO_4)_6[PO_3(OH)]$, Merrillite-(Ca)* $(Ca,[ ])_{19}Mg_2(PO_4)_{14}$, Merrillite $Ca_{18}Na_2Mg_2(PO_4)_{14}$, Merrillite-(Y)* $Ca_{16}Y_2Mg_2(PO_4)_{14}$, Ferromerrillite $Ca_9NaFe(PO_4)_7$, Tu-ite $Ca_3(PO_4)_2$, Bobdownsite $Ca_9Mg(PO_3F)(PO_4)_6$

  - Xanthiosite $Ni_3(AsO_4)_2$, Graftonite $(Fe^{2+},Mn,Ca)_3(PO_4)_2$, Beusite $(Mn^{2+},Fe^{2+},Ca,Mg)_3(PO_4)_2$, Stanfieldite $Ca_4(Mg,Fe^{2+},Mn)_5(PO_4)_6$, Hurlbutite $CaBe_2(PO_4)_2$, Stranskiite $Zn_2Cu^{2+}(AsO_4)_2$, Keyite $(Cu^{2+})_3(Zn,Cu)_4Cd_2(AsO_4)_6 \cdot 2H_2O$, Lammerite $Cu_3[(As,P)O_4]_2$, Mc-birneyite $Cu_3(VO_4)_2$, Tillmannsite $(Ag_3Hg)(V,As)O_4$, IMA2009-002 $Cu_3(AsO_4)_2$

"Monazite" Supergroup

- Anhydrous phosphates, etc. $A^+ XO_4$

- Berlinite group

  - Berlinite $AlPO_4$, Alarsite $AlAsO_4$, Rodolicoite $Fe^{3+}PO_4$

- Monazite group (Monoclinic: P21/n)

  - Monazite-(Ce) $(Ce,La,Nd,Th)PO_4$, Monazite-(La) $(La,Ce,Nd)PO_4$, Cheralite-(Ce)?

(Ce,Ca,Th)(P,Si)O_4, Brabantite? CaTh(PO_4)$_2$, Monazite-(Nd) (Nd,Ce,La)(P,Si)O_4, Gasparite-(Ce) CeAsO_4, Monazite-(Sm) SmPO_4

- Lithiophosphate group

  - Lithiophosphate Li$_3$PO_4, Olympite LiNa$_5$(PO_4)$_2$, Nalipoite NaLi$_2$PO_4

- Zenotime group (Tetragonal: I41/amd)

  - Xenotime-(Y) YPO_4, Chernovite-(Y) YAsO_4, Wakefieldite-(Y) YVO_4, Wakefieldite-(Ce) (Ce^{3+},Pb^{2+},Pb^{4+})VO_4, Pretulite ScPO_4, Xenotime-(Yb) YbPO_4, Wakefieldite-(La) LaVO_4, Wakefieldite-(Nd) NdVO_4

- Heterosite Fe^{3+}PO_4, Purpurite Mn^{3+}PO_4, Rooseveltite BiAsO_4, Tetrarooseveltite BiAsO_4, Pucherite BiVO_4, Clinobisvanite BiVO_4, Dreyerite BiVO_4, Ximengite BiPO_4, Kosnarite K(Zr^{4+})$_2$(PO_4)$_3$, Petewilliamsite (Ni,Co,Cu)$_{30}$(As$_2$O$_7$)$_{15}$

"Adelite" Supergroup

- Anhydrous phosphates, etc. containing hydroxyl or halogen where (A B)$_2$ (XO_4) Zq

- Adelite group

  - Adelite CaMg(AsO_4)(OH), Conichalcite CaCu(AsO_4)(OH), Austinite CaZn(AsO_4)(OH), Duftite-beta? PbCu(AsO_4)(OH), Gabrielsonite PbFe^{2+}(AsO_4)(OH), Tangeite CaCu(VO_4)(OH), Nickelaustinite Ca(Ni,Zn)(AsO_4)(OH), Cobaltaustinite CaCo(AsO_4)(OH), Arsendescloizite PbZn(AsO_4)(OH), Gottlobite CaMg(VO_4,AsO_4)(OH)

- Descloizite group

  - Descloizite PbZn(VO_4)(OH), Mottramite PbCu(VO_4)(OH), Pyrobelonite PbMn(VO_4)(OH), Cechite Pb(Fe^{2+},Mn)(VO_4)(OH), Duftite-alpha PbCu(AsO_4)(OH)

- Herderite group

  - Herderite CaBe(PO_4)F, Hydroxylherderite CaBe(PO_4)(OH), Vayrynenite MnBe(PO_4)(OH,F), Bergslagite CaBe(AsO_4)(OH)

- Lacroixite group

  - Lacroixite NaAl(PO_4)F, Durangite NaAl(AsO_4)F, Maxwellite NaFe^{3+}(AsO_4)F

- Tilasite group

  - Tilasite CaMg(AsO_4)F, Isokite CaMg(PO_4)F, Panasqueiraite CaMg(PO_4)(OH,F)

- Amblygonite group

  - Amblygonite (Li,Na)Al(PO_4)(F,OH), Montebrasite? LiAl(PO_4)(OH,F), Natromontebrasite? (Na,Li)Al(PO_4)(OH,F)

- Dussertite group

  - Dussertite Ba(Fe^{3+})$_3$(AsO_4)$_2$(OH)$_5$, Florencite-(Ce) CeAl$_3$(PO_4)$_2$(OH)$_6$, Florencite-(La) (La,Ce)Al$_3$(PO_4)$_2$(OH)$_6$, Florencite-(Nd) (Nd,Ce)Al$_3$(PO_4)$_2$(OH)$_6$

- Arsenoflorencite group

  - Arsenoflorencite-(Ce) $(Ce,La)Al_3(AsO_4)_2(OH)_6$, Arsenoflorencite-(Nd)* (Nd,La,Ce,Ba) $(Al,Fe^{3+})_3(AsO_4,PO_4)_2(OH)_6$, Arsenoflorencite-(La)* $(La,Sr)Al_3(AsO_4,SO_4,PO_4)_2(OH)_6$, Graulichite-(Ce) $Ce(Fe^{3+})_3(AsO_4)_2(OH)_6$

- Waylandite group

  - Waylandite $BiAl_3(PO_4)_2(OH)_6$, Eylettersite $(Th,Pb)_{(1-x)}Al_3(PO_4,SiO_4)_2(OH)_6$ (?), Zairite $Bi(Fe^{3+},Al)_3(PO_4)_2(OH)_6$, Arsenogorceixite $BaAl_3AsO_3(OH)(AsO_4,PO_4)(OH,F)_6$

- Babefphite $BaBe(PO_4)(F,O)$, Brazilianite $NaAl_3(PO_4)_2(OH)_4$, Tavorite $LiFe^{3+}(PO_4)(OH)$, Vesignieite $Cu_3Ba(VO_4)_2(OH)_2$, Bayldonite $(Cu,Zn)_3Pb(AsO_3OH)_2(OH)_2$, Curetonite $Ba_4Al_3Ti(PO_4)_4(O,OH)_6$, Thadeuite $(Ca,Mn^{2+})(Mg,Fe^{2+},Mn^{3+})_3(PO_4)_2(OH,F)_2$, Leningradite $Pb(Cu^{2+})_3(VO_4)_2Cl_2$, Arctite $Na_2Ca_4(PO_4)_3F$, Wilhelmkleinite $Zn(Fe^{3+})_3(AsO_4)_2(OH)_2$, Artsmithite $Hg^+_4Al(PO_4)_{1.74}(OH)_{1.78}$

"Olivenite" Supergroup

- Anhydrous phosphates, etc. containing hydroxyl or halogen where $(A)_2(XO_4)$ Zq

- Zwieselite group

  - Zwieselite $(Fe^{2+},Mn)_2(PO_4)F$, Triplite $(Mn,Fe^{2+},Mg,Ca)_2(PO_4)(F,OH)$, Magniotriplite? $(Mg,Fe^{2+},Mn)_2(PO_4)F$

- Wagnerite group

  - Wagnerite $(Mg,Fe^{2+})_2(PO_4)F$, Hydroxylwagnerite $Mg_2(PO_4)(OH)$

- Wolfeite group

  - Wolfeite $(Fe^{2+},Mn^{2+})_2(PO_4)(OH)$, Triploidite $(Mn,Fe^{2+})_2(PO_4)(OH)$, Sarkinite $(Mn^{2+})_2(AsO_4)(OH)$, Stanekite $Fe^{3+}(Mn,Fe^{2+},Mg)(PO_4)O$, Joosteite $(Mn^{2+},Mn^{3+},Fe^{3+})_2(PO_4)O$

- Satterlyite group

  - Satterlyite $(Fe^{2+},Mg)_2(PO_4)(OH)$, Holtedahlite $Mg_{12}(PO_3OH,CO_3)(PO_4)_5(OH,O)_6$

- Olivenite group

  - Olivenite subgroup

    - Adamite $Zn_2(AsO_4)(OH)$, Eveite $(Mn^{2+})_2[OH|AsO_4]$, Libethenite Cu_2PO_4OH, Olivenite $Cu_2[OH|AsO_4]$, Zincolivenite $CuZn(AsO_4)(OH)$, Zincolibethenite $CuZn(PO_4)OH$

  - Tarbuttite subgroup

    - Tarbuttite Zn_2PO_4OH, Paradamite $Zn_2[OH|AsO_4]$

- Althausite $Mg_2(PO_4)(OH,F,O)$, Augelite $Al_2(PO_4)(OH)_3$, Arsenobismite? $Bi_2(AsO_4)(OH)_3$, Angelellite $(Fe^{3+})_4(AsO_4)_2O_3$, Spodiosite? $Ca_2(PO_4)F$

"Arrojadite" Supergroup

- Anhydrous phosphates, etc. containing hydroxyl or halogen where $(A\ B)_m\ (XO_4)_4\ Zq$

  - Palermoite group

    - Palermoite $(Sr,Ca)(Li,Na)_2Al_4(PO_4)_4(OH)_4$, Bertossaite $Li_2CaAl_4(PO_4)_4(OH)_4$

  - Arrojadite group (Arrojadite subgroup) (Al in Al site, OH in W site, Fe in M site)

    - Arrojadite $KNa_4Ca(Mn^{2+})_4(Fe^{2+})_{10}Al(PO_4)_{12}(OH,F)_2$, Arrojadite-(KNa) $KNa_4Ca(Fe,Mn,Mg)_{13}Al(PO_4)_{11}(PO_3OH)(OH,F)_2$, Arrojadite-(KFe) $KNa_2CaNa_2(Fe^{2+},Mn,Mg)_{13}Al(PO_4)_{11}(PO_3OH)(OH,F)_2$, Arrojadite-(NaFe) $NaNa_2CaNa_2(Fe^{2+},Mn,Mg)_{13}Al(PO_4)_{11}(PO_3OH)(OH,F)_2$, Arrojadite-(BaNa) $BaFe^{2+}Na_2Ca(Fe^{2+},Mn,Mg)_{13}Al(PO_4)_{11}(PO_3OH)(OH,F)_2$, Arrojadite-(BaFe) $(Ba,K,Pb)Na_3(Ca,Sr)(Fe^{2+},Mg,Mn)_{14}Al(PO_4)_{11}(PO_3OH)(OH,F)_2$, Arrojadite-(SrFe) $SrFe^{2+}Na_2Ca(Fe^{2+},Mn,Mg)_{13}Al(PO_4)_{11}(PO_3OH)(OH,F)_2$, Arrojadite-(PbFe) $PbFe^{2+}Na_2Ca(Fe^{2+},Mn,Mg)_{13}Al(PO_4)_{11}(PO_3OH)(OH,F)_2$

  - Arrojadite group (Fluorarrojadite subgroup) (Al in Al site, F in W site, Fe in M site)

    - Fluorarrojadite-(KNa) $KNa_4Ca(Fe,Mn,Mg)_{13}Al(PO_4)_{11}(PO_3OH)(F,OH)_2$, Fluorarrojadite-(BaNa) $BaFe^{2+}Na_2Ca(Fe^{2+},Mn,Mg)_{13}Al(PO_4)_{11}(PO_3OH)(F,OH)_2$, Fluorarrojadite-(BaFe) $(Ba,K,Pb)Na_3(Ca,Sr)(Fe^{2+},Mg,Mn)_{14}Al(PO_4)_{11}(PO_3OH)(F,OH)_2$

  - Arrojadite group (Dickinsonite subgroup) (Fe in Al site, OH in W site, Fe in M site)

    - Dickinsonite? $KNa_4Ca(Mn^{2+},Fe^{2+})_{14}Al(PO_4)_{12}(OH)_2$, Dickinsonite-(KMnNa) $KNaMnNa_3Ca(Mn,Fe,Mg)_{13}Al(PO_4)_{11}(PO_4)(OH,F)_2$, Dickinsonite-(KNaNa) $KNaNa_4Ca(Mn,Fe,Mg)_{13}Al(PO_4)_{11}(PO_4)(OH,F)_2$, Dickinsonite-(K-Na) $KNa_4Ca(Mn,Fe,Mg)_{13}Al(PO_4)_{11}(PO_4)(OH,F)_2$, Dickinsonite-(NaNa) $Na_2Na_4Ca(Mn,Fe,Mg)_{13}Al(PO_4)_{11}(PO_4)(OH,F)_2$

  - Ferri-arrojadite-(BaNa) $BaFe^{2+}Na_2Ca(Fe^{2+},Mn,Mg)_{13}Al(PO_4)_{11}(PO_3OH)(F,OH)_2$

  - Lulzacite $Sr_2Fe^{2+}(Fe^{2+},Mg)_2Al_4(PO_4)_4(OH)_{10}$

"Apatite" Supergroup

- Anhydrous phosphates, etc. containing hydroxyl or halogen where $(A)_5\ (XO_4)_3\ Zq$

  - Morelandite group

    - Morelandite $(Ba,Ca,Pb)_5(AsO_4,PO_4)_3Cl$, Alforsite $Ba_5(PO_4)_3Cl$

  - Clinomimetite group

    - Clinomimetite $Pb_5(AsO_4)_3Cl$, Apatite-(CaOH)-M $(Ca,Na)_5[(P,S)O_4]_3(OH,Cl)$

  - Apatite group

    - Apatite* $Ca_5(PO_4)_3(OH,F,Cl)$, Apatite-(CaF) $Ca_5(PO_4)_3F$, Apatite-(CaCl) $Ca_5(PO_4)_3Cl$, Apatite-(CaOH) $Ca_5(PO_4)_3(OH)$, Carbonate-fluorapatite? $Ca_5(PO_4,CO_3)_3F$, Carbonate-hydroxylapatite? $Ca_5(PO_4,CO_3)_3(OH)$, Belovite-(Ce) $(Sr,Ce,Na,-$

$Ca)_5(PO_4)_3(OH)$, Belovite-(La) $(Sr,La,Ce,Ca)_5(PO_4)_3(F,OH)$, Kuannersuite-(Ce) $Ba_6Na_2REE_2(PO_4)_6FCl$, Apatite-(SrOH) $(Sr,Ca)_5(PO_4)_3(F,OH)$, Fluorcaphite (Ca,Sr,Ce,Na)$_5(PO_4)_3F$, Deloneite-(Ce) $NaCa_2SrCe(PO_4)_3F$, Phosphohedyphane $Ca_2Pb_3(PO_4)_3Cl$, IMA2008-009 $Sr_5(PO_4)_3F$, IMA2008-068 $Ca_2Pb_3(PO_4)_3F$

- Svabite group

  - Svabite $Ca_5(AsO_4)_3F$, Turneaureite $Ca_5[(As,P)O_4]_3Cl$, Johnbaumite $Ca_5(AsO_4)_3(OH)$, Fermorite $(Ca,Sr)_5(AsO_4,PO_4)_3(OH)$

- Hedyphane $Ca_2Pb_3(AsO_4)_3Cl$, Phosphohedyphane $Ca_2Pb_3(PO_4)_3Cl$

- Pyromorphite group

  - Pyromorphite $Pb_5(PO_4)_3Cl$, Mimetite $Pb_5(AsO_4)_3Cl$, Vanadinite $Pb_5(VO_4)_3Cl$, Hydroxylpyromorphite $Pb_5(PO_4)_3OH$

"Rockbridgeite" Supergroup

- Anhydrous phosphates, etc. containing hydroxyl or halogen where $(A\ B)_5\ (XO_4)_3\ Zq$

  - Kulanite group

    - Kulanite $Ba(Fe^{2+},Mn,Mg)_2Al_2(PO_4)_3(OH)_3$, Penikisite $BaMg_2Al_2(PO_4)_3(OH)_3$, Bjarebyite $(Ba,Sr)(Mn^{2+},Fe^{2+},Mg)_2Al_2(PO_4)_3(OH)_3$, Perloffite $Ba(Mn,Fe^{2+})_2(Fe^{3+})_2(PO_4)_3(OH)_3$, Johntomaite $Ba(Fe^{2+},Ca,Mn^{2+})_2(Fe^{3+})_2(PO_4)_3(OH)_3$

  - Rockbridgeite group

    - Rockbridgeite $(Fe^{2+},Mn)(Fe^{3+})_4(PO_4)_3(OH)_5$, Frondelite $Mn^{2+}(Fe^{3+})_4(PO_4)_3(OH)_5$, Plimerite $Zn(Fe^{3+})_4(PO_4)_3(OH)_5$

  - Griphite $Ca(Mn,Na,Li)_6Fe^{2+}Al_2(PO_4)_6(F,OH)_2$

"Lazulite" Supergroup

- Anhydrous phosphates, etc. containing hydroxyl or halogen where $(A^{2+}\ B^{2+})_3\ (XO_4)_2\ Zq$

- Lazulite group

  - Lazulite $MgAl_2(PO_4)_2(OH)_2$, Scorzalite $(Fe^{2+},Mg)Al_2(PO_4)_2(OH)_2$, Hentschelite $Cu^{2+}(Fe^{3+})_2(PO_4)_2(OH)_2$, Barbosalite $Fe^{2+}(Fe^{3+})_2(PO_4)_2(OH)_2$

- Lipscombite group

  - Lipscombite $(Fe^{2+},Mn)(Fe^{3+})_2(PO_4)_2(OH)_2$, Zinclipscombite $Zn(Fe^{3+})_2(PO_4)_2(OH)$

- Goedkenite group

  - Goedkenite $(Sr,Ca)_2Al(PO_4)_2(OH)$, Bearthite $Ca_2Al(PO_4)_2(OH)$, Gamagarite $Ba_2(Fe^{3+},Mn^{3+})(VO_4)_2(OH)$, Tokyoite $Ba_2Mn(VO_4)_2(OH)$

- Carminite group

- Carminite $Pb(Fe^{3+})_2(AsO_4)_2(OH)_2$, Sewardite $Ca(Fe^{3+})_2(AsO_4)_2(OH)_2$

- Mounanaite group

  - Mounanaite $Pb(Fe^{3+})_2(VO_4)_2(OH)_2$, Krettnichite $Pb(Mn^{3+})_2(VO_4)_2(OH)_2$

- Preisingerite group

  - Preisingerite $Bi_3(AsO_4)_2O(OH)$, Schumacherite $Bi_3[(V,As,P)O_4]_2O(OH)$

- Jagowerite $BaAl_2(PO_4)_2(OH)_2$, Melonjosephite $CaFe^{2+}Fe^{3+}(PO_4)_2(OH)$, Samuelsonite $(Ca,Ba)Ca_8(Fe^{2+},Mn)_4Al_2(PO_4)_{10}(OH)_2$, Petitjeanite $(Bi^{3+})_3(PO_4)_2O(OH)$, Drugmanite $Pb_2(Fe^{3+},Al)H(PO_4)_2(OH)_2$

Class: Hydrated Phosphates

- Hureaulite group

  - Hureaulite $Mn_5(PO_3OH)_2(PO_4)_2 \bullet 4H_2O$, Sainfeldite $Ca_5(AsO_3OH)_2(AsO_4)_2 \bullet 4H_2O$, Villyaellenite $(Mn^{2+})_5(AsO_3OH)_2(AsO_4)_2 \bullet 4H_2O$, IMA2008-047 $Cd_3Zn_2(AsO_3OH)_2(AsO_4)_2 \bullet 4H_2O$, IMA2008-066 $Mn_5(H_2O)_4(AsO_3OH)_2(AsO_4)_2$

- Lindackerite group

  - Lindackerite $CuCu_4(AsO_4)_2(AsO_3OH)_2 \bullet {\sim}9H_2O$, Braithwaiteite Na-$Cu_5(Ti,Sb)_2O_2(AsO_4)_4[AsO_3(OH)]_2 \bullet 8H_2O$, Veselovskyite $(Zn,Cu,Co)Cu_4(AsO_4)_2(AsO_3OH)_2 \bullet 9H_2O$, IMA2008-010 $CaCu_4(AsO4)_2(AsO_3OH)_2 \bullet 10H_2O$

- Struvite group

  - Struvite $(NH_4)MgPO_4 \bullet 6H_2O$, Struvite-(K) $KMg(PO_4) \bullet 6H_2O$, Hazenite $KNaMg_2(PO_4)_2 \bullet 14H_2O$

- Autunite group

- Formula: $A(UO_2)_2(XO_4)_2 \cdot (10\text{-}12)H_2O$

  - A = Cu, Ca, Ba, or Mg; X = P or As.

  - Autunite $Ca(UO_2)_2(PO_4)_2 \bullet (10\text{-}12)H_2O$, Heinrichite $Ba(UO_2)_2(AsO_4)_2 \bullet (10\text{-}12)H_2O$, Kahlerite $Fe^{2+}(UO_2)_2(AsO_4)_2 \bullet (10\text{-}12)H_2O$, Novacekite-I $Mg(UO_2)_2(AsO_4)_2 \bullet 12H_2O$, Sabugalite $HAl(UO_2)_4(PO_4)_4 \bullet 16H_2O$, Saleeite $Mg(UO_2)_2(PO_4)_2 \bullet 10H_2O$, Torbernite $Cu(UO_2)_2(PO_4)_2 \bullet (8\text{-}12)H_2O$, Uranocircite $Ba(UO_2)_2(PO_4)_2 \bullet 12H_2O$, Uranospinite $Ca(UO_2)_2(AsO_4)_2 \bullet 10H_2O$, Zeunerite $Cu(UO_2)_2(AsO_4)_2 \bullet (10\text{-}16)H_2O$

- Meta-autunite group

- Formula: $A(UO_2)_2(XO_4)_2 \cdot nH_2O$ (n = 6, 7 or 8)

  - A = Cu, Ca, Ba, or Mg and X = P or As.

  - Abernathyite $K_2(UO_2)_2(AsO_4)_2 \cdot 6H_2O$, Bassetite $Fe^{2+}(UO_2)_2(PO_4)_2 \cdot 8H_2O$, Chernikovite $(H_3O)_2(UO_2)_2(PO_4)_2 \cdot 6H_2O$, Lehnerite $Mn^{2+}(UO_2)_2(PO_4)_2 \cdot 8H_2O$, Meta-ankoleite $K_2(UO_2)_2(PO_4)_2 \cdot 6H_2O$, Meta-autunite $Ca(UO_2)_2(PO_4)_2 \cdot (6\text{-}8)H_2O$, Metakahlerite

$Fe^{2+}(UO_2)_2(AsO_4)_2 \cdot 8H_2O$, Metakirchheimerite $Co(UO_2)_2(AsO_4)_2 \cdot 8H_2O$, Metalode-vite $Zn(UO_2)_2(AsO_4)_2 \cdot 10H_2O$, Metanovacekite $Mg(UO_2)_2(AsO_4)_2 \cdot (4-8)H_2O$, Meta-torbernite $Cu(UO_2)_2(PO_4)_2 \cdot 8H_2O$, Metauranocircite $Ba(UO_2)_2(PO_4)_2 \cdot (6-8)H_2O$, Metauranospinite $Ca(UO_2)_2(AsO_4)_2 \cdot 8H_2O$, Metazeunerite $Cu(UO_2)_2(AsO_4)_2 \cdot 8H_2O$, Natrouranospinite $(Na_2,Ca)(UO_2)_2(AsO_4)_2 \cdot 5H_2O$, Sodium Meta-autunite $Na_2(UO_2)_2(PO_4)_2 \cdot (6-8)H_2O$, Uramarsite $(NH_4,H_3O)_2(UO_2)_2(AsO_4,PO_4)_2 \cdot 6H_2O$, Uramphite $(NH_4)(UO_2)(PO_4) \cdot 3H_2O$

- Vivianite group

  - Vivianite $(Fe^{2+})_3(PO_4)_2 \cdot 8H_2O$, Baricite $(Mg,Fe^{2+})_3(PO_4)_2 \cdot 8H_2O$, Erythrite $Co_3(AsO_4)_2 \cdot 8H_2O$, Annabergite $Ni_3(AsO_4)_2 \cdot 8H_2O$, Kottigite $Zn_3(AsO_4)_2 \cdot 8H_2O$, Parasymplesite $(Fe^{2+})_3(AsO_4)_2 \cdot 8H_2O$, Hornesite $Mg_3(AsO_4)_2 \cdot 8H_2O$, Arupite $(Ni,Fe^{2+})_3(PO_4)_2 \cdot 8H_2O$, Pakhomovskyite $Co_3(PO_4)_2 \cdot 8H_2O$

- Walpurgite group

  - Walpurgite $Bi_4(UO_2)(AsO_4)_2O_4 \cdot 2H_2O$, Orthowalpurgite $(UO_2)Bi_4O_4(AsO_4)_2 \cdot 2H_2O$, Phosphowalpurgite $(UO_2)Bi_4(PO_4)O_4 \cdot 2H_2O$

- Roscherite group

  - Roscherite $Ca(Mn^{2+},Fe^{2+})_5Be_4(PO_4)_6(OH)_4 \cdot 6H_2O$, Zanazziite $Ca_2(Mg,Fe^{2+})(Mg,Fe^{2+},Al,Mn,Fe^{3+})_4Be_4(PO_4)_6(OH)_4 \cdot 6H_2O$, Greifensteinite $Ca_2Be_4(Fe^{2+},Mn)_5(PO_4)_6(OH)_4 \cdot 6H_2O$, Atencioite $Ca_2Fe^{2+}[ ]Mg_2(Fe^{2+})_2Be_4(PO_4)_6(OH)_4 \cdot 6H_2O$, Guimaraesite $Ca_2(Zn,Mg,Fe)_5Be_4(PO_4)_6(OH)_4 \cdot 6H_2O$, Footemineite $Ca_2Mn^{2+}(Mn^{2+})_2(Mn^{2+})_2Be_4(PO_4)_6(OH)_4 \cdot 6H_2O$, Ruifrancoite $Ca_2([ ],Mn)_2(Fe^{3+},Mn,Mg)_4Be_4(PO_4)_6(OH)_4 \cdot 6H_2O$

- Pharmacosiderite group

  - Pharmacosiderite $K(Fe^{3+})_4(AsO_4)_3(OH)_4 \cdot (6-7H)_2O$, Alumopharmacosiderite $KAl_4(AsO_4)_3(OH)_4 \cdot 6 \cdot 5H_2O$, Bariopharmacosiderite $Ba(Fe^{3+})_4(AsO_4)_3(OH)_5 \cdot 5H_2O$, Barium-alumopharmacosiderite? $BaAl_4(AsO_4)_3(OH)_5 \cdot 5H_2O$, Natropharmacosiderite $(Na,K)_2(Fe^{3+})_4(AsO_4)_3(OH)_5 \cdot 7H_2O$

- 08.CE.75 group

  - Malhmoodite $FeZr(PO_4)_2 \cdot 4H_2O$, Zigrasite $ZnZr(PO_4)_2 \cdot 4H_2O$, Unnamed (Ca-analogue of zigrasite) $CaZr(PO_4)_2 \cdot 4H_2O$

- "Variscite" supergroup

- Hydrated phosphates, etc. where $A^{3+} XO_4 \cdot xH_2O$

  - Variscite group

    - Variscite $AlPO_4 \cdot 2H_2O$, Strengite $Fe^{3+}PO_4 \cdot 2H_2O$, Scorodite $Fe^{3+}AsO_4 \cdot 2H_2O$, Mansfieldite $AlAsO_4 \cdot 2H_2O$, Yanomamite $In(AsO_4) \cdot 2H_2O$

  - Metavariscite group

    - Metavariscite $AlPO_4 \cdot 2H_2O$, Phosphosiderite $Fe^{3+}PO_4 \cdot 2H_2O$, Kolbeckite $ScPO_4 \cdot 2H_2O$

- Rhabdophane group

  - Rhabdophane-(Ce) $(Ce,La)PO_4 \cdot H_2O$, Rhabdophane-(La) $(La,Ce)PO_4 \cdot H_2O$, Rhabdophane-(Nd) $(Nd,Ce,La)PO_4 \cdot H_2O$, Grayite $(Th,Pb,Ca)PO_4 \cdot H_2O$, Brockite $(Ca,Th,Ce)(PO_4) \cdot H_2O$, Tristramite $(Ca,U^{4+},Fe^{3+})(PO_4,SO_4) \cdot 2H_2O$

- Ningyoite group

  - Ningyoite $(U,Ca,Ce)_2(PO_4)_2 \cdot (1-2)H_2O$, Lermontovite $U^{4+}(PO_4)(OH) \cdot H_2O$ (?), Vyacheslavite $U^{4+}(PO_4)(OH) \cdot 2.5H_2O$

- Koninckite $Fe^{3+}PO_4 \cdot 3H_2O$ (?), Kankite $Fe^{3+}AsO_4 \cdot 3.5H_2O$, Steigerite $AlVO_4 \cdot 3H_2O$, Churchite-(Y) $YPO_4 \cdot 2H_2O$, Churchite-(Nd) $Nd(PO_4) \cdot 2H_2O$, Parascorodite $Fe^{3+}AsO_4 \cdot 2H_2O$, Serrabrancaite $MnPO_4 \cdot H_2O$

- "Mixite" supergroup

- Hydrated phosphates, etc., containing hydroxyl or halogen where $(A)_2 (XO_4) Zq \cdot xH_2O$

  - Mixite group (Arsenate series)

    - Mixite $BiCu_6(AsO_4)_3(OH)_6 \cdot 3H_2O$, Agardite-(Y) $(Y,Ca)Cu_6(AsO_4)_3(OH)_6 \cdot 3H_2O$, Agardite-(La) $(La,Ca)Cu_6(AsO_4)_3(OH)_6 \cdot 3H_2O$, Agardite-(Nd) $(Pb,Nd,Y,La,-Ca)Cu_6(AsO_4)_3(OH)_6 \cdot 3H_2O$, Agardite-(Dy) $(Dy,La,Ca)Cu_6(AsO_4)_3(OH)_6 \cdot 3H_2O$, Agardite-(Ca) $CaCu_6(AsO_4)_3(OH)_6 \cdot 3H_2O$, Agardite-(Ce) $(Ce,Ca)Cu_6(AsO_4)_3(OH)_6 \cdot 3H_2O$, Goudeyite $(Al,Y)Cu_6(AsO_4)_3(OH)_6 \cdot 3H_2O$, Zalesiite $(Ca,Y)Cu_6[(AsO_4)_2(AsO_3OH)(OH)_6] \cdot 3H_2O$, Plumboagardite $(Pb,REE,Ca)Cu_6(AsO_4)_3(OH)_6 \cdot 3H_2O$

  - Mixite group (Phosphate series)

    - Petersite-(Y) $(Y,Ce,Nd,Ca)Cu_6(PO_4)_3(OH)_6 \cdot 3H_2O$, Calciopetersite $CaCu_6[(PO_4)_2(PO_3OH)(OH)_6] \cdot 3H_2O$

- Zapatalite $Cu_3Al_4(PO_4)_3(OH)_9 \cdot 4H_2O$, Juanitaite $(Cu,Ca,Fe)_{10}Bi(AsO_4)_4(OH)_{11} \cdot 2H_2O$

"Brackebushite" Supergroup

- Hydrated phosphates, etc. where $A^{2+} (B^{2+})_2 (XO_4) \cdot xH_2O$

  - Fairfieldite subgroup

    - Fairfieldite $Ca_2(Mn,Fe^{2+})(PO_4)_2 \cdot 2H_2O$, Messelite $Ca_2(Fe^{2+},Mn)(PO_4)_2 \cdot 2H_2O$, Collinsite $Ca_2(Mg,Fe^{2+})(PO_4)_2 \cdot 2H_2O$, Cassidyite $Ca_2(Ni,Mg)(PO_4)_2 \cdot 2H_2O$, Talmessite $Ca_2Mg(AsO_4)_2 \cdot 2H_2O$, Gaitite $Ca_2Zn(AsO_4)_2 \cdot 2H_2O$, Roselite-beta $Ca_2(Co,Mg)(AsO_4)_2 \cdot 2H_2O$, Parabrandtite $Ca_2Mn^{2+}(AsO_4) \cdot 2H_2O$, Hillite $Ca_2(Zn, Mg)[PO_4]_2 \cdot 2H_2O$, Nickeltalmessite $Ca_2Ni(AsO_4)_2 \cdot 2H_2O$

  - Roselite subgroup

    - Roselite $Ca_2(Co,Mg)(AsO_4)_2 \cdot 2H_2O$, Brandtite $Ca_2(Mn,Mg)(AsO_4)_2 \cdot 2H_2O$, Zincroselite $Ca_2Zn(AsO_4)_2 \cdot 2H_2O$, Wendwilsonite $Ca_2(Mg,Co)(AsO_4)_2 \cdot 2H_2O$, Manganlotharmeyerite $Ca(Mn^{3+},Mg,)_2(AsO_4)_2(OH,H_2O)_2$

- Brackebushite group

  - Brackebuschite $Pb_2(Mn,Fe_{2+})(VO_4)_2(OH)$, Arsenbrackebuschite $Pb_2(Fe^{2+},Zn)(AsO_4)_2 \cdot H_2O$, Feinglosite $Pb_2(Zn,Fe)[(As,S)O_4]_2 \cdot H_2O$, Calderonite $Pb_2Fe^{3+}(VO_4)_2(OH)$, Bushmakinite $Pb_2Al(PO_4)(VO_4)(OH)$

- Helmutwinklerite subgroup

  - Tsumcorite $PbZnFe^{2+}(AsO_4)_2 \cdot H_2O$, Helmutwinklerite $PbZn_2(AsO_4)_2 \cdot 2H_2O$, Thometzekite $Pb(Cu,Zn)_2(AsO_4)_2 \cdot 2H_2O$, Mawbyite $Pb(Fe^{3+}Zn)_2(AsO_4)_2(OH,H_2O)_2$, Rappoldite $Pb(Co,Ni,Zn,)_2(AsO_4)_2 \cdot 2H_2O$, Schneebergite $Bi(Co,Ni)_2(AsO_4)_2(OH,H_2O)_2$, Nickelschneebergite $Bi(Ni,Co)_2(AsO_4)_2(OH,H_2O)_2$, Cobalttsumcorite $Pb(Co,Fe)_2(AsO_4)_2(OH,H_2O)_2$

- Unnamed group

  - Wicksite $NaCa_2(Fe^{2+},Mn^{2+})_4MgFe^{3+}(PO_4)_6 \cdot 2H_2O$, Bederite $([\ ],Na)Ca_2(Mn^{2+},Mg,Fe^{2+})_2(Fe^{3+},Mg^{2+},Al)_2Mn^{2+}_2(PO_4)_6 \cdot 2H_2O$, Tassieite $(Na,[\ ])Ca_2(Mg,Fe^{2+},Fe^{3+})_2(Fe^{3+},Mg)_2(Fe^{2+},Mg)_2(PO_4)_6 \cdot 2H_2O$

- Anapaite $Ca_2Fe^{2+}(PO_4)_2 \cdot 4H_2O$, Prosperite $CaZn_2(AsO_4)_2 \cdot H_2O$, Parascholzite $CaZn_2(PO_4)_2 \cdot 2H_2O$, Scholzite $CaZn_2(PO_4)_2 \cdot 2H_2O$, Phosphophyllite $Zn_2(Fe^{2+},Mn)(PO_4)_2 \cdot 4H_2O$, Cabalzarite $Ca(Mg,Al,Fe^{2+})_2(AsO_4)_2(H_2O,OH)_2$, Grischunite $NaCa_2(Mn^{2+})_5Fe^{3+}(AsO_4)_6 \cdot 2H_2O$

"Turquoise" Supergroup

- Hydrated phosphates, etc., containing hydroxyl or halogen where $(A)_3(XO_4)_2 Zq \cdot xH_2O$

  - Burangaite group

    - Burangaite $(Na,Ca)_2(Fe^{2+},Mg)_2Al_{10}(PO_4)_8(OH,O)_{12} \cdot 4H_2O$, Dufrenite $Fe_{2+}(Fe^{3+})_4(PO_4)_3(OH)_5 \cdot 2H_2O$, Natrodufrenite $Na(Fe^{3+},Fe^{2+})(Fe^{3+},Al)_5(PO_4)_4(OH)_6 \cdot 2H_2O$, Matioliite $NaMgAl_5(PO_4)_4(OH)_6 \cdot 2H_2O$, IMA2008-056 $NaMn^{2+}(Fe^{3+})_5(PO_4)_4(OH)_6 \cdot 2H_2O$

  - Souzalite group

    - Souzalite $(Mg,Fe^{2+})_3(Al,Fe^{3+})_4(PO_4)_4(OH)_6 \cdot 2H_2O$, Gormanite $Fe^{2+}_3Al_4(PO_4)_4(OH)_6 \cdot 2H_2O$, Andyrobertsite $KCdCu_5(AsO_4)_4[As(OH)_2O_2] \cdot 2H_2O$, Calcioandyrobertsite-1M $KCaCu_5(AsO_4)_4[As(OH)_2O_2] \cdot 2H_2O$, Calcioandyrobertsite-2O $KCaCu_5(AsO_4)_4[As(OH)_2O_2] \cdot 2H_2O$

  - Turquoise group

    - Aheylite $(Fe^{2+},Zn)Al_6[(OH)_4|(PO_4)_2]_2 \cdot 4H_2O$, Chalcosiderite $Cu(Fe^{3+},Al)_6[(OH)_4|(PO_4)_2]_2 \cdot 4H_2O$, Faustite $(Zn,Cu)Al_6[(OH)_4|(PO_4)_2]_2 \cdot 4H_2O$, Planerite $Al_6[(OH)_4|HPO_4|PO_4]_2 \cdot 4H_2O$, Turquoise $Cu(Al,Fe^{3+})_6[(OH)_4|(PO_4)_2]_2 \cdot 4H_2O$

  - Unnamed group

    - Sampleite $NaCaCu_5(PO_4)_4Cl \cdot 5H_2O$, Lavendulan $NaCaCu_5(AsO_4)_4Cl \cdot 5H_2O$,

Zdenekite $NaPb(Cu^{2+})_5(AsO_4)_4Cl•5H_2O$, Mahnertite $(Na,Ca)(Cu^{2+})_3(AsO_4)_2Cl•5H_2O$, Lemanskiite $NaCaCu_5(AsO_4)_4Cl•5H_2O$

- Duhamelite? $Pb_2Cu_4Bi(VO_4)_4(OH)_3•8H_2O$, Santafeite $(Mn,Fe,Al,Mg)_2(Mn^{4+},Mn^{2+})_2(Ca,Sr,Na)_3(VO_4,AsO_4)_4(OH)_3•2H_2O$, Ogdensburgite $Ca_2(Zn,Mn)(Fe^{3+})_4(AsO_4)_4(OH)_6•6H_2O$, Dewindtite $Pb_3[H(UO_2)_3O_2(PO_4)_2]_2•12H_2O$

"Overite" Supergroup

- Hydrated phosphates, etc., containing hydroxyl or halogen where $(AB)_4(XO_4)_3 Zq \cdot xH_2O$

  - Overite group

    - Overite $CaMgAl(PO_4)_2(OH)•4H_2O$, Segelerite $CaMgFe^{3+}(PO_4)_2(OH)•4H_2O$, Manganosegelerite $(Mn,Ca)(Mn,Fe^{2+},Mg)Fe^{3+}(PO_4)_2(OH)•4H_2O$, Lunokite $(Mn,Ca)(Mg,Fe^{2+},Mn)Al(PO_4)_2(OH)•4H_2O$, Wilhelmvierlingite $CaMn^{2+}Fe^{3+}(PO_4)_2(OH)•2H_2O$, Kaluginite* $(Mn^{2+},Ca)MgFe^{3+}(PO_4)_2(OH)•4H_2O$, Juonniite $CaMgSc(PO_4)_2(OH)•4H_2O$

  - Jahnsite group

    - Jahnsite-(CaMnMg) $CaMnMg_2(Fe^{3+})_2(PO_4)_4(OH)_2•8H_2O$, Jahnsite-(CaMnFe) $CaMn^{2+}(Fe^{2+})_2(Fe^{3+})_2(PO_4)_4(OH)_2•8H_2O$, Jahnsite-(CaMnMn) $CaMn^{2+}(Mn^{2+})_2(Fe^{3+})_2(PO_4)_4(OH)_2•8H_2O$, Jahnsite-(MnMnMn)* $MnMnMn_2(Fe^{3+})_2(PO_4)_4(OH)_2•8H_2O$

  - Whiteite group

    - Whiteite-(CaFeMg) $Ca(Fe^{2+},Mn^{2+})Mg_2Al_2(PO_4)_4(OH)_2•8H_2O$, Whiteite-(MnFeMg) $(Mn^{2+},Ca)(Fe^{2+},Mn^{2+})Mg_2Al_2(PO_4)_4(OH)_2•8H_2O$, Whiteite-(CaMnMg) $CaMn^{2+}Mg_2Al_2(PO_4)_4(OH)_2•8H_2O$, Rittmannite $Mn^{2+}Mn^{2+}Fe^{2+}Al_2(OH)_2(PO_4)_4•8H_2O$, Jahnsite-(CaFeFe) $(Ca,Mn)(Fe^{2+},Mn^{2+})(Fe^{2+})_2(Fe^{3+})_2(PO_4)_4(OH)_2•8H_2O$, Jahnsite-(NaFeMg) $NaFe^{3+}Mg_2(Fe^{3+})_2(PO_4)_4(OH)_2•8H_2O$, Jahnsite-(CaMgMg) $CaMgMg_2(Fe^{3+})_2(PO_4)_4(OH)_2•8H_2O$, Jahnsite-(NaMnMg) $NaMnMg_2(Fe^{3+})_2(PO_4)_4(OH)_2•8H_2O$

  - Leucophosphite group

  - Leucophosphite $K(Fe^{3+})_2(PO_4)_2(OH)•2H_2O$, Tinsleyite $KAl_2(PO_4)_2(OH)•2H_2O$, Spheniscidite $(NH_4,K)(Fe^{3+},Al)_2(PO_4)_2(OH)•2H_2O$

  - Montgomeryite group

    - Montgomeryite $Ca_4MgAl_4(PO_4)_6(OH)_4•12H_2O$, Kingsmountite $(Ca,Mn^{2+})_4(Fe^{2+},Mn^{2+})Al_4(PO_4)_6(OH)_4•12H_2O$, Calcioferrite $Ca_4Fe^{2+}(Fe^{3+},Al)_4(PO_4)_6(OH)_4•12H_2O$, Zodacite $Ca_4Mn^{2+}(Fe^{3+})_4(PO_4)_6(OH)_4•12H_2O$, Angastonite $CaMgAl_2(PO_4)_2(OH)_4•7H_2O$

  - Strunzite group

    - Strunzite $Mn^{2+}(Fe^{3+})_2(PO_4)_2(OH)_2•6H_2O$, Ferrostrunzite $Fe^{2+}(Fe^{3+})_2(PO_4)_2(OH)_2•6H_2O$, Ferristrunzite $Fe^{3+}(Fe^{3+})_2(PO_4)_2(OH)_3•5H_2O$

- Laueite group

  - Laueite $Mn^{2+}(Fe^{3+})_2(PO_4)_2(OH)_2 \cdot 8H_2O$, Stewartite $Mn^{2+}(Fe^{3+})_2(PO_4)_2(OH)_2 \cdot 8H_2O$, Pseudolaueite $Mn^{2+}(Fe^{3+})2(PO_4)_2(OH)_2 \cdot (7\text{-}8)H_2O$, Ushkovite $Mg(Fe^{3+})_2(PO_4)_2(OH)_2 \cdot 8H_2O$, Ferrolaueite $Fe^{2+}(Fe^{3+})_2(PO4)_2(OH)_2 \cdot 8H_2O$

- Gatumbaite group

  - Gatumbaite $CaAl_2(PO_4)_2(OH)_2 \cdot H_2O$, Kleemanite $ZnAl_2(PO_4)_2(OH)_2 \cdot 3H_2O$

- Vanuralite group

  - Vanuralite $Al(UO_2)_2(VO_4)_2(OH) \cdot 11H_2O$, Metavanuralite $Al(UO_2)_2(VO_4)_2(OH) \cdot 8H_2O$, Threadgoldite $Al(UO_2)_2(PO_4)_2(OH) \cdot 8H_2O$, Chistyakovaite $Al(UO_2)_2(AsO_4)_2(F,OH) \cdot 6.5H_2O$

- Vauxite group

  - Vauxite $Fe^{2+}Al_2(PO_4)_2(OH)_2 \cdot 6H_2O$, Paravauxite $Fe^{2+}Al_2(PO_4)_2(OH)_2 \cdot 8H_2O$, Sigloite $Fe^{3+}Al_2(PO_4)_2(OH)_3 \cdot 5H_2O$, Gordonite $MgAl_2(PO_4)_2(OH)_2 \cdot 8H_2O$, Mangangordonite $(Mn^{2+},Fe^{2+},Mg)Al_2(PO_4)_2(OH)_2 \cdot 8H_2O$, Kastningite $(Mn^{2+},Fe^{2+},Mg)Al_2(PO_4)_2(OH)_2 \cdot 8H_2O$, Maghrebite $MgAl_2(AsO_4)_2(OH)_2 \cdot 8H_2O$

- Bermanite group

  - Bermanite $Mn^{2+}(Mn^{3+})_2(PO_4)_2(OH)_2 \cdot 4H_2O$, Ercitite $Na_2(H_2O)_4[(Mn^{3+})_2(OH)_2(PO_4)_2]$

- Arthurite group/ Whitmoreite group

  - Whitmoreite $Fe^{2+}(Fe^{3+})_2(PO_4)_2(OH)_2 \cdot 4H_2O$, Arthurite $Cu(Fe^{3+})_2(AsO_4,PO_4,SO_4)_2(O,OH)_2 \cdot 4H_2O$, Ojuelaite $Zn(Fe^{3+})_2(AsO_4)^2(OH)_2 \cdot 4H_2O$, Earlshannonite $(Mn,Fe^{2+})(Fe^{3+})_2(PO_4)_2(OH)_2 \cdot 4H_2O$, Gladiusite $(Fe^{2+})_2(Fe^{3+},Mg)_4(PO_4)(OH)_{13} \cdot H_2O$, Cobaltarthurite $Co(Fe^{3+})_2(AsO_4)_2(OH)_2 \cdot 4H_2O$, Kunatite $Cu(Fe^{3+})_2(PO_4)_2(OH)_2 \cdot 4H_2O$, Bendadaite $Fe^{2+}(Fe^{3+})_2(AsO_4)_2(OH)_2 \cdot 4H_2O$

- Sincosite group

  - Sincosite $Ca(V^{4+}O)_2(PO_4)_2 \cdot 5H_2O$, Phosphovanadylite $(Ba,Ca,K,Na)_x[(V,Al)_4P_2(O,OH)_{16}] \cdot 12H_2O$ x~0.66, Bariosincosite $Ba(V^{4+}O)_2(PO_4)_2 \cdot 4H_2O$

- Paulkerrite group

  - Paulkerrite $K(Mg,Mn)_2(Fe^{3+},Al)_2Ti(PO_4)_4(OH)_3 \cdot 15H_2O$, Mantienneite $KMg_2Al_2Ti(PO_4)_4(OH)_3 \cdot 15H_2O$, Matveevite? $KTiMn_2(Fe^{3+})_2(PO_4)_4(OH)_3 \cdot 15H_2O$, Benyacarite $(H_2O,K)_2Ti(Mn^{2+},Fe^{2+})_2(Fe^{3+},Ti)_2Ti(PO_4)_4(O,F)_2 \cdot 14H_2O$

- Keckite $Ca(Mn,Zn)_2(Fe^{3+})_3(PO_4)_4(OH)_3 \cdot 2H_2O$, Minyulite $KAl_2(PO_4)_2(OH,F) \cdot 4H_2O$, Giniite $Fe^{2+}(Fe^{3+})_4(PO_4)_4(OH)_2 \cdot 2H_2O$, Metavauxite $Fe^{2+}Al_2(PO_4)_2(OH)_2 \cdot 8H_2O$, Metavauxite $Fe^{2+}Al_2(PO_4)_2(OH)_2 \cdot 8H_2O$, Xanthoxenite $Ca_4(Fe^{3+})_2(PO_4)_4(OH)_2 \cdot 3H_2O$, Beraunite $Fe^{2+}(Fe^{3+})_5(PO_4)_4(OH)_5 \cdot 4H_2O$, Furongite $Al_2(UO_2)(PO_4)_3(OH)_2 \cdot 8H_2O$, Mcauslanite $H(Fe^{2+})_3Al_2(PO_4)_4F \cdot 18H_2O$, Vochtenite $(Fe^{2+},Mg)Fe^{3+}[(UO_2)(PO_4)]_4(OH) \cdot (12\text{-}13)H_2O$

Alunite Supergroup - Part II

- Beudantite group, $AB_3(XO_4)(SO_4)(OH)_6$

  - Beudantite $PbFe_3[(OH)_6|SO_4|AsO_4]$, Corkite $PbFe_3(PO_4)(SO_4)(OH)_6$, Gallobeudantite $PbGa_3(AsO_4)(SO_4)(OH)_6$, Hidalgoite $PbAl_3(AsO_4)(SO_4)(OH)_6$, Hinsdalite $PbAl_3(PO_4)(SO_4)(OH)_6$, Kemmlitzite $(Sr,Ce)Al_3(AsO_4)(SO_4)(OH)_6$, Svanbergite $SrAl_3(PO_4)(SO_4)(OH)_6$, Weilerite $BaAl_3H[(As,P)O_4]_2(OH)_6$, Woodhouseite $CaAl_3(PO_4)(SO_4)(OH)_6$

- Dussertite group/ Arsenocrandallite group

  - Arsenocrandallite $(Ca,Sr)Al_3[(As,P)O_4]_2(OH)_5•H_2O$, Arsenoflorencite-(Ce) $(Ce,La)Al_3(AsO_4)_2(OH)$, Arsenogorceixite $BaAl_3AsO_3(OH)(AsO_4,PO_4)(OH,F)_6$, Arsenogoyazite $(Sr,Ca,Ba)Al_3(AsO_4,PO_4)_2(OH,F)_5•H_2O$, Dussertite $Ba(Fe^{3+})_3(AsO_4)_2(OH)_5$, Graulichite-(Ce) $Ce(Fe^{3+})_3(AsO_4)_2(OH)_6$, Philipsbornite $PbAl_3(AsO_4)_2(OH)_5•H_2O$, Segnitite $Pb(Fe^{3+})_3H(AsO_4)_2(OH)_6$

- Plumbogummite group/ Crandallite group

  - Benauite $HSr(Fe^{3+})_3(PO_4)_2(OH)_6$, Crandallite $CaAl_3[(OH)_5|(PO_4)_2]·H_2O$, Eylettersite $(Th,Pb)_{(1-x)}Al_3(PO_4,SiO_4)_2(OH)_6(?)$, Florencite-(Ce) $CeAl_3(PO_4)_2(OH)_6$, Florencite-(La) $LaAl_3(PO_4)_2(OH)_6$, Florencite-(Nd) $(Nd,La,Ce)Al_3(PO_4)_2(OH)_6$, Gorceixite $BaAl_3(PO_4)(PO_3OH)(OH)_6$, Goyazite $SrAl_3(PO_4)_2(OH)_5•H_2O$, Kintoreite $Pb(Fe^{3+})_3(PO_4)_2(OH,H_2O)_6$, Plumbogummite $PbAl_3(PO_4)_2(OH)_5•H_2O$, Springcreekite $BaV_3(PO_4)_2(OH,H_2O)_6$, Waylandite $BiAl_3(PO_4)_2(OH)_6$, Zairite $Bi(Fe^{3+},Al)_3(PO_4)_2(OH)_6$, Mills et al. (2009)

Class: Non Simple Phosphates

- Stibiconite group

  - Stibiconite $Sb^{3+}(Sb^{5+})_2O_6(OH)$, Bindheimite $Pb_2Sb_2O_6(O,OH)$, Romeite $(Ca,Fe^{2+},Mn,Na)_2(Sb,Ti)_2O_6(O,OH,F)$, Hydroxycalcioroméite (Lewisite) $(Ca,Fe^{2+},Na)_2(Sb,Ti)_2O_7$, Monimolite $(Pb,Ca)_2Sb_2O_7$, Stetefeldtite $Ag_2Sb_2O_6(O,OH)$, Bismutostibiconite $Bi(Sb^{5+},Fe^{3+})_2O_7$, Partzite $Cu_2Sb_2(O,OH)_7$ (?)

- Rossite group

  - Rossite $CaV_2O_6•4H_2O$, Metarossite $CaV_2O_6•2H_2O$, Ansermetite $MnV_2O_6•4H_2O$

- Pascoite group

  - Pascoite $Ca_3V_{10}O_{28}•17H_2O$, Magnesiopascoite $Ca_2Mg[V_{10}O_{28}]•16H_2O$

- Vanadium oxysalts (Hydrated)

  - Hewettite group

    - Hewettite $CaV_6O_{16}•9H_2O$, Metahewettite $CaV_6O_{16}•3H_2O$, Barnesite $Na_2V_6O_{16}•3H_2O$, Hendersonite $Ca_{1.3}V_6O_{16}•6H_2O$, Grantsite $Na_4Ca_x(V^{4+})_{2x}V^{5+}_{(12-2x)}O_{32}•8H_2O$

  - Straczekite group

- Straczekite $(Ca,K,Ba)(V^{5+},V^{4+})_8O_{20}\cdot 3H_2O$, Corvusite $(Na,Ca,K)V_8O_{20}\cdot 4H_2O)$, Fernandinite $CaV_8O_{20}\cdot 4H_2O$, Bariandite $Al_{0.6}V_8O_{20}\cdot 9H_2O$, Bokite $(Al,Fe^{3+})_{1.3}(V^{4+},Fe)_8O_{20}\cdot 4.7H_2O$, Kazakhstanite $(Fe^{3+})_5(V^{4+})_3(V^{5+})_{12}O_{39}(OH)_9\cdot 9H_2O$
  - Schubnelite $Fe^{2+}(V^{5+}O4)H_2O$, Fervanite $(Fe^{3+})_4(VO_4)_4\cdot 5H_2O$, Bannermanite $(Na,K)_{0.7}(V^{5+})_6O_{15}$, Melanovanadite $Ca(V^{4+})_2(V^{5+})_2O_{10}\cdot 5H_2O$
- Anhydrous Molybdates and Tungstates where A XO_4
  - Wolframite series
    - Wolframite* $(Fe,Mn)WO_4$, Hubnerite $MnWO_4$, Ferberite $Fe^{2+}WO_4$, Sanmartinite $(Zn,Fe^{2+})WO_4$, Heftetjernite $ScTaO_4$
  - Scheelite series
    - Scheelite $CaWO_4$, Powellite $CaMoO_4$
  - Wulfenite Series
    - Wulfenite $PbMoO_4$, Stolzite $PbWO_4$
  - Raspite $PbWO_4$

Category 10

- Organic Compounds
  - Category:Coal
  - Category:Oil shale

Class: Organic Minerals

- Category:Oxalate minerals

Extras

- Rocks, ores and other mixtures of minerals
  - Lapislazuli*, Psilomelane*, Olivine* (Fayalite-Forsterite Series)
- Ice
- Liquids: Water, Mercury Hg, Asphaltum*
- Amorphous solids: Polycrase, Pyrobitumen*, Amber*
  - Vitreous (melts by heating): Tektite, Obsydian

Nickel–Strunz Classification

Nickel–Strunz classification is a scheme for categorizing minerals based upon their chemical composition, introduced by German mineralogist Karl Hugo Strunz (24 February 1910 – 19 April 2006) in his *Mineralogische Tabellen* (1941). The 4th and the 5th edition was edited by Christel Tennyson too (1966). It was followed by A.S. Povarennykh with a modified classification (1966 in Russian, 1972 in English).

As curator of the Mineralogical Museum of Friedrich-Wilhelms-Universität (now known as the Humboldt University of Berlin), Strunz had been tasked with sorting the museum's geological collection according to crystal-chemical properties. His *Mineralogical Tables*, has been through a number of modifications; the most recent edition, published in 2001, is the ninth (Mineralogical Tables by Hugo Strunz and Ernest H. Nickel (31 August 1925 – 18 July 2009)). James A. Ferraiolo was responsible for it at Mindat.org. The IMA/CNMNC supports the Nickel–Strunz database.

Classifications

Nickel–Strunz mineral classes

The current scheme divides minerals into ten classes, which are further divided into divisions, families and groups according to chemical composition and crystal structure.

- elements
- sulfides and sulfosalts
- halides
- oxides, hydroxides and arsenites
- carbonates and nitrates
- borates
- sulfates, chromates, molybdates and tungstates
- phosphates, arsenates and vanadates
- silicates
- Organic compounds

IMA/CNMNC mineral classes

IMA/CNMNC proposes a new hierarchical scheme (Mills et al. 2009), using the Nickel–Strunz classes (10 ed) this gives:

- Classification of minerals (non silicates)
  - Nickel–Strunz class 01: Native Elements
    - Class: native elements
  - Nickel–Strunz class 02: Sulfides and Sulfosalts
    - Class 02.A – 02.G: sulfides, selenides, tellurides; arsenides, antimonides, bismuthides

- Class 02.H – 02.M: sulfosalts; sulfarsenites, sulfantimonites, sulfbismuthites, etc.
- Nickel–Strunz class 03: Halogenides
  - Class: halides
- Nickel–Strunz class 04: Oxides
  - Class: oxides
  - Class: hydroxides
  - Class: arsenites (including antimonites, bismuthites, sulfites, selenites and tellurites)
- Nickel–Strunz class 05: Carbonates and Nitrates
  - Class: carbonates
  - Class: nitrates
- Nickel–Strunz class 06: Borates
  - Class: borates
    - Subclass: nesoborates
    - Subclass: soroborates
    - Subclass: cycloborates
    - Subclass: inoborates
    - Subclass: phylloborates
    - Subclass: tectoborates
- Nickel–Strunz class 07: Sulfates, Selenates, Tellurates
  - Class: sulfates, selenates, tellurates
  - Class: chromates
  - Class: molybdates
  - Class: tungstates
  - Nickel–Strunz class 08: Phosphates, Arsenates, Vanadates
  - Class: phosphates
  - Class: arsenates
  - Class: vanadates
- Nickel–Strunz class 10: Organic Compounds
  - Class: organic compounds

- Classification of minerals (silicates)

  - Nickel–Strunz class 09: Silicates and Germanates

    - Class: silicates

    - Subclass 09.A: nesosilicates

    - Subclass 09.B: sorosilicates

    - Subclass 09.C: cyclosilicates

    - Subclass 09.D: inosilicates

    - Subclass 09.E: phyllosilicates

    - Subclass: tectosilicates

      - 09.F: without zeolitic H_2O

      - 09.G: with zeolitic H_2O; zeolite family

    - Subclass 09.J: germanates

References

- Klein, Cornelis and Cornelius Hurlbut, Jr., Manual of Mineralogy, Wiley, 20th ed., 1985 pp. 320 - 325 ISBN 0-471-80580-7

- Klein, Cornelis and Cornelius S. Hurlbut, 1985, Manual of Mineralogy, 20th ed., John Wiley and Sons, New York, pp. 347-354 ISBN 0-471-80580-7

- Hurlbut, Cornelius S.; Klein, Cornelis, 1985, Manual of Mineralogy, 20th ed., John Wiley and Sons, New York ISBN 0-471-80580-7

- Knobloch, Eberhard (2003). The shoulders on which We stand/Wegbereiter der Wissenschaft (in German and English). Springer. pp. 170–173. ISBN 3-540-20557-8.

- Ernest H. Nickel and Monte C. Nichols (2008-05-22). "IMA/CNMNC List of Mineral Name based on the database MINERAL, which Materials Data, Inc. (MDI) makes available" (PDF). Retrieved 2011-01-31.

- Dyar, Gunter, and Tasa (2007). Mineralogy and Optical Mineralogy. Mineralogical Society of America. pp. 2–4. ISBN 978-0939950812.

- Bindi, L.; Paul J. Steinhardt; Nan Yao; Peter J. Lu (2011). "Icosahedrite, $Al_{63}Cu_{24}Fe_{13}$, the first natural quasicrystal" (PDF). American Mineralogist. 96: 928–931. doi:10.2138/am.2011.3758.

Theories and Concepts of Mineralogy

This chapter elucidates the crucial theories and concepts of minerals. Biomineralization refers to the process by which living organisms produce minerals often to stiffen and harden tissues. Biocrystallization is the formation of crystals from macromolecules by living organisms. The content explores the topics of biocrystallization, biomineralization, mineral alteration, crystal growth, evaporite etc.

Concretion

Concretions on Bowling Ball Beach (Mendocino County, California) weathered out of steeply tilted Cenozoic mudstone

A concretion is a hard, compact mass of matter formed by the precipitation of mineral cement within the spaces between particles, and is found in sedimentary rock or soil. Concretions are often ovoid or spherical in shape, although irregular shapes also occur. The word 'concretion' is derived from the Latin *con* meaning 'together' and *crescere* meaning 'to grow'. Concretions form within layers of sedimentary strata that have already been deposited. They usually form early in the burial history of the sediment, before the rest of the sediment is hardened into rock. This concretionary cement often makes the concretion harder and more resistant to weathering than the host stratum.

There is an important distinction to draw between concretions and nodules. Concretions are formed from mineral precipitation around some kind of nucleus while a nodule is a replacement body.

Descriptions dating from the 18th century attest to the fact that concretions have long been regarded as geological curiosities. Because of the variety of unusual shapes, sizes and compositions, concretions have been interpreted to be dinosaur eggs, animal and plant fossils (called pseudofossils), extraterrestrial debris or human artifacts.

Origins

Detailed studies (i.e., Boles *et al.*, 1985; Thyne and Boles, 1989; Scotchman, 1991; Mozley and Burns, 1993; McBride *et al.*, 2003; Chan *et al.*, 2005; Mozley and Davis, 2005) published in peer-reviewed journals have demonstrated that concretions form subsequent to burial during diagenesis. They quite often form by the precipitation of a considerable amount of cementing material around a nucleus, often organic, such as a leaf, tooth, piece of shell or fossil. For this reason, fossil collectors commonly break open concretions in their search for fossil animal and plant specimens. Some of the most unusual concretion nuclei, as documented by Al-Agha *et al.* (1995), are World War II military shells, bombs, and shrapnel, which are found inside siderite concretions found in an English coastal salt marsh.

Depending on the environmental conditions present at the time of their formation, concretions can be created by either concentric or pervasive growth (Mozley, 1996; Raiswell and Fisher, 2000). In concentric growth, the concretion grows as successive layers of mineral accrete to its surface. This process results in the radius of the concretion growing with time. In case of pervasive growth, cementation of the host sediments, by infilling of its pore space by precipitated minerals, occurs simultaneously throughout the volume of the area, which in time becomes a concretion.

Appearance

Samples of small rock concretions found at Hells Hollow State Park in Pennsylvania.

Concretions vary in shape, hardness and size, ranging from objects that require a magnifying lens to be clearly visible to huge bodies three meters in diameter and weighing several thousand pounds. The giant, red concretions occurring in Theodore Roosevelt National Park, in North Dakota, are almost 3 m (9.8 ft) in diameter. Spheroidal concretions, as large as 9 m (30 ft) in diameter, have been found eroding out of the Qasr El Sagha Formation within the Faiyum depression of Egypt. Concretions are usually similar in color to the rock in which they are found. Concretions occur in a wide variety of shapes, including spheres, disks, tubes, and grape-like or soap bubble-like aggregates.

Composition

They are commonly composed of a carbonate mineral such as calcite; an amorphous or microcrystalline form of silica such as chert, flint, or jasper; or an iron oxide or hydroxide such as goethite and hematite. They can also be composed of other minerals that include dolomite, ankerite, siderite, pyrite, marcasite, barite and gypsum.

Although concretions often consist of a single dominant mineral, other minerals can be present depending on the environmental conditions which created them. For example, carbonate concretions, which form in response to the reduction of sulfates by bacteria, often contain minor percentages of pyrite. Other concretions, which formed as a result of microbial sulfate reduction, consist of a mixture of calcite, barite, and pyrite.

Occurrence

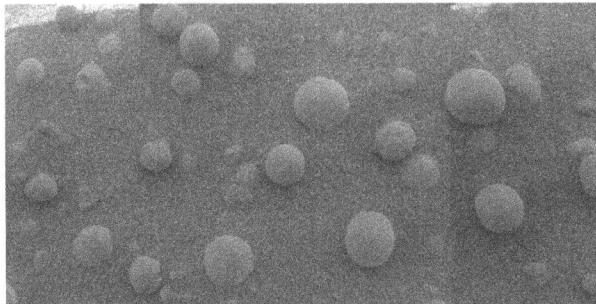

Mosaic shows spherules, some partly embedded, spread over (smaller) soil grains on the Martian surface.

Concretions are found in a variety of rocks, but are particularly common in shales, siltstones, and sandstones. They often outwardly resemble fossils or rocks that look as if they do not belong to the stratum in which they were found. Occasionally, concretions contain a fossil, either as its nucleus or as a component that was incorporated during its growth but concretions are not fossils themselves. They appear in nodular patches, concentrated along bedding planes, protruding from weathered cliffsides, randomly distributed over mudhills or perched on soft pedestals.

Small hematite concretions, dubbed "blueberries" due to their resemblance to blueberries in a muffin, have been observed by the *Opportunity* rover in the Eagle Crater on Mars.

Types of Concretion

Concretions vary considerably in their compositions, shapes, sizes and modes of origin.

Septarian Concretions

Septarian concretions or septarian nodules, are concretions containing angular cavities or cracks, called "septaria". The word comes from the Latin word *septum*; "partition", and refers to the cracks/separations in this kind of rock. There is an incorrect explanation that it comes from the Latin word for "seven", *septem*, referring to the number of cracks that commonly occur. Cracks are highly variable in shape and volume, as well as the degree of shrinkage they indicate. Although it has commonly been assumed that concretions grew incrementally from the inside outwards, the fact that radially oriented cracks taper towards the margins of septarian concretions is taken as evidence that in these cases the periphery was stiffer while the inside was softer, presumably due to a gradient in the amount of cement precipitated.

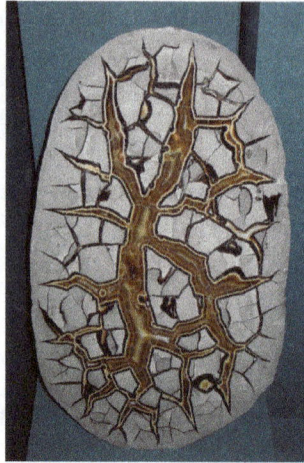

A slice of a septarian nodule

The process that created the septaria, which characterize septarian concretions, remains a mystery. A number of mechanisms, *e.g.* the dehydration of clay-rich, gel-rich, or organic-rich cores; shrinkage of the concretion's center; expansion of gases produced by the decay of organic matter; brittle fracturing or shrinkage of the concretion interior by either earthquakes or compaction; and others, have been proposed for the formation of septaria (Pratt 2001). At this time, it is uncertain, which, if any, of these and other proposed mechanisms is responsible for the formation of septaria in septarian concretions (McBride *et al.* 2003). Septaria usually contain crystals precipitated from circulating solutions, usually of calcite. Siderite or pyrite coatings are also occasionally observed on the wall of the cavities present in the septaria, giving rise respectively to a panoply of bright reddish and golden colors. Some septaria may also contain small calcite stalactites and well-shaped millimetric pyrite single crystals.

Moeraki Boulders, New Zealand

A spectacular example of septarian concretions, which are as much as 3 meters (9.8 feet) in diameter, are the Moeraki Boulders. These concretions are found eroding out of Paleocene mudstone of the Moeraki Formation exposed along the coast near Moeraki, South Island, New Zealand. They are composed of calcite-cemented mud with septarian veins of calcite and rare late-stage quartz and ferrous dolomite (Boles *et al.* 1985, Thyne and Boles 1989). Very similar concretions, which are as much as 3 meters (9.8 feet) in diameter and called "Koutu Boulders", litter the beach between Koutu and Kauwhare points along the south shore of the Hokianga Harbour of Hokianga, North Island, New Zealand. The much smaller septarian concretions found in the Kimmeridge Clay exposed in cliffs along the Wessex Coast of England are more typical examples of septarian concretions (Scotchman 1991).

Cannonball Concretions

Cannonball concretions are large spherical concretions, which resemble cannonballs. These are found along the Cannonball River within Morton and Sioux Counties, North Dakota, and can reach 3 m (9.8 ft) in diameter. They were created by early cementation of sand and silt by calcite. Similar cannonball concretions, which are as much as 4 to 6 m (13 to 20 ft) in diameter, are found associated with sandstone outcrops of the Frontier Formation in northeast Utah and central Wyoming. They formed by the early cementation of sand by calcite (McBride *et al.* 2003). Somewhat weathered and eroded giant cannonball concretions, as large as 6 meters (20 feet) in diameter, occur in abundance at "Rock City" in Ottawa County, Kansas. Large and spherical boulders are also found along Koekohe beach near Moeraki on the east coast of the South Island of New Zealand. The Moeraki Boulders and Koutu Boulders of New Zealand are examples of septarian concretions, which are also cannonball concretions. Large spherical rocks, which are found on the shore of Lake Huron near Kettle Point, Ontario, and locally known as "kettles", are typical cannonball concretions. Cannonball concretions have also been reported from Van Mijenfjorden, Spitsbergen; near Haines Junction, Yukon Territory, Canada; Jameson Land, East Greenland; near Mecevici, Ozimici, and Zavidovici in Bosnia-Herzegovina; in Alaska in the Kenai Peninsula Captain Cook State Park on north of Cook Inlet beach. Reports of cannonball concretions have also come from Bandeng and Zhanlong hills near Gongxi Town, Hunan Province, China.

Hiatus Concretions

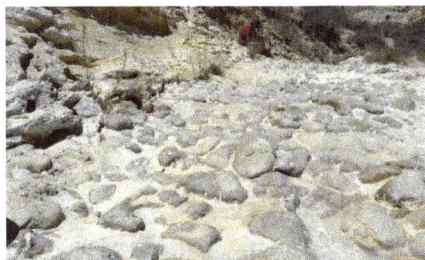

Hiatus concretions at the base of the Menuha Formation (Upper Cretaceous), the Negev, southern Israel.

Hiatus concretion encrusted by bryozoans (thin, branching forms) and an edrioasteroid; Kope Formation (Upper Ordovician), northern Kentucky.

Hiatus concretions are distinguished by their stratigraphic history of exhumation, exposure and reburial. They are found where submarine erosion has concentrated early diagenetic concretions as lag surfaces by washing away surrounding fine-grained sediments (Zaton 2010). Their significance for stratigraphy, sedimentology and paleontology was first noted by Voigt (1968) who referred to them as *Hiatus-Konkretionen*. "Hiatus" refers to the break in sedimentation that allowed this erosion and exposure. They are found throughout the fossil record but are most common during periods in which calcite sea conditions prevailed, such as the Ordovician, Jurassic and Cretaceous (Zaton 2010). Most are formed from the cemented infillings of burrow systems in siliciclastic or carbonate sediments.

A distinctive feature of hiatus concretions separating them from other types is that they were often encrusted by marine organisms including bryozoans, echinoderms and tube worms in the Paleozoic (e.g., Wilson 1985) and bryozoans, oysters and tube worms in the Mesozoic and Cenozoic (e.g., Taylor and Wilson 2001). Hiatus concretions are also often significantly bored by worms and bivalves (Taylor and Wilson 2001).

Elongate Concretions

Elongate concretions form parallel to sedimentary strata and have been studied extensively due to the inferred influence of phreatic (saturated) zone groundwater flow direction on the orientation of the axis of elongation (e.g., Johnson, 1989; McBride *et al.*, 1994; Mozley and Goodwin, 1995; Mozley and Davis, 2005). In addition to providing information about the orientation of past fluid flow in the host rock, elongate concretions can provide insight into local permeability trends (i.e., permeability correlation structure; Mozley and Davis, 1996), variation in groundwater velocity (Davis, 1999), and the types of geological features that influence flow.

Elongate concretions are well known in the Kimmeridge Clay formation of northwest Europe. In outcrops, where they have acquired the name "doggers", they are typically only a few metres across, but in the subsurface they can be seen to penetrate up to tens of metres of along-hole dimension. Unlike limestone beds, however, it is impossible to consistently correlate them between even closely spaced wells.

Moqui Marbles

Moqui Marbles, hematite, goethite concretions, from the Navajo Sandstone of southeast Utah. The cube at the top is one cubic centimeter in size.

Moqui Marbles, also called Moqui balls, and "Moki marbles", are iron oxide concretions which can be found eroding in great abundance out of outcrops of the Navajo Sandstone within south-central and southeastern Utah. These concretions range in shape from spheres to discs, buttons, spiked balls, cylindrical forms, and other odd shapes. They range from pea-size to baseball-size. They were created by the precipitation of iron, which was dissolved in groundwater. They are further described by (Chan and Parry 2002, Chan *et al.* 2005, Loope et al. 2010, 2011).

Kansas Pop Rocks

Kansas pop rocks are concretions of either iron sulfide, *i.e.* pyrite and marcasite, or in some cases jarosite, which are found in outcrops of the Smoky Hill Chalk Member of the Niobrara Formation within Gove County, Kansas. They are typically associated with thin layers of altered volcanic ash, called bentonite, that occur within the chalk comprising the Smoky Hill Chalk Member. A few of these concretions enclose, at least in part, large flattened valves of inoceramid bivalves. These concretions range in size from a few millimeters to as much as 0.7 m (2.3 ft) in length and 12 cm (0.39 ft) in thickness. Most of these concretions are oblate spheroids. Other "pop rocks" are small polycuboidal pyrite concretions, which are as much as 7 cm (0.23 ft) in diameter (Hattin 1982). These concretions are called "pop rocks" because they explode if thrown in a fire. Also, when they are either cut or hammered, they produce sparks and a burning sulfur smell. Contrary to what has been published on the Internet, none of the iron sulfide concretions, which are found in the Smoky Hill Chalk Member were created by either the replacement of fossils or by metamorphic processes. In fact, metamorphic rocks are completely absent from the Smoky Hill Chalk Member (Hattin 1982). Instead, all of these iron sulfide concretions were created by the precipitation of iron sulfides within anoxic marine calcareous ooze after it had accumulated and before it had lithified into chalk.

Iron sulfide concretions, such as the Kansas Pop rocks, consisting of either pyrite and marcasite, are nonmagnetic (Hobbs and Hafner 1999). On the other hand, iron sulfide concretions, which either are composed of or contain either pyrrhotite or smythite, will be magnetic to varying degrees (Hoffmann, 1993). Prolonged heating of either a pyrite or marcasite concretion will convert portions of either mineral into pyrrhotite causing the concretion to become slightly magnetic.

Calcium Carbonate Disc Concretions

Marleka fairy stone from Stensö in Sweden

These so-called fairy stones consist of single or multiple discs, usually 6–10 cm in diameter and often with concentric grooves on their surfaces. They form in Quaternary clay as calcium carbonate migrates to some small fossil or pebble. Fairy stones are particularly common in the Harricana River valley in the Abitibi-Témiscamingue administrative region of Quebec, and in Östergötland county, Sweden.

Crystal Growth

A crystal is a solid material whose constituent atoms, molecules, or ions are arranged in an orderly repeating pattern extending in all three spatial dimensions. Crystal growth is a major stage of a crystallization process, and consists in the addition of new atoms, ions, or polymer strings into the characteristic arrangement of a crystalline Bravais lattice. The growth typically follows an initial stage of either homogeneous or heterogeneous (surface catalyzed) nucleation, unless a "seed" crystal, purposely added to start the growth, was already present.

Quartz is one of the several thermodynamically stable crystalline forms of silica, SiO_2

The action of crystal growth yields a crystalline solid whose atoms or molecules are typically close packed, with fixed positions in space relative to each other. The crystalline state of matter is characterized by a distinct structural rigidity and virtual resistance to deformation (i.e. changes of shape and/or volume). Most crystalline solids have high values both of Young's modulus and of the shear modulus of elasticity. This contrasts with most liquids or fluids, which have a low shear modulus, and typically exhibit the capacity for macroscopic viscous flow.

Introduction

Crystalline solids are typically formed by cooling and solidification from the liquid state. According to the Ehrenfest classification of first-order phase transitions, there is a discontinuous change in volume (and thus a discontinuity in the slope or first derivative with respect to temperature, dV/dT) at the melting point. Within this context, the crystal and melt are distinct phases with an interfacial discontinuity having a surface of tension with a positive surface energy. Thus, a metastable parent phase is always stable with respect to the nucleation of small embryos or droplets from a daughter phase, provided it has a positive surface of tension. Such first-order transitions must proceed by the advancement of an interfacial region whose structure and properties vary discontinuously from the parent phase.

The process of nucleation and growth generally occurs in two different stages. In the first nucleation stage, a small nucleus containing the newly forming crystal is created. Nucleation occurs relatively slowly as the initial crystal components must impinge on each other in the correct orientation and placement for them to adhere and form the crystal. After crystal nucleation, the second stage of growth rapidly ensues. Crystal growth spreads outwards from the nucleating site. In this faster process, the elements which form the motif add to the growing crystal in a prearranged system, the crystal lattice, started in crystal nucleation. As first pointed out by Frank, perfect crystals would only grow exceedingly slowly. Real crystals grow comparatively rapidly because they contain dislocations (and other defects), which provide the necessary growth points, thus providing the necessary catalyst for structural transformation and long-range order formation.

Discontinuity

The conditions of a homogeneous environment are often approximated to but rarely ever realized. Crystal growth always involves some form of transport of matter or heat (or both). And homogeneous conditions for the transport process can only exist for spherical, cylindrical, or infinite plane surfaces. A polyhedral crystal cannot grow (remaining polyhedral) with uniform levels of supersaturation (or supercooling) over its faces. In general, the supersaturation is greatest at its corners. This refutes the assumption that the growth rate is a function of orientation and local supersaturation.

Thus, the crystal face must grow as a whole. The growth rate of the entire face is determined by the driving force (level of supersaturation) at the point of emergence of the predominant point of growth (e.g. a dislocation, a foreign particle acting as catalyst, or crystal twin). The defect-free habit face can thus resist a finite level of supersaturation without any growth at all.

Josiah Willard Gibbs was the first to point out that in the growth of a perfect crystal, the first derivative of the free energy with respect to mass becomes periodically undefinable — at each time that an additional layer on the crystal face is completed. There is discontinuity in the chemical potential at each such point.

In one sense, the crystal can then be in equilibrium with environments having a range of chemical potentials. In another sense, it is not in equilibrium. There are available states of lower free energy. But any free energy barrier must be passed by a fluctuation, or nucleation process, in order to access it. The fundamental thermodynamic effect of a screw dislocation is to eliminate this discontinuity in the chemical potential, by making it impossible to ever complete a single crystal face.

Nucleation

Silver crystal growing on a ceramic substrate.

Nucleation can be either homogeneous, without the influence of foreign particles, or heterogeneous, with the influence of foreign particles. Generally, heterogeneous nucleation takes place more quickly since the foreign particles act as a scaffold for the crystal to grow on, thus eliminating the necessity of creating a new surface and the incipient surface energy requirements.

Heterogeneous nucleation can take place by several methods. Some of the most typical are small inclusions, or cuts, in the container the crystal is being grown on. This includes scratches on the sides and bottom of glassware. A common practice in crystal growing is to add a foreign substance, such as a string or a rock, to the solution, thereby providing nucleation sites for facilitating crystal growth and reducing the time to fully crystallize.

The number of nucleating sites can also be controlled in this manner. If a brand-new piece of glassware or a plastic container is used, crystals may not form because the container surface is too smooth to allow heterogeneous nucleation. On the other hand, a badly scratched container will result in many lines of small crystals. To achieve a moderate number of medium-sized crystals, a container which has a few scratches works best. Likewise, adding small previously made crystals, or seed crystals, to a crystal growing project will provide nucleating sites to the solution. The addition of only one seed crystal should result in a larger single crystal.

Some important features during growth are the arrangement, the origin of growth, the interface form (important for the driving force), and the final size. When origin of growth is only in one direction for all the crystals, it can result in the material becoming very anisotropic (different properties in different directions). The interface form determines the additional free energy for each volume of crystal growth.

Lattice arrangement in metals often takes the structure of body centered cubic, face centered cubic, or hexagonal close packed. The final size of the crystal is important for mechanical properties of materials. (For example, in metals it is widely acknowledged that large crystals can stretch further due to the longer deformation path and thus lower internal stresses.).

Mechanisms of Growth

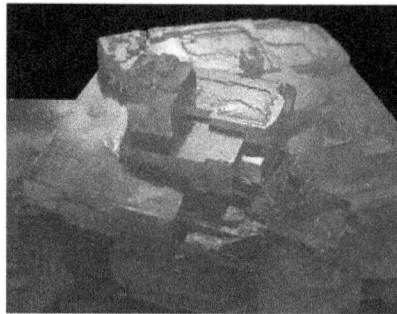

An example of the cubic crystals typical of the rock-salt structure.

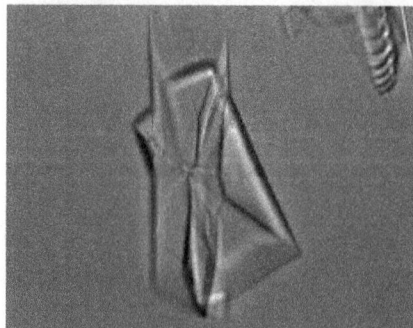

Time-lapse of growth of a citric acid crystal. The video covers an area of 2.0 by 1.5 mm and was captured over 7.2 min.

The interface between a crystal and its vapor can be molecularly sharp at temperatures well below the melting point. An ideal crystalline surface grows by the spreading of single layers, or equivalently, by the lateral advance of the growth steps bounding the layers. For perceptible growth rates, this mechanism requires a finite driving force (or degree of supercooling) in order to lower the nucleation barrier sufficiently for nucleation to occur by means of thermal fluctuations. In the theory of crystal growth from the melt, Burton and Cabrera have distinguished between two major mechanisms:

Non-uniform lateral growth. The surface advances by the lateral motion of steps which are one interplanar spacing in height (or some integral multiple thereof). An element of surface undergoes no change and does not advance normal to itself except during the passage of a step, and then it advances by the step height. It is useful to consider the step as the transition between two adjacent regions of a surface which are parallel to each other and thus identical in configuration — displaced from each other by an integral number of lattice planes. Note here the distinct possibility of a step in a diffuse surface, even though the step height would be much smaller than the thickness of the diffuse surface.

Uniform normal growth. The surface advances normal to itself without the necessity of a stepwise growth mechanism. This means that in the presence of a sufficient thermodynamic driving force, every element of surface is capable of a continuous change contributing to the advancement of the interface. For a sharp or discontinuous surface, this continuous change may be more or less uniform over large areas each successive new layer. For a more diffuse surface, a continuous growth mechanism may require change over several successive layers simultaneously.

Non-uniform lateral growth is a geometrical motion of steps — as opposed to motion of the entire surface normal to itself. Alternatively, uniform normal growth is based on the time sequence of an element of surface. In this mode, there is no motion or change except when a step passes via a continual change. The prediction of which mechanism will be operative under any set of given conditions is fundamental to the understanding of crystal growth. Two criteria have been used to make this prediction:

- Whether or not the surface is diffuse. A diffuse surface is one in which the change from one phase to another is continuous, occurring over several atomic planes. This is in contrast to a sharp surface for which the major change in property (e.g. density or composition) is discontinuous, and is generally confined to a depth of one interplanar distance.

- Whether or not the surface is singular. A singular surface is one in which the surface tension as a function of orientation has a pointed minimum. Growth of singular surfaces is known to requires steps, whereas it is generally held that non-singular surfaces can continuously advance normal to themselves.

Driving Force

Consider next the necessary requirements for the appearance of lateral growth. It is evident that the lateral growth mechanism will be found when any area in the surface can reach a metastable equilibrium in the presence of a driving force. It will then tend to remain in such an equilibrium configuration until the passage of a step. Afterward, the configuration will be identical except that each part of the step but will have advanced by the step height. If the surface cannot reach equilibrium in the presence of a driving force, then it will continue to advance without waiting for the lateral motion of steps.

Thus, Cahn concluded that the distinguishing feature is the ability of the surface to reach an equilibrium state in the presence of the driving force. He also concluded that for every surface or interface in a crystalline medium, there exists a critical driving force, which, if exceeded, will enable the surface or interface to advance normal to itself, and, if not exceeded, will require the lateral growth mechanism.

Thus, for sufficiently large driving forces, the interface can move uniformly without the benefit of either a heterogeneous nucleation or screw dislocation mechanism. What constitutes a sufficiently large driving force depends upon the diffuseness of the interface, so that for extremely diffuse interfaces, this critical driving force will be so small that any measurable driving force will exceed it. Alternatively, for sharp interfaces, the critical driving force will be very large, and most growth will occur by the lateral step mechanism.

Note that in a typical solidification or crystallization process, the thermodynamic driving force is dictated by the degree of supercooling.

Morphology

Silver sulfide whiskers growing out of surface-mount resistors.

It is generally believed that the mechanical and other properties of the crystal are also pertinent to the subject matter, and that crystal morphology provides the missing link between growth kinetics and physical properties. The necessary thermodynamic apparatus was provided by Josiah Willard Gibbs'study of heterogeneous equilibrium. He provided a clear definition of surface energy, by which the concept of surface tension is made applicable to solids as well as liquids. He also appreciated that *an anisotropic surface free energy implied a non-spherical equilibrium shape*, which should be thermodynamically defined as *the shape which minimizes the total surface free energy*.

It may be instructional to note that whisker growth provides the link between the mechanical phenomenon of high strength in whiskers and the various growth mechanisms which are responsible for their fibrous morphologies. (Prior to the discovery of carbon nanotubes, single-crystal whiskers had the highest tensile strength of any materials known). Some mechanisms produce defect-free whiskers, while others may have single screw dislocations along the main axis of growth — producing high strength whiskers.

The mechanism behind whisker growth is not well understood, but seems to be encouraged by compressive mechanical stresses including mechanically induced stresses, stresses induced by diffusion of different elements, and thermally induced stresses. Metal whiskers differ from metallic dendrites in several respects. Dendrites are fern-shaped like the branches of a tree, and grow across the surface of the metal. In contrast, whiskers are fibrous and project at a right angle to the surface of growth, or substrate.

Diffusion-Control

NASA animation of dendrite formation in microgravity.

Manganese dendrites on a limestone bedding plane from Solnhofen, Germany. Scale in mm.

Very commonly when the supersaturation (or degree of supercooling) is high, and sometimes even when it is not high, growth kinetics may be diffusion-controlled. Under such conditions, the polyhedral crystal form will be unstable, it will sprout protrusions at its corners and edges where the degree of supersaturation is at its highest level. The tips of these protrusions will clearly be the points of highest supersaturation. It is generally believed that the protrusion will become longer (and thinner at the tip) until the effect of interfacial free energy in raising the chemical potential slows the tip growth and maintains a constant value for the tip thickness.

In the subsequent tip-thickening process, there should be a corresponding instability of shape. Minor bumps or "bulges" should be exaggerated — and develop into rapidly growing side branches. In such an unstable (or metastable) situation, minor degrees of anisotropy should be sufficient to determine directions of significant branching and growth. The most appealing aspect of this argument, of course, is that it yields the primary morphological features of dendritic growth.

Evaporite

Evaporite is a name for a water-soluble mineral sediment that results from concentration and crystallization by evaporation from an aqueous solution. There are two types of evaporite deposits:

marine, which can also be described as ocean deposits, and non-marine, which are found in standing bodies of water such as lakes. Evaporites are considered sedimentary rocks and are formed by chemical sediments.

Cobble encrusted with halite evaporated from the Dead Sea, Israel

Formation of Evaporite Rocks

Although all water bodies on the surface and in aquifers contain dissolved salts, the water must evaporate into the atmosphere for the minerals to precipitate. For this to happen, the water body must enter a restricted environment where water input into this environment remains below the net rate of evaporation. This is usually an arid environment with a small basin fed by a limited input of water. When evaporation occurs, the remaining water is enriched in salts, and they precipitate when the water becomes supersaturated.

Evaporite Depositional Environments

Marine Evaporites

Anhydrite

Marine evaporites tend to have thicker deposits and are usually the focus of more extensive research. They also have a system of evaporation. When scientists evaporate ocean water in a laboratory, the minerals are deposited in a defined order that was first demonstrated by Usiglio in 1884. The first phase of the experiment begins when about 50% of the original water depth remains. At this point, minor carbonates begin to form. The next phase in the sequence comes when the experiment is left with about 20% of its original level. At this point, the mineral gypsum begins to form, which is then followed by halite at 10%, excluding carbonate minerals that tend not to be evaporates. The most common minerals that are generally considered to be the most representative of marine evaporates are calcite, gypsum and anhydrite, halite, sylvite, carnallite, langbeinite, polyhalite, and kainite. Kieserite ($MgSO_4$) may also be included, which often will make up less than four percent of the overall content. However, there are approximately 80 different minerals that have been reported found in evaporite deposits (Stewart,1963;Warren,1999), though only about a dozen are common enough to be considered important rock formers.

Non-Marine Evaporites

Non-marine evaporites are usually composed of minerals that are not common in marine environments, because in general the water from which non-marine evaporite precipitates have proportions of chemical elements different from those found in the marine environments. Common minerals that are found in these deposits include blödite, borax, epsomite, gaylussite, glauberite, mirabilite, thenardite and trona. Non-marine deposits may also contain halite, gypsum, and anhydrite, and may in some cases even be dominated by these minerals, although they did not come from ocean deposits. This, however, does not make non-marine deposits any less important; these deposits often help to paint a picture into past Earth climates. Some particular deposits even show important tectonic and climatic changes. These deposits also may contain important minerals that help in today's economy. Thick non-marine deposits that accumulate tend to form where evaporation rates will exceed the inflow rate, and where there is sufficient soluble supplies. The inflow also has to occur in a closed basin, or one with restricted outflow, so that the sediment has time to pool and form in a lake or other standing body of water. Primary examples of this are called "saline lake deposits". Saline lakes includes things such as perennial lakes, which are lakes that are there year-round, playa lakes, which are lakes that appear only during certain seasons, or any other terms that are used to define places that hold standing bodies of water intermittently or year-round. Examples of modern non-marine depositional environments include the Great Salt Lake in Utah and the Dead Sea, which lies between Jordan and Israel.

Evaporite depositional environments that meet the above conditions include:

- Graben areas and half-grabens within continental rift environments fed by limited riverine drainage, usually in subtropical or tropical environments
    - Example environments at the present that match this is the Denakil Depression, Ethiopia; Death Valley, California
- Graben environments in oceanic rift environments fed by limited oceanic input, leading to eventual isolation and evaporation
    - Examples include the Red Sea, and the Dead Sea in Jordan and Israel
- Internal drainage basins in arid to semi-arid temperate to tropical environments fed by ephemeral drainage
    - Example environments at the present include the Simpson Desert, Western Australia, the Great Salt Lake in Utah
- Non-basin areas fed exclusively by groundwater seepage from artesian waters

- Example environments include the seep-mounds of the Victoria Desert, fed by the Great Artesian Basin, Australia
- Restricted coastal plains in regressive sea environments
  - Examples include the sabkha deposits of Iran, Saudi Arabia, and the Red Sea; the Garabogazköl of the Caspian Sea
- Drainage basins feeding into extremely arid environments
  - Examples include the Chilean deserts, certain parts of the Sahara, and the Namib

The most significant known evaporite depositions happened during the Messinian salinity crisis in the basin of the Mediterranean.

Evaporitic Formations

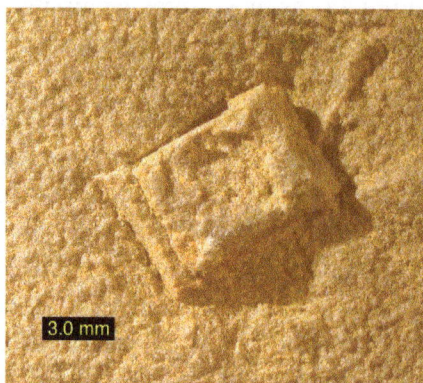

Hopper crystal cast of halite in a Jurassic rock, Carmel Formation, southwestern Utah

Evaporite formations need not be composed entirely of halite salt. In fact, most evaporite formations do not contain more than a few percent of evaporite minerals, the remainder being composed of the more typical detrital clastic rocks and carbonates. Examples of evaporate formations include occurrences of evaporite sulfur in Eastern Europe and West Asia.

For a formation to be recognised as evaporitic it may simply require recognition of halite pseudomorphs, sequences composed of some proportion of evaporite minerals, and recognition of mud crack textures or other textures.

Economic Importance of Evaporites

Evaporites are important economically because of their mineralogy, their physical properties in-situ, and their behaviour within the subsurface.

Evaporite minerals, especially nitrate minerals, are economically important in Peru and Chile. Nitrate minerals are often mined for use in the production on fertilizer and explosives.

Thick halite deposits are expected to become an important location for the disposal of nuclear waste because of their geologic stability, predictable engineering and physical behaviour, and imperviousness to groundwater.

Halite formations are famous for their ability to form diapirs, which produce ideal locations for trapping petroleum deposits.

Major Groups of Evaporite Minerals

Calcite

This is a chart that shows minerals that form the marine evaporite rocks, they are usually the most common minerals that appear in this kind of deposit.

Mineral class	Mineral name	Chemical Composition
Chlorides	Halite	NaCl
	Sylvite	KCl
	Carnallite	$KMgCl_3 * 6H_2O$
	Kainite	$KMg(SO_4)Cl * 3H_2O$
Sulfates	Anhydrite	$CaSO_4$
	Gypsum	$CaSO_4 * 2H_2O$
	Kieserite	$MgSO_4 * H_2O$
	Langbeinite	$K_2Mg_2(SO_4)_3$
	Polyhalite	$K_2Ca_2Mg(SO_4)_6 * H_2O$
Carbonates	Dolomite	$CaMg(CO_3)_2$
	Calcite	$CaCO_3$
	Magnesite	$MgCO_3$

- Halides: halite, sylvite (KCl), and fluorite

- Sulfates: such as gypsum, barite, and anhydrite

- Nitrates: nitratine (soda niter) and niter

- Borates: typically found in arid-salt-lake deposits plentiful in the southwestern US. A common borate is borax, which has been used in soaps as a surfactant.

- Carbonates: such as trona, formed in inland brine lakes.

    - Some evaporite minerals, such as Hanksite, are from multiple groups.

Evaporite minerals start to precipitate when their concentration in water reaches such a level that they can no longer exist as solutes.

The minerals precipitate out of solution in the reverse order of their solubilities, such that the order of precipitation from sea water is :

1. Calcite ($CaCO_3$) and dolomite ($CaMg(CO_3)_2$)

2. Gypsum ($CaSO_4 \cdot 2H_2O$) and anhydrite ($CaSO_4$).

3. Halite (i.e. common salt, NaCl)

4. Potassium and magnesium salts

The abundance of rocks formed by seawater precipitation is in the same order as the precipitation given above. Thus, limestone (calcite) and dolomite are more common than gypsum, which is more common than halite, which is more common than potassium and magnesium salts.

Evaporites can also be easily recrystallized in laboratories in order to investigate the conditions and characteristics of their formation.

Salt Dome

Astronaut photography of Jashak salt dome in the Zagros Mountains in Bushehr Province, Iran (the white area in the middle)

Salt dome in Fars Province, Iran

A salt dome is a type of structural dome formed when a thick bed of evaporite minerals (mainly salt, or halite) found at depth intrudes vertically into surrounding rock strata, forming a diapir. It is important in petroleum geology because salt structures are impermeable and can lead to the formation of a stratigraphic trap.

Formation

The formation of a salt dome begins with the deposition of salt in a restricted marine basin. Because the flow of salt-rich seawater into the basin is not balanced by outflow, much to all water lost from the basin is via evaporation, resulting in the precipitation and deposition of salt evaporites. The rate of sedimentation of salt is significantly larger than the rate of sedimentation of clastics, but it is recognized that a single evaporation event is rarely enough to produce the vast quantities of salt needed to form a layer thick enough for salt diapirs to be formed. This indicates that a sustained period of episodic flooding and evaporation of the basin must occur, as can be seen from the example of the Mediterranean Messinian salinity crisis. At the present day, evaporite deposits can be seen accumulating in basins that merely have restricted access but do not completely dry out; they provide an analogue to some deposits recognized in the geologic record, such as the Garabogazköl basin in Turkmenistan.

Over time, the layer of salt is covered with deposited sediment, becoming buried under an increasingly large overburden. The overlying sediment will undergo compaction, causing an increase in density and therefore a decrease in buoyancy. Unlike clastics, pressure has a significantly smaller effect on the density of salt due to its crystal structure and this eventually leads to it becoming more buoyant than the sediment above it. The ductility of salt initially allows it to plastically deform and flow laterally, decoupling the overlying sediment from the underlying sediment. Since the salt has a larger buoyancy than the sediment above - and if a significant faulting event affects the lower surface of the salt - the salt can begin to flow vertically, forming a salt pillow. The vertical growth of these salt pillows creates pressure on the upward surface, causing extension and faulting.

Possible forces that drive the flow of salt are differential loading on the source layer and density contrasts in the overburdening sediment. Forces that resist this flow are the mass of the roof block and the block's inherent resistance to faulting, aka strength. To accommodate common density contrast between the overburden sediment and the salt, beginning active diapirism, the diapir height must be more than two-thirds to three-quarters the thickness of the overburden. If the diapir is narrow its height must be greater.

Eventually, over millions of years, the salt will pierce and break through the overlying sediment, first as a dome-shaped and then a mushroom-shaped - fully formed salt diapir. If the rising salt diapir breaches the surface, it can become a flowing salt glacier. In cross section, these large domes may be anywhere from 1 to 10 kilometres (0.62 to 6.21 mi) across, and extend as deep as 6.5 kilometres (4.0 mi).

Structure

Typical structures of active diapirism are a central crestal graben flanked by flaps that rotate upward and outward. Reverse faults can separate the flaps from the overburden. Normal faults create the crestal graben and propagate downward. New faults form farther outward as the dome arch becomes more intense. These structures occur beneath the surface and are not necessarily associated with the dome at the surface. Emergence of the dome will not occur unless the dome is very wide or tall relative to the overburden's thickness.

Recognizing Salt Domes in Seismic Data

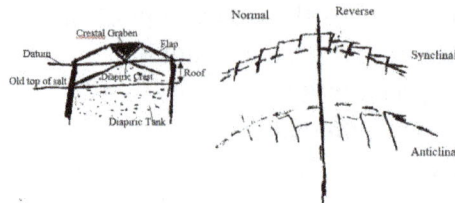

Left shows a generalized structure of a salt dome. Right shows two fault regimes that make up the crestal graben, the dotted line indicates the top of the salt.

If a salt dome has not pierced the surface they can found located beneath the surface in various ways. The unique surficial structures can be observed as indicating the salt dome beneath the surface. Salt domes can also be interpreted from seismic reflection where the stark density contrast between the salt and surrounding sediments outlines the salt structures. Salt domes can also be associated with sulfur springs and natural gas vents.

Examples

The first salt dome was discovered in 1890 when an exploratory oil well was drilled into Spindletop Hill near Beaumont TX.

Major occurrences of salt domes are found along the Gulf Coast of the USA in Texas and Louisiana. One example of an island formed by a salt dome is Avery Island in Louisiana. At present ocean levels it is no longer surrounded by the sea but it is surrounded by bayous on all sides. The Gulf Coast is home to over 500 currently discovered salt domes.

Another example of an emergent salt dome is at Onion Creek, Utah / Fisher Towers near Moab, Utah, U.S. These two images show a Cretaceous age salt body that has risen as a ridge through several hundred meters of overburden, predominantly sandstone. As the salt body rose, the overburden formed an anticline (arching upward along its center line) which fractured and eroded to expose the salt body.

End-on view of emergent salt dome between remnants of displaced overburden

Lateral view of emergent salt dome from ridge of remnant of displaced overburden

The term "salt dome" is also sometimes inaccurately used to refer to dome-shaped silos used to store rock salt for melting snow on highways. These domes are actually called monolithic domes and are used to store a variety of bulk goods.

Commercial Uses

The rock salt that is found in salt domes is mostly impermeable. As the salt moves up towards the surface, it can penetrate and/or bend strata of existing rock with it. As these strata are penetrated, they are generally bent slightly upwards at the point of contact with the dome, and can form pockets where petroleum and natural gas can collect between impermeable strata of rock and the salt. The strata immediately above the dome that are not penetrated are pushed upward, creating a dome-like reservoir above the salt where petroleum can also gather. These oil pools can eventually be extracted, and indeed form a major source of the petroleum produced along the coast of the Gulf of Mexico.

The caprock above the salt domes is sometimes the site of deposits of native sulfur, which is recovered by the Frasch process.

Other uses include storing oil, natural gas, hydrogen gas, or even hazardous waste in large caverns formed after salt mining, as well as excavating the domes themselves for uses in everything from table salt to the granular material used to prevent roadways from icing over.

Mineral Alteration

Mineral alteration refers to the various natural processes that alter a mineral's chemical composition or crystallography.

Mineral alteration is essentially governed by the laws of thermodynamics related to energy conservation, relevant to environmental conditions, often in presence of catalysts, the most common and influential being water (H_2O).

The degree and scales of time in which different minerals alter vary depending on the initial product and its physical properties and susceptibility to alteration. Some minerals such as quartz and zircon are highly resistant to alteration under normal weathering conditions. Yet quartz may alter to stishovite with intense pressure, and zircon to *crytolite* (a metamict zircon) with amount of radioactive components and time.

In some circumstances, a mineral alters while maintaining its outer form known as a pseudomorph.

Mineral alteration is distinctly different than the rock alteration process metamorphism. It also differs from weathering. However, these processes assist in mineral alteration. Some minerals are members of a solid solution series and are samples of a range of compositional changes in a continuum, and thus are not 'mineral alteration' products.

Examples of Mineral Alterations

Oxidation

A common oxidation example is when a natural ferrous iron mineral such as pyrite is oxidized to form goethite or other ferric iron hydroxides or sulfates.

Hydration and n

The common mineral gypsum is a hydrous sulfate mineral that readily alters to the anhydrous sulfate aptly named anhydrite with prolonged desiccation.

$$CaSO_4 \cdot 2H_2O <=> CaSO_4$$

This is a reversible reaction.

Kaolinization

Kaolinization refers to the alteration of alkali feldspar into the clay mineral kaolinite in the presence of slightly acidic solutions. Rain readily dissolves carbon dioxide (CO_2) from the atmosphere, promoting weathering of granitic rocks. As demonstrated in the following reaction, in the presence of carbonic acid and water, potassium feldspar is altered to kaolinite, with potassium ion, bicarbonate, and silica in solution as byproducts.

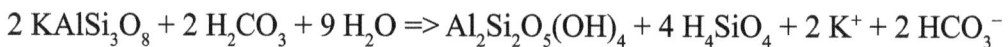

$$2\ KAlSi_3O_8 + 2\ H_2CO_3 + 9\ H_2O => Al_2Si_2O_5(OH)_4 + 4\ H_4SiO_4 + 2\ K^+ + 2\ HCO_3^-$$

Epidotization

Epidotization is the alteration process in which plagioclase feldspars convert into the epidote group minerals.

Chloritization

Chloritization is the alteration of pyroxene or amphibole minerals into the chlorite group minerals. Chloritization is a common process in metamorphic transitions to the greenschist facies, and amphibolite facies retrograde metamorphism.

Shock Induced Alteration

As observed in and around astroblems such as impact craters, ordinary silica or quartz crystals may alter to the minerals stishovite and coesite as a result of meteorite impacts producing an extreme pressure and high temperature environment.

Radioactive Decay

A common example of a radioactive decay alteration is when a radioactive element bearing zircon or allanite crystal becomes metamict or amorphous due to structural damage.

Serpentinization

Serpentinization is the mineral alteration process that results in the formation of serpentine group of minerals mainly from the olivine group, with hydration and changes in pressure as major factors.

Dolomitization

Dolomitization refers to the varied suggested manners in which a predominantly calcite rich calcium bearing sedimentary rock such as limestone may alter into the magnesian dolomite rich dolostone. Diagenesis is a likely culprit that involves volumes of water and fairly low heat, as an ionic exchange catalyst. The reaction is as follows:

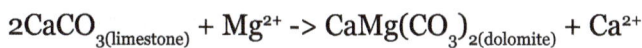

$$2CaCO_{3(limestone)} + Mg^{2+} \rightarrow CaMg(CO_3)_{2(dolomite)} + Ca^{2+}$$

Pyritization

Pyritization involves the ionic replacement by iron and sulfur atoms that combine to form the mineral pyrite.

Opalization

Opalization is the alteration of amorphous silica, often as organic remains of siliceous microfossils in lithified sedimentary rocks, into the mineraloid opal.

Perovskite (Structure)

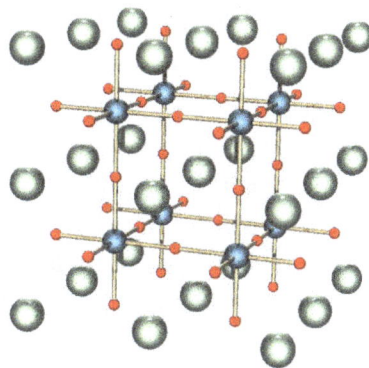

Structure of a perovskite with a chemical formula ABX_3. The red spheres are X atoms (usually oxygens), the blue spheres are B-atoms (a smaller metal cation, such as Ti^{4+}), and the green spheres are the A-atoms (a larger metal cation, such as Ca^{2+}). Pictured is the undistorted cubic structure; the symmetry is lowered to orthorhombic, tetragonal or trigonal in many perovskites.

A Perovskite mineral (calcium titanate) from Kusa, Russia. Taken at the Harvard Museum of Natural History.

A perovskite is any material with the same type of crystal structure as calcium titanium oxide ($CaTiO_3$), known as the *perovskite structure*, or $^{XII}A^{2+VI}B^{4+}X^{2-}_3$ with the oxygen in the face centers. Perovskites take their name from the mineral, which was first discovered in the Ural mountains of Russia by Gustav Rose in 1839 and is named after Russian mineralogist L. A. Perovski (1792–1856). The general chemical formula for perovskite compounds is ABX_3, where 'A' and 'B' are two cations of very different sizes, and X is an anion that bonds to both. The 'A' atoms are larger than the 'B' atoms. The ideal cubic-symmetry structure has the B cation in 6-fold coordination, surrounded by an octahedron of anions, and the A cation in 12-fold cuboctahedral coordination. The relative ion size requirements for stability of the cubic structure are quite stringent, so slight buckling and distortion can produce several lower-symmetry distorted versions, in which the co-ordination numbers of A cations, B cations or both are reduced.

Natural compounds with this structure are perovskite, loparite, and the silicate perovskite bridgmanite.

Structure

The perovskite structure is adopted by many oxides that have the chemical formula ABO_3.

In the idealized cubic unit cell of such a compound, type 'A' atom sits at cube corner positions (0, 0, 0), type 'B' atom sits at body centre position (1/2, 1/2, 1/2) and oxygen atoms sit at face centred positions (1/2, 1/2, 0). (The diagram shows edges for an equivalent unit cell with A in body centre, B at the corners, and O in mid-edge).

The relative ion size requirements for stability of the cubic structure are quite stringent, so slight buckling and distortion can produce several lower-symmetry distorted versions, in which the coordination numbers of A cations, B cations or both are reduced. Tilting of the BO_6 octahedra reduces the coordination of an undersized A cation from 12 to as low as 8. Conversely, off-centering of an undersized B cation within its octahedron allows it to attain a stable bonding pattern. The resulting electric dipole is responsible for the property of ferroelectricity and shown by perovskites such as $BaTiO_3$ that distort in this fashion.

The orthorhombic and tetragonal phases are most common non-cubic variants.

Complex perovskite structures contain two different B-site cations. This results in the possibility of ordered and disordered variants.

Common Occurrence

The most common mineral in the Earth is bridgmanite, a magnesium-rich silicate which adopts the perovskite structure at high pressure. As pressure increases, the SiO_4^{4-} tetrahedral units in the dominant silica-bearing minerals become unstable compared with SiO_6^{8-} octahedral units. At the pressure and temperature conditions of the lower mantle, the most abundant material is a perovskite-structured mineral with the formula $(Mg,Fe)SiO_3$, with the second most abundant material likely the rocksalt-structured $(Mg,Fe)O$ oxide, periclase.

At the high pressure conditions of the Earth's lower mantle, the pyroxene enstatite, $MgSiO_3$, transforms into a denser perovskite-structured polymorph; this phase may be the most common mineral in the Earth. This phase has the orthorhombically distorted perovskite structure ($GdFeO_3$-type structure) that is stable at pressures from ~24 GPa to ~110 GPa. However, it cannot be transported from depths of several hundred km to the Earth's surface without transforming back into less dense materials. At higher pressures, $MgSiO_3$ perovskite transforms to post-perovskite.

Although the most common perovskite compounds contain oxygen, there are a few perovskite compounds that form without oxygen. Fluoride perovskites such as $NaMgF_3$ are well known. A large family of metallic perovskite compounds can be represented by RT_3M (R: rare-earth or other relatively large ion, T: transition metal ion and M: light metalloids). The metalloids occupy the octahedrally coordinated "B" sites in these compounds. RPd_3B, RRh_3B and $CeRu_3C$ are examples. $MgCNi_3$ is a metallic perovskite compound and has received lot of attention because of its superconducting properties. An even more exotic type of perovskite is represented by the mixed oxide-aurides of Cs and Rb, such as Cs_3AuO, which contain large alkali cations in the traditional "anion" sites, bonded to O^{2-} and Au^- anions.

Material Properties

Perovskite materials exhibit many interesting and intriguing properties from both the theoretical and the application point of view. Colossal magnetoresistance, ferroelectricity, superconductivity, charge ordering, spin dependent transport, high thermopower and the interplay of structural, magnetic and transport properties are commonly observed features in this family. These compounds are used as sensors and catalyst electrodes in certain types of fuel cells and are candidates for memory devices and spintronics applications.

Many superconducting ceramic materials (the high temperature superconductors) have perovskite-like structures, often with 3 or more metals including copper, and some oxygen positions left vacant. One prime example is yttrium barium copper oxide which can be insulating or superconducting depending on the oxygen content.

Chemical engineers are considering a cobalt-based perovskite material as a replacement for platinum in catalytic converters in diesel vehicles.

Applications

Physical properties of interest to materials science among perovskites include superconductivity, magnetoresistance, ionic conductivity, and a multitude of dielectric properties, which are of great importance in microelectronics and telecommunication. Because of the flexibility of bond angles inherent in the perovskite structure there are many different types of distortions which can occur from the ideal structure. These include tilting of the octahedra, displacements of the cations out of the centers of their coordination polyhedra, and distortions of the octahedra driven by electronic factors (Jahn-Teller distortions).

Photovoltaics

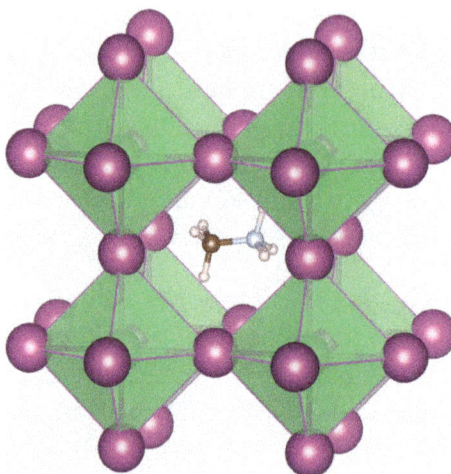

Crystal structure of $CH_3NH_3PbX_3$ perovskites (X=I, Br and/or Cl). The methylammonium cation ($CH_3NH_3^+$) is surrounded by PbX_6 octahedra.

Synthetic perovskites have been identified as possible inexpensive base materials for high-efficiency commercial photovoltaics – they showed a conversion efficiency of up to 15% and can be manufactured using the same thin-film manufacturing techniques as that used for thin film silicon solar cells. Methylammonium tin halides and methylammonium lead halides are of interest for use in dye-sensitized solar cells. In 2016, power conversion efficiency have reached 21%.In July 2016, a team of researchers led by Dr. Alexander Weber-Bargioni demonstrated that perovskite PV cells could reach a theoretical peak efficiency of 31%.

Among the methylammonium halides studied so far the most common is the methylammonium lead triiodide (CH3NH3PbI3). It has a high charge carrier mobility and charge carrier lifetime that allow light-generated electrons and holes to move far enough to be extracted as current, instead of losing their energy as heat within the cell. CH3NH3PbI3 effective diffusion lengths are some 100 nm for both electrons and holes.

Methylammonium halides are deposited by low-temperature solution methods (typically spin-coating). Other low-temperature (below 100 °C) solution-processed films tend to have considerably smaller diffusion lengths. Stranks et al. described nanostructured cells using a mixed methylammonium lead halide ($CH_3NH_3PbI_{3-x}Cl_x$) and demonstrated one amorphous thin-film solar cell with an 11.4% conversion efficiency, and another that reached 15.4% using vacuum evaporation. The film thickness of about 500 to 600 nm implies that the electron and hole diffusion lengths were at least of this order. They measured values of the diffusion length exceeding 1 µm for the mixed perovskite, an order of magnitude greater than the 100 nm for the pure iodide. They also showed that carrier lifetimes in the mixed perovskite are longer than in the pure iodide.

For CH

3NH

3PbI

3, open-circuit voltage (V_{OC}) typically approaches 1 V, while for CH

3NH

3PbI(I,Cl)

3 with low Cl content, $V_{OC} > 1.1$ V has been reported. Because the band gaps (E_g) of both are 1.55 eV, V_{OC}-to-E_g ratios are higher than usually observed for similar third-generation cells. With wider bandgap perovskites, V_{OC} up to 1.3 V has been demonstrated.

The technique offers the potential of low cost because of the low temperature solution methods and the absence of rare elements. Cell durability is currently insufficient for commercial use.

Planar heterojunction perovskite solar cells can be manufactured in simplified device architectures (without complex nanostructures) using only vapor deposition. This technique produces 15% solar-to-electrical power conversion as measured under simulated full sunlight.

Lasers

Also in 2008 researchers demonstrated that perovskite can generate laser light. $LaAlO_3$ doped with neodymium gave laser emission at 1080 nm. In 2014 it was shown that mixed methylammonium lead halide ($CH_3NH_3PbI_{3-x}Cl_x$) cells fashioned into optically pumped vertical-cavity surface-emitting lasers (VCSELs) convert visible pump light to near-IR laser light with a 70% efficiency.

Light Emitting Diodes

Due to their high photoluminesence quantum efficiencies, perovskites may be good candidates for use in light-emitting diodes (LEDs). However, the propensity for radiative recombination has mostly been observed at liquid nitrogen temperatures.

Photoelectrolysis

In September 2014, researchers of EPFL in Lausanne, Switzerland reported achieving water electrolysis at 12.3% efficiency in a highly efficient and low-cost water-splitting cell using perovskite photovoltaics.

Inclusion (Mineral)

Dark inclusions of aegerine in light-green apatite

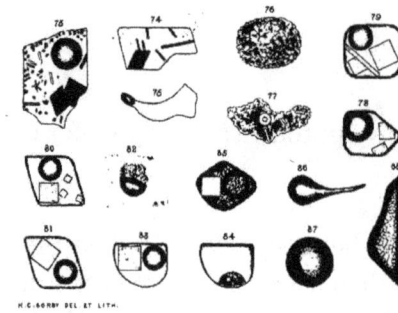

Sketch showing different shapes of inclusions

In mineralogy, an inclusion is any material that is trapped inside a mineral during its formation.

In gemology, an inclusion is a characteristic enclosed within a gemstone, or reaching its surface from the interior.

According to Hutton's law of inclusions, fragments included in a host rock are older than the host rock itself.

Mineralogy

Inclusions are usually other minerals or rocks, but may also be water, gas or petroleum. Liquid or vapor inclusions are known as fluid inclusions. In the case of amber it is possible to find insects and plants as inclusions.

The analysis of atmospheric gas bubbles as inclusions in ice cores is an important tool in the study of climate change.

A xenolith is a pre-existing rock which has been picked up by a lava flow. Melt inclusions form when bits of melt become trapped inside crystals as they form in the melt.

Gemology

Inclusions are one of the most important factors when it comes to gem valuation. In many gemstones, such as diamonds, inclusions affect the clarity of the stone, diminishing the stone's value. In some stones, however, such as star sapphires, the inclusion actually increases the value of the stone.

Many colored gemstones, such as amethyst, emerald, and sapphire, are expected to have inclusions, and the inclusions do not greatly affect the stone's value. Colored gemstones are categorized into three types as follows:

- Type I colored stones include stones with very little or no inclusions. They include aquamarines, topaz and zircon.

- Type II colored stones include stones that often have a few inclusions. They include sapphire, ruby, garnet and spinel.

- Type III colored stones include stones that almost always have inclusions. Stones in this category include emerald and tourmaline.

- Clear gemstone with metallic inclusion.

- Peridot with milky inclusion.

- Natural Ruby with inclusions.

Metallurgy

The term "inclusion" is also used in the context of metallurgy and metals processing. During the melt stage of processing hard particles such as oxides can enter or form in the liquid metal which are subsequently trapped when the melt solidifies. The term is usually used negatively such as when the particle could act as a fatigue crack nucleator or as an area of high stress intensity.

Post-Perovskite

Post-perovskite (pPv) is a high-pressure phase of magnesium silicate ($MgSiO_3$). It is composed of the prime oxide constituents of the Earth's rocky mantle (MgO and SiO_2), and its pressure and temperature for stability imply that it is likely to occur in portions of the lowermost few hundred km of Earth's mantle.

The post-perovskite phase has implications for the D''-layer that influences the convective mixing in the mantle responsible for plate tectonics.

Post-perovskite has the same crystal structure as the synthetic solid compound $CaIrO_3$, and is often referred to as the "$CaIrO_3$-type phase of $MgSiO_3$" in the literature. The crystal system of post-perovskite is orthorhombic, its space group is *Cmcm*, and its structure is a stacked SiO_6-octahedral sheet along the *b* axis. The name "post-perovskite" derives from silicate perovskite, the stable phase of $MgSiO_3$ throughout most of Earth's mantle, which has the perovskite structure.

The prefix "post-" refers to the fact that it occurs after perovskite structured $MgSiO_3$ as pressure increases (and historically, the progression of high pressure mineral physics). At upper mantle pressures, nearest Earth's surface, $MgSiO_3$ persists as the silicate mineral enstatite, a pyroxene rock forming mineral found in igneous and metamorphic rocks of the crust.

History

The $CaIrO_3$-type phase of $MgSiO_3$ phase was discovered in 2004 using the laser-heated diamond anvil cell (LHDAC) technique by a group at the Tokyo Institute of Technology and, independently, by researchers from the Swiss Federal Institute of Technology (ETH Zurich) and Japan Agency for Marine-Earth Science and Technology who used a combination of quantum-mechanical simulations and LHDAC experiments. The TIT group's paper appeared in the journal *Science*. The ETH/JAM-EST collaborative paper and TIT group's second paper appeared two months later in the journal *Nature*. This simultaneous discovery was preceded by S. Ono's experimental discovery of a similar phase, possessing exactly the same structure, in Fe_2O_3.

Importance in Earth's Mantle

Post-perovskite phase is stable above 120 GPa at 2500 K, and exhibits a positive Clapeyron slope such that the transformation pressure increases with temperature. Because these conditions correspond to a depth of about 2600 km and the D" seismic discontinuity occurs at similar depths, the perovskite to post-perovskite phase change is considered to be the origin of such seismic discontinuities in this region. Post-perovskite also holds great promise for mapping experimentally determined information regarding the temperatures and pressures of its transformation into direct information regarding temperature variations in the D" layer once the seismic discontinuities attributed to this transformation have been sufficiently mapped out. Such information can be used, for example, to:

1) better constrain the amount of heat leaving Earth's core

2) determine whether or not subducted slabs of oceanic lithosphere reach the base of the mantle

3) help delineate the degree of chemical heterogeneity in the lower mantle

4) find out whether or not the lowermost mantle is unstable to convective instabilities that result in upwelling hot thermal plumes of rock which rise up and possibly trace out volcanic hot spot tracks at Earth's surface.

For these reasons the finding of the $MgSiO_3$-post-perovskite phase transition is considered by many geophysicists to be the most important discovery in deep Earth science in several decades, and was only made possible by the concerted efforts of mineral physics scientists around the world as they sought to increase the range and quality of LHDAC experiments and as *ab initio* calculations attained predictive power.

Physical Properties

The sheet structure of post-perovskite makes the compressibility of the *b* axis higher than that of the *a* or *c* axis. This anisotropy may yield the morphology of a platy crystal habit parallel to the (010) plane; the seismic anisotropy observed in the D" region might qualitatively (but not quantitatively) be explained by this characteristic. Theory predicted the (110) slip associated with particularly favorable stacking faults and confirmed by later experiments. Some theorists predicted other slip systems, which await experimental confirmation. In 2005 and 2006 Ono and Oganov published two papers predicting that post-perovskite should have high electrical conductivity, perhaps two orders of magnitude higher than perovskite's conductivity. In 2008 Hirose's group published an experimental report confirming this prediction. A highly conductive post-perovskite layer provides an explanation for the observed decadal variations of the length of day.

Chemical Properties

Another potentially important effect that needs to be better characterized for the post-perovskite phase transition is the influence of other chemical components that are known to be present to some degree in Earth's lowermost mantle. The phase transition pressure (characterized by a two-phase loop in this system), was initially thought to decrease as the FeO content increases, but some recent experiments suggest the opposite. However, it is possible that the effect of Fe_2O_3 is more relevant as most of iron in post-perovskite is likely to be trivalent (ferric). Such components as Al_2O_3 or the more oxidized Fe_2O_3 also affect the phase transition pressure, and might have strong mutual interactions with one another. The influence of variable chemistry present in the Earth's lowermost mantle upon the post-perovskite phase transition raises the issue of both thermal and chemical modulation of its possible appearance (along with any associated discontinuities) in the D" layer.

Summary

Experimental and theoretical work on the perovskite/post-perovskite phase transition continues, while many important features of this phase transition remain ill-constrained. For example, the Clapeyron slope (characterized by the Clausius–Clapeyron relation) describing the increase in the pressure of the phase transition with increasing temperature is known to be relatively high in comparison to other solid-solid phase transitions in the Earth's mantle, however, the experimentally determined value varies from about 5 MPa/K to as high as 13 MPa/K. *Ab initio* calculations give a tighter range, between 7.5 MPa/K and 9.6 MPa/K, and are probably the most reliable estimates available today. The difference between experimental estimates arises primarily because different materials were used as pressure standards in LHDAC experiments. A well-characterized equation of state for the pressure standard, when combined with high energy synchrotron generated X-ray diffraction patterns of the pressure standard (which is mixed in with the experimental sample material), yields information on the pressure-temperature conditions of the experiment. However, as these extreme pressures and temperatures have not been sufficiently explored in experiments, the equations of state for many popular pressure standards are not yet well characterized and often yield different results. Another source of uncertainty in LHDAC experiments is the measurement of temperature from a sample's thermal radiation, which is required to obtain the pressure from the equation of state of the pressure standard. In laser-heated experiments at such high pressures (over 1 million atmospheres), the samples are necessarily small and numerous approximations (e.g., gray body) are required to obtain estimates of the temperature.

Solid Solution

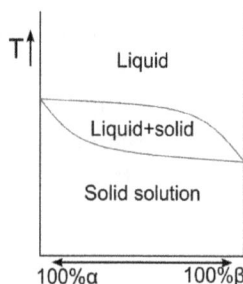

Fig. 1 A binary phase diagram displaying solid solutions over the full range of relative concentrations.

IUPAC definition

Solid in which components are compatible and form a unique phase.

Note 1: The definition "crystal containing a second constituent which fits into and is distributed in the lattice of the host crystal" given in refs., is not general and, thus, is not recommended.

Note 2: The expression is to be used to describe a solid phase containing more than one substance when, for convenience, one (or more) of the substances,called the solvent, is treated differently from the other substances, called solutes.

Note 3: One or several of the components can be *macromolecules*. Some of the other components can then act as plasticizers, i.e., as molecularly dispersed substances that decrease the glass-transition temperature at which the amorphous phase of a *polymer* is converted between glassy and rubbery states.

Note 4: In pharmaceutical preparations, the concept of solid solution is often applied to the case of mixtures of *drug* and *polymer*.

Note 5: The number of drug molecules that do behave as solvent (plasticizer) of polymers is small.

A solid solution is a solid-state solution of one or more solutes in a solvent. Such a mixture is considered a solution rather than a compound when the crystal structure of the solvent remains unchanged by addition of the solutes, and when the mixture remains in a single homogeneous phase. This often happens when the two elements (generally metals) involved are close together on the periodic table; conversely, a chemical compound generally results when two metals involved are not near each other on the periodic table.

The solid solution needs to be distinguished from a mechanical mixtures of powdered solids like two salts, sugar and salt, etc. The mechanical mixtures have total or partial miscibility gap in solid state. Examples of solid solutions include crystallized salts from their liquid mixture, metal alloys, moist solids. In the case of metal alloys intermetallic compounds occur frequently.

Details

The solute may incorporate into the solvent crystal lattice *substitutionally*, by replacing a solvent particle in the lattice, or *interstitially*, by fitting into the space between solvent particles. Both of these types of solid solution affect the properties of the material by distorting the crystal lattice and disrupting the physical and electrical homogeneity of the solvent material.

Some mixtures will readily form solid solutions over a range of concentrations, while other mixtures will not form solid solutions at all. The propensity for any two substances to form a solid solution is a complicated matter involving the chemical, crystallographic, and quantum properties of the substances in question. Substitutional solid solutions, in accordance with the Hume-Rothery rules, may form if the solute and solvent have:

- Similar atomic radii (15% or less difference)
- Same crystal structure
- Similar electronegativities
- Similar valency

The phase diagram in *Fig. 1* displays an alloy of two metals which forms a solid solution at all relative concentrations of the two species. In this case, the pure phase of each element is of the same crystal structure, and the similar properties of the two elements allow for unbiased substitution through the full range of relative concentrations.

Solid solutions have important commercial and industrial applications, as such mixtures often have superior properties to pure materials. Many metal alloys are solid solutions. Even small amounts of solute can affect the electrical and physical properties of the solvent.

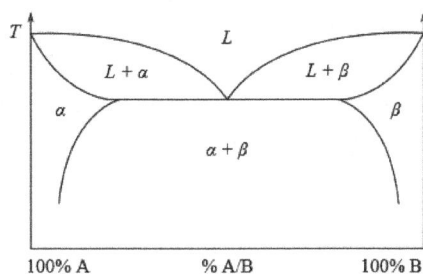

Fig. 2 This binary phase diagram shows two solid solutions: and .

The binary phase diagram in *Fig. 2* shows the phases of a mixture of two substances in varying concentrations, A and B. The region labeled "α" is a solid solution, with B acting as the solute in a matrix of A. On the other end of the concentration scale, the region labeled "β" is also a solid solution, with acting as the solute in a matrix of . The large solid region in between the and solid solutions, labeled "α + β", is *not* a solid solution. Instead, an examination of the microstructure of a mixture in this range would reveal two phases — solid solution A -in- B and solid solution B -in- A would form separate phases, perhaps lamella or grains.

Application

In the phase diagram, at three different concentrations, the material will be solid until it's heated to its melting point, and then (after adding the heat of fusion) become liquid at that same temperature:

- the unalloyed extreme left

- the unalloyed extreme right

- the dip in the center (the eutectic composition).

At other proportions, the material will enter a mushy or pasty phase until it warms up to being completely melted.

The mixture at the dip point of the diagram is called a eutectic alloy. Lead-tin mixtures formulated at that point (37/63 mixture) are useful when soldering electronic components, particularly if done manually, since the solid phase is quickly entered as the solder cools. In contrast, when lead-tin mixtures were used to solder seams in automobile bodies a pasty state enabled a shape to be formed with a wooden paddle or tool, so a 70-30 lead to tin ratio was used. (Lead is being removed from such applications owing to its toxicity and consequent difficulty in recycling devices and components that include lead.)

Exsolution

When a solid solution becomes unstable — due to a lower temperature, for example — exsolution occurs and the two phases separate into distinct microscopic to megascopic lamellae. This is mainly caused by difference in cation size. Cations which have a large difference in radii are not likely to readily substitute.

Take the alkali feldspar minerals for example, whose end members are albite, $NaAlSi_3O_8$ and microcline, $KAlSi_3O_8$. At high temperatures Na^+ and K^+ readily substitute for each other and so the minerals will form a solid solution, yet at low temperatures albite can only substitute a small amount of K^+ and the same applies for Na^+ in the microcline. This leads to exsolution where they will separate into two separate phases. In the case of the alkali feldspar minerals, thin white albite layers will alternate between typically pink microcline.

Biomineralization

Calcitic skeletal parts of belemnites (Jurassic of Wyoming)

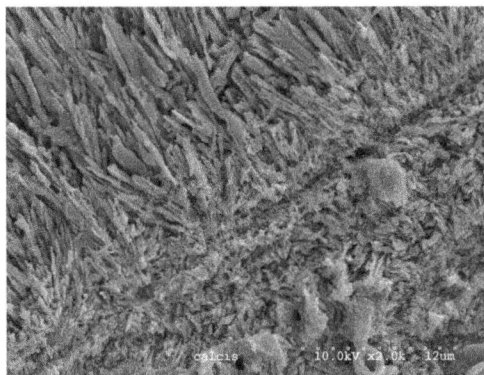

Glomerula piloseta (Sabellidae), longitudinal section of the tube, aragonitic spherulitic prismatic structure

IUPAC definition

Mineralization caused by cell-mediated phenomena.

Biomineralization is the process by which living organisms produce minerals, often to harden or stiffen existing tissues. Such tissues are called mineralized tissues. It is an extremely widespread phenomenon; all six taxonomic kingdoms contain members that are able to form minerals, and over 60 different minerals have been identified in organisms. Examples include silicates in algae and diatoms, carbonates in invertebrates, and calcium phosphates and carbonates in vertebrates. These minerals often form structural features such as sea shells and the bone in mammals and birds. Organisms have been producing mineralised skeletons for the past 550 million years. Other examples include copper, iron and gold deposits involving bacteria. Biologically-formed minerals often have special uses such as magnetic sensors in magnetotactic bacteria (Fe_3O_4), gravity sensing devices ($CaCO_3$, $CaSO_4$, $BaSO_4$) and iron storage and mobilization ($Fe_2O_3 \cdot H_2O$ in the protein ferritin).

In terms of taxonomic distribution, the most common biominerals are the phosphate and carbonate salts of calcium that are used in conjunction with organic polymers such as collagen and chitin to give structural support to bones and shells. The structures of these biocomposite materials are highly controlled from the nanometer to the macroscopic level, resulting in complex architectures that provide multifunctional properties. Because this range of control over mineral growth is desirable for materials engineering applications, there is significant interest in understanding and elucidating the mechanisms of biologically controlled biomineralization.

Biological Roles

Biominerals perform a variety of roles in organisms, the most important being support, defense and feeding.

Biology

If present on a super-cellular scale, biominerals are usually deposited by a dedicated organ, which is often defined very early in the embryological development. This organ will contain an organic matrix that facilitates and directs the deposition of crystals. The matrix may be collagen, as in deuterostomes, or based on chitin or other polysaccharides, as in molluscs.

Shell Formation in Molluscs

The mollusc shell is a biogenic composite material that has been the subject of much interest in materials science because of its unusual properties and its model character for biomineralization. Molluscan shells consist of 95–99% calcium carbonate by weight, while an organic component makes up the remaining 1–5%. The resulting composite has a fracture toughness ~3000 times greater than that of the crystals themselves. In the biomineralization of the mollusc shell, specialized proteins are responsible for directing crystal nucleation, phase, morphology, and growths dynamics and ultimately give the shell its remarkable mechanical strength. The application of biomimetic principles elucidated from mollusc shell assembly and structure may help in fabricating new composite materials with enhanced optical, electronic, or structural properties.

Chemistry

Because extracellular iron is strongly involved in inducing calcification, its control is essential in developing shells; the protein ferritin plays an important role in controlling the distribution of iron. The most common mineral present in biomineralization is hydroxyapatite (HA), which is a naturally occurring mineral form of calcium apatite with the formula $Ca_{10}(PO_4)_6(OH)_2$. Hydroxyapatite crystals are found in many biological materials including bones, fish scales, and cartilage. Each material has a mineral content which corresponds with the required mechanical properties, where increasing HA content typically leads to increased stiffness but reduced extensibility.

Evolution

The first evidence of biomineralization dates to some 750 million years ago, and sponge-grade organisms may have formed calcite skeletons 630 million years ago. But in most lineages, biomineralization first occurred in the Cambrian or Ordovician periods. Organisms used whichever form of calcium carbonate was more stable in the water column at the point in time when they became biomineralized, and stuck with that form for the remainder of their biological history. The stability is dependent on the Ca/Mg ratio of seawater, which is thought to be controlled primarily by the rate of sea floor spreading, although atmospheric CO_2 levels may also play a role.

Biomineralization evolved multiple times, independently, and most animal lineages first expressed biomineralized components in the Cambrian period. Interestingly, many of the same processes are used in unrelated lineages, which suggests that biomineralization machinery was assembled from pre-existing "off-the-shelf" components already used for other purposes in the organism. Although the biomachinery facilitating biomineralization is complex – involving signalling transmitters, inhibitors, and transcription factors – many elements of this 'toolkit' are shared between phyla as diverse as corals, molluscs, and vertebrates. The shared components tend to perform quite fundamental tasks, such as designating that cells will be used to create the minerals, whereas genes controlling more finely tuned aspects that occur later in the biomineralization process – such as the precise alignment and structure of the crystals produced – tend to be uniquely evolved in different lineages. This suggests that Precambrian organisms were employing the same elements, albeit for a different purpose — perhaps to *avoid* the inadvertent precipitation of calcium carbonate from the supersaturated Proterozoic oceans. Forms of mucus that are involved in inducing mineralization in most metazoan lineages appear to have performed such an anticalcifatory function in the ancestral state. Further, certain proteins that would originally have been involved in maintaining calcium concentrations within cells are homologous to all metazoans, and appear to have been co-opted into biomineralization after the divergence of the metazoan lineages. The *galaxins* are one probable example of a gene being co-opted from a different ancestral purpose into controlling biomineralization, in this case being 'switched' to this purpose in the Triassic scleractinian corals; the role performed appears to be functionally identical to the unrelated pearlin gene in molluscs. Carbonic anhydrase serves a role in mineralization in sponges, as well as metazoans, implying an ancestral role. Far from being a rare trait that evolved a few times and remained stagnant, biomineralization pathways in fact evolved many times and are still evolving rapidly today; even within a single genus it is possible to detect great variation within a single gene family.

The homology of biomineralization pathways is underlined by a remarkable experiment whereby the nacreous layer of a molluscan shell was implanted into a human tooth, and rather than experiencing an immune response, the molluscan nacre was incorporated into the host bone matrix. This points to the exaptation of an original biomineralization pathway.

The most ancient example of biomineralization, dating back 2 billion years, is the deposition of magnetite, which is observed in some bacteria, as well as the teeth of chitons and the brains of vertebrates; it is possible that this pathway, which performed a magnetosensory role in the common ancestor of all bilaterians, was duplicated and modified in the Cambrian to form the basis for calcium-based biomineralization pathways. Iron is stored in close proximity to magnetite-coated chiton teeth, so that the teeth can be renewed as they wear. Not only is there a marked similarity between the magnetite deposition process and enamel deposition in vertebrates but some vertebrates even have comparable iron storage facilities near their teeth.

Type of mineralization	Examples of organisms
Calcium carbonate (calcite or aragonite)	<ul><li>foraminifera</li><li>coccolithophores</li><li>calcareous sponge spicules</li><li>corals</li><li>Archaeocyatha</li><li>bryozoans</li><li>brachiopod and mollusc shells</li><li>Echinoderms</li><li>Serpulidae</li></ul>
Silica	<ul><li>radiolarians</li><li>diatoms</li><li>most sponge spicules</li></ul>

Apatite (phosphate carbonate)	• enamel (Vertebrate teeth)
	• Vertebrate bone
	• conodonts

Astrobiology

It has been suggested that biominerals could be important indicators of extraterrestrial life and thus could play an important role in the search for past or present life on Mars. Furthermore, organic components (biosignatures) that are often associated with biominerals are believed to play crucial roles in both pre-biotic and biotic reactions.

On January 24, 2014, NASA reported that current studies by the *Curiosity* and *Opportunity* rovers on the planet Mars will now be searching for evidence of ancient life, including a biosphere based on autotrophic, chemotrophic and/or chemolithoautotrophic microorganisms, as well as ancient water, including fluvio-lacustrine environments (plains related to ancient rivers or lakes) that may have been habitable. The search for evidence of habitability, taphonomy (related to fossils), and organic carbon on the planet Mars is now a primary NASA objective.

Potential Applications

Most traditional approaches to synthesis of nanoscale materials are energy inefficient, requiring stringent conditions (e.g., high temperature, pressure or pH) and often produce toxic byproducts. Furthermore, the quantities produced are small, and the resultant material is usually irreproducible because of the difficulties in controlling agglomeration. In contrast, materials produced by organisms have properties that usually surpass those of analogous synthetically manufactured materials with similar phase composition. Biological materials are assembled in aqueous environments under mild conditions by using macromolecules. Organic macromolecules collect and transport raw materials and assemble these substrates and into short- and long-range ordered composites with consistency and uniformity. The aim of biomimetics is to mimic the natural way of producing minerals such as apatites. Many man-made crystals require elevated temperatures and strong chemical solutions, whereas the organisms have long been able to lay down elaborate mineral structures at ambient temperatures. Often, the mineral phases are not pure but are made as composites that entail an organic part, often protein, which takes part in and controls the biomineralisation. These composites are often not only as hard as the pure mineral but also tougher, as the micro-environment controls biomineralisation.

Biomineralization of Uranium Contaminants in Groundwater

Biomineralization may be used to remediate groundwater contaminated with uranium. The biomineralization of uranium primarily involves the precipitation of uranium phosphate minerals associated with the release of phosphate by microorganisms. Negatively charged ligands at the surface of the cells attract the positively charged uranyl ion (UO_2^{2+}). If the concentrations of phosphate and UO_2^{2+} are sufficiently high, minerals such as autunite ($Ca(UO_2)_2(PO_4)_2 \cdot 10\text{-}12H_2O$) or polycrystalline HUO_2PO_4 may form thus reducing the mobility of UO_2^{2+}. Compared to the direct addition of inorganic phosphate to contaminated groundwater, biomineralization has the advantage that the ligands produced by microbes will target uranium compounds more specifically rather than react actively with all aqueous metals. Stimulating bacterial phosphatase activity to liberate phosphate under controlled conditions limits the rate of bacterial hydrolysis of organophosphate and the release of phosphate to the system, thus avoiding clogging of the injection location with metal phosphate minerals. The high concentration of ligands near the cell surface also provides nucleation foci for precipitation, which leads to higher efficiency than chemical precipitation.

Biocrystallization

Biocrystallization is the formation of crystals from organic macromolecules by living organisms. This may be a stress response, a normal part of metabolism such as processes that dispose of waste compounds, or a pathology. Template mediated crystallization is qualitatively different from *in vitro* crystallization. Inhibitors of biocrystallization are of interest in drug design efforts against lithiasis and against pathogens that feed on blood, since many of these organisms use this process to safely dispose of heme.

DNA

Under severe stress conditions the bacteria *Escherichia coli* protects its DNA from damage by sequestering it within a crystalline structure. This process is mediated by the stress response protein Dps and allows the bacteria to survive varied assaults such as oxidative stress, heat shock, ultraviolet light, gamma radiation and extremes of pH.

Heme

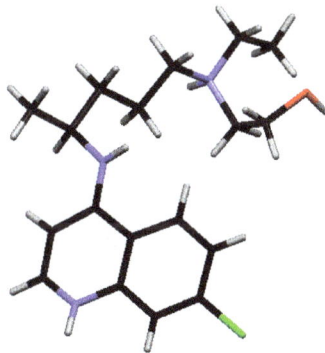

The biocrystallization inhibitor chloroquine was developed in Germany in the 1930s. For 20 years Chloroquine was a "magic bullet".

Blood feeding organisms digest hemoglobin and release high quantities of free toxic heme. To avoid destruction by this molecule, the parasite biocrystallizes heme to form hemozoin. To date, the only definitively characterized product of hematin disposal is the pigment hemozoin. Hemozoin is *per definitionem* not a mineral and therefore not formed by biomineralization. Heme biocrystallization has been found in blood feeding organisms of great medical importance including *Plasmodium*, *Rhodnius* and *Schistosoma*. Heme biocrystallization is inhibited by quinoline antimalarials such as Chloroquine.

Targeting heme biocrystallization remains one of the most promising avenues for antimalarial drug development because the drug target is highly specific to the malarial parasite, and outside the genetic control of the parasite.

Lithiasis

Lithiasis (formation of stones) is a global human health problem. Stones can form in both urinary and gastrointestinal tracts. Related to the formation of stones is the formation of crystals; this can occur in joints (e.g. gout) and in the viscera.

References

- Wenk, Hans-Rudolf; Bulakh, Andrei (2004). Minerals: Their Constitution and Origin. New York, NY: Cambridge University Press. ISBN 978-0-521-52958-7.

- John Lloyd; John Mitchinson. "What's the commonest material in the world". QI: The Book of General Ignorance. Faber & Faber. ISBN 0-571-23368-6.

- Callister Jr., William D. (2006). Materials Science and Engineering: An Introduction (7th ed.). John Wiley & Sons. ISBN 0-471-35446-5.

- Nesse, William D. (2000). Introduction to Mineralogy. New York: Oxford University Press. p91-92. ISBN 978-0-19-510691-6

- "Harvesting hydrogen fuel from the Sun using Earth-abundant materials". Phys.org. Sep 25, 2014. Retrieved 26 September 2014.

- Bullis, Kevin (8 August 2013). "A Material That Could Make Solar Power "Dirt Cheap"". MIT Technology Review. Retrieved 8 August 2013.

Study of the Properties of Minerals

The chapter lists and studies the density and hardness of various minerals. It introduces the reader to the Rosiwal and Mohs scales of mineral hardness. The content also studies the tenacity, cleavage, fracture, crystal habit and lustre of minerals. These characteristics form the distinguishing characteristics that help classify minerals.

Density

The density, or more precisely, the volumetric mass density, of a substance is its mass per unit volume. The symbol most often used for density is ρ (the lower case Greek letter rho), although the Latin letter D can also be used. Mathematically, density is defined as mass divided by volume:

A graduated cylinder containing various coloured liquids with different densities.

where ρ is the density, m is the mass, and V is the volume. In some cases (for instance, in the United States oil and gas industry), density is loosely defined as its weight per unit volume, although this is scientifically inaccurate – this quantity is more specifically called specific weight.

For a pure substance the density has the same numerical value as its mass concentration. Different materials usually have different densities, and density may be relevant to buoyancy, purity and packaging. Osmium and iridium are the densest known elements at standard conditions for temperature and pressure but certain chemical compounds may be denser.

To simplify comparisons of density across different systems of units, it is sometimes replaced by the dimensionless quantity "relative density" or "specific gravity", i.e. the ratio of the density of the material to that of a standard material, usually water. Thus a relative density less than one means that the substance floats in water.

The density of a material varies with temperature and pressure. This variation is typically small for solids and liquids but much greater for gases. Increasing the pressure on an object decreases the volume of the object and thus increases its density. Increasing the temperature of a substance (with a few exceptions) decreases its density by increasing its volume. In most materials, heating the bottom of a fluid results in convection of the heat from the bottom to the top, due to the decrease in the density of the heated fluid. This causes it to rise relative to more dense unheated material.

The reciprocal of the density of a substance is occasionally called its specific volume, a term sometimes used in thermodynamics. Density is an intensive property in that increasing the amount of a substance does not increase its density; rather it increases its mass.

History

In a well-known but probably apocryphal tale, Archimedes was given the task of determining whether King Hiero's goldsmith was embezzling gold during the manufacture of a golden wreath dedicated to the gods and replacing it with another, cheaper alloy. Archimedes knew that the irregularly shaped wreath could be crushed into a cube whose volume could be calculated easily and compared with the mass; but the king did not approve of this. Baffled, Archimedes is said to have taken an immersion bath and observed from the rise of the water upon entering that he could calculate the volume of the gold wreath through the displacement of the water. Upon this discovery, he leapt from his bath and ran naked through the streets shouting, "Eureka! Eureka!". As a result, the term "eureka" entered common parlance and is used today to indicate a moment of enlightenment.

The story first appeared in written form in Vitruvius' *books of architecture*, two centuries after it supposedly took place. Some scholars have doubted the accuracy of this tale, saying among other things that the method would have required precise measurements that would have been difficult to make at the time.

From the equation for density ($\rho = m / V$), mass density has units of mass divided by volume. As there are many units of mass and volume covering many different magnitudes there are a large number of units for mass density in use. The SI unit of kilogram per cubic metre (kg/m^3) and the cgs unit of gram per cubic centimetre (g/cm^3) are probably the most commonly used units for density.1 kg/m^3 equals 1,000 g/cm^3. (The cubic centimeter can be alternately called a *millilitre* or a *cc*.) In industry, other larger or smaller units of mass and or volume are often more practical and US customary units may be used.

Measurement of Density

Homogeneous Materials

The density at all points of a homogeneous object equals its total mass divided by its total volume. The mass is normally measured with a scale or balance; the volume may be measured directly (from the geometry of the object) or by the displacement of a fluid. To determine the density of a liquid or a gas, a hydrometer, a dasymeter or a Coriolis flow meter may be used, respectively. Similarly, hydrostatic weighing uses the displacement of water due to a submerged object to determine the density of the object.

Heterogeneous Materials

If the body is not homogeneous, then its density varies between different regions of the object. In that case the density around any given location is determined by calculating the density of a small volume around that location. In the limit of an infinitesimal volume the density of an inhomogeneous object at a point becomes: $\rho(\vec{r}) = dm/dV$, where dV is an elementary volume at position r. The mass of the body then can be expressed as

$$m = \int_V \rho(\vec{r}) dV.$$

Non-Compact Materials

In practice, bulk materials such as sugar, sand, or snow contain voids. Many materials exist in nature as flakes, pellets, or granules.

Voids are regions which contain something other than the considered material. Commonly the void is air, but it could also be vacuum, liquid, solid, or a different gas or gaseous mixture.

The bulk volume of a material—inclusive of the void fraction—is often obtained by a simple measurement (e.g. with a calibrated measuring cup) or geometrically from known dimensions.

Mass divided by *bulk* volume determines bulk density. This is not the same thing as volumetric mass density.

To determine volumetric mass density, one must first discount the volume of the void fraction. Sometimes this can be determined by geometrical reasoning. For the close-packing of equal spheres the non-void fraction can be at most about 74%. It can also be determined empirically. Some bulk materials, however, such as sand, have a *variable* void fraction which depends on how the material is agitated or poured. It might be loose or compact, with more or less air space depending on handling.

In practice, the void fraction is not necessarily air, or even gaseous. In the case of sand, it could be water, which can be advantageous for measurement as the void fraction for sand saturated in water—once any air bubbles are thoroughly driven out—is potentially more consistent than dry sand measured with an air void.

In the case of non-compact materials, one must also take care in determining the mass of the material sample. If the material is under pressure (commonly ambient air pressure at the earth's surface) the determination of mass from a measured sample weight might need to account for buoyancy effects due to the density of the void constituent, depending on how the measurement was conducted. In the case of dry sand, sand is so much denser than air that the buoyancy effect is commonly neglected (less than one part in one thousand).

Mass change upon displacing one void material with another while maintaining constant volume can be used to estimate the void fraction, if the difference in density of the two voids materials is reliably known.

Changes of Density

In general, density can be changed by changing either the pressure or the temperature. Increasing the pressure always increases the density of a material. Increasing the temperature generally decreases the density, but there are notable exceptions to this generalization. For example, the density of water increases between its melting point at 0 °C and 4 °C; similar behavior is observed in silicon at low temperatures.

The effect of pressure and temperature on the densities of liquids and solids is small. The compressibility for a typical liquid or solid is 10^{-6} bar^{-1} (1 bar = 0.1 MPa) and a typical thermal expansivity is 10^{-5} K^{-1}. This roughly translates into needing around ten thousand times atmospheric pressure to reduce the volume of a substance by one percent. (Although the pressures needed may be around a thousand times smaller for sandy soil and some clays.) A one percent expansion of volume typically requires a temperature increase on the order of thousands of degrees Celsius.

In contrast, the density of gases is strongly affected by pressure. The density of an ideal gas is $\rho = \dfrac{MP}{RT}$,

where M is the molar mass, P is the pressure, R is the universal gas constant, and T is the absolute temperature. This means that the density of an ideal gas can be doubled by doubling the pressure, or by halving the absolute temperature.

In the case of volumic thermal expansion at constant pressure and small intervals of temperature the temperature dependence of density is : $\rho = \dfrac{\rho_{T_0}}{1+\alpha \cdot \Delta T}$

where is the density at a reference temperature, is the thermal expansion coefficient of the material at temperatures close to .

Density of Solutions

The density of a solution is the sum of mass (massic) concentrations of the components of that solution.

Mass (massic) concentration of each given component ρ_i in a solution sums to density of the solution. $\rho = \sum_i \varrho_i$

Expressed as a function of the densities of pure components of the mixture and their volume participation, it allows the determination of excess molar volumes:

$$\rho = \sum_i \rho_i \frac{V_i}{V} = \sum_i \rho_i \varphi_i = \sum_i \rho_i \frac{V_i}{\sum_i V_i + \sum_i V^E{}_i}$$

provided that there is no interaction between the components.

Knowing the relation between excess volumes and activity coefficients of the components, one can determine the activity coefficients. $\overline{V^E}_i = RT \dfrac{\partial\left(ln(\gamma_i)\right)}{\partial P}$

Densities

Water

Density of liquid water at 1 atm pressure	
Temp. (°C)	Density (kg/m³)
−30	983.854
−20	993.547
−10	998.117
0	999.8395
4	999.9720
10	999.7026
15	999.1026
20	998.2071
22	997.7735
25	997.0479
30	995.6502
40	992.2
60	983.2
80	971.8
100	958.4
Notes: Values below 0 °C refer to supercooled water.	

Air

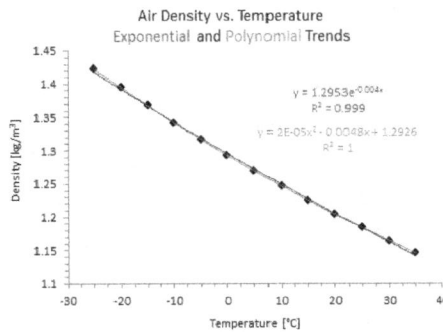

Air density vs. temperature

Density of air at 1 atm pressure	
T (°C)	ρ (kg/m³)
−25	1.423
−20	1.395

−15	1.368
−10	1.342
−5	1.316
0	1.293
5	1.269
10	1.247
15	1.225
20	1.204
25	1.184
30	1.164
35	1.146

Various Materials

Densities of various materials covering a range of values		
Material	ρ (kg/m³)	Notes
Helium	0.179	
Aerographite	0.2	[note 2]
Metallic microlattice	0.9	[note 2]
Aerogel	1.0	[note 2]
Air	1.2	At sea level
Tungsten hexafluoride	12.4	One of the heaviest known gases at standard conditions
Liquid hydrogen	70	At approx. −255 °C
Styrofoam	75	Approx.
Cork	240	Approx.
Pine	373	
Lithium	535	
Wood	700	Seasoned, typical
Oak	710	
Potassium	860	
Sodium	970	
Ice	916.7	At temperature < 0 °C
Water (fresh)	1,000	At 4 °C, the temperature of its maximum density
Water (salt)	1,030	
Nylon	1,150	
Plastics	1,175	Approx.; for polypropylene and PETE/PVC
Tetrachloroethene	1,622	
Magnesium	1,740	
Beryllium	1,850	
Glycerol	1,261	
Concrete	2,000	
Silicon	2,330	
Aluminium	2,700	

Diiodomethane	3,325	Liquid at room temperature
Diamond	3,500	
Titanium	4,540	
Selenium	4,800	
Vanadium	6,100	
Antimony	6,690	
Zinc	7,000	
Chromium	7,200	
Tin	7,310	
Manganese	7,325	Approx.
Iron	7,870	
Niobium	8,570	
Brass	8,600	
Cadmium	8,650	
Cobalt	8,900	
Nickel	8,900	
Copper	8,940	
Bismuth	9,750	
Molybdenum	10,220	
Silver	10,500	
Lead	11,340	
Thorium	11,700	
Rhodium	12,410	
Mercury	13,546	
Tantalum	16,600	
Uranium	18,800	
Tungsten	19,300	
Gold	19,320	
Plutonium	19,840	
Platinum	21,450	
Iridium	22,420	
Osmium	22,570	

Notes: Unless otherwise noted, all densities given are at standard conditions for temperature and pressure, that is, 273.15 K (0.00 °C) and 100 kPa (0.987 atm).

1.   Air contained in material excluded when calculating density

Others

Entity	ρ (kg/m³)	Notes
Interstellar medium	1×10^{-19}	Assuming 90% H, 10% He; variable T
The Earth	5,515	Mean density.
The inner core of the Earth	13,000	Approx., as listed in Earth.
The core of the Sun	33,000–160,000	Approx.
Super-massive black hole	9×10^5	Density of a 4.5-million-solar-mass black hole Event horizon radius is 13.5 million km.

White dwarf star	2.1×10^9	Approx.
Atomic nuclei	2.3×10^{17}	Does not depend strongly on size of nucleus
Neutron star	1×10^{18}	
Stellar-mass black hole	1×10^{18}	Density of a 4-solar-mass black hole Event horizon radius is 12 km.

Common Units

The SI unit for density is:

- kilograms per cubic meter (kg/m³)

Litres and metric tons are not part of the SI, but are acceptable for use with it, leading to the following units:

- kilograms per liter (kg/L)
- grams per milliliter (g/mL)
- metric tons per cubic meter (t/m³)

Densities using the following metric units all have exactly the same numerical value, one thousandth of the value in (kg/m³). Liquid water has a density of about 1 kg/dm³, making any of these SI units numerically convenient to use as most solids and liquids have densities between 0.1 and 20 kg/dm³.

- kilograms per cubic decimetre (kg/dm³)
- grams per cubic centimetre (g/cm³)
    - 1 gram/cm³ = 1000 kg/m³
- megagrams (metric tons) per cubic metre (Mg/m³)

In US customary units density can be stated in:

- Avoirdupois ounces per cubic inch (1 g/cc ≈ 0.578036672 oz/cu in)
- Avoirdupois ounces per fluid ounce (1 g/cc ≈ 1.04317556 oz/fl. oz = 1.04317556 lbs/pint)
- Avoirdupois pounds per cubic inch (1 g/cc ≈ 0.036127292 lb/cu in)
- pounds per cubic foot (1 g/cc ≈ 62.427961 lb/cu ft)
- pounds per cubic yard (1 g/cc ≈ 1685.5549 lb/cu yd)
- pounds per US liquid gallon (1 g/cc ≈ 8.34540445 lb/gal)
- pounds per US bushel (1 g/cc ≈ 77.6888513 lb/bu)
- slugs per cubic foot

Imperial units differing from the above (as the Imperial gallon and bushel differ from the US units) in practice are rarely used, though found in older documents. The Imperial gallon was based on the concept that an Imperial fluid ounce of water would have a mass of one Avoirdupois ounce, and indeed 1 g/cc ≈ 1.00224129 ounces per Imperial fluid ounce = 10.0224129 lbs per Imperial gallon. The density of precious metals could conceivably be based on Troy ounces and pounds, a possible cause of confusion.

Hardness

Hardness is a measure of how resistant solid matter is to various kinds of permanent shape change when a compressive force is applied. Some materials (e.g. metals) are harder than others (e.g. plastics). Macroscopic hardness is generally characterized by strong intermolecular bonds, but the behavior of solid materials under force is complex; therefore, there are different measurements of hardness: *scratch hardness*, *indentation hardness*, and *rebound hardness*.

Hardness is dependent on ductility, elastic stiffness, plasticity, strain, strength, toughness, visco-elasticity, and viscosity.

Common examples of hard matter are ceramics, concrete, certain metals, and superhard materials, which can be contrasted with soft matter.

Measuring Hardness

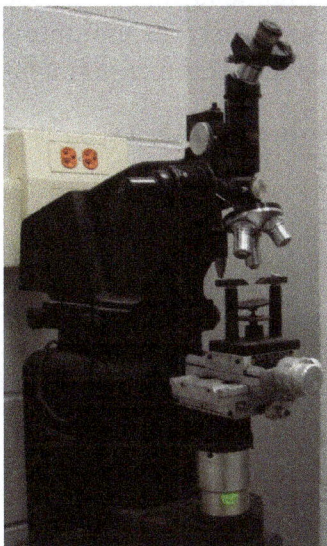

A Vickers hardness tester

There are three main types of hardness measurements: *scratch*, *indentation*, and *rebound*. Within each of these classes of measurement there are individual measurement scales. For practical reasons conversion tables are used to convert between one scale and another.

Scratch Hardness

Scratch hardness is the measure of how resistant a sample is to fracture or permanent plastic deformation due to friction from a sharp object. The principle is that an object made of a harder material will scratch an object made of a softer material. When testing coatings, scratch hardness refers to the force necessary to cut through the film to the substrate. The most common test is Mohs scale, which is used in mineralogy. One tool to make this measurement is the sclerometer.

Another tool used to make these tests is the pocket hardness tester. This tool consists of a scale arm with graduated markings attached to a four-wheeled carriage. A scratch tool with a sharp rim is mounted at a predetermined angle to the testing surface. In order to use it a weight of known mass is added to the scale arm at one of the graduated markings, the tool is then drawn across the test surface. The use of the weight and markings allows a known pressure to be applied without the need for complicated machinery.

Indentation Hardness

Indentation hardness measures the resistance of a sample to material deformation due to a constant compression load from a sharp object; they are primarily used in engineering and metallurgy fields. The tests work on the basic premise of measuring the critical dimensions of an indentation left by a specifically dimensioned and loaded indenter.

Common indentation hardness scales are Rockwell, Vickers, Shore, and Brinell.

Rebound Hardness

Rebound hardness, also known as *dynamic hardness*, measures the height of the "bounce" of a diamond-tipped hammer dropped from a fixed height onto a material. This type of hardness is related to elasticity. The device used to take this measurement is known as a scleroscope.

Two scales that measures rebound hardness are the Leeb rebound hardness test and Bennett hardness scale.

Hardening

There are five hardening processes: Hall-Petch strengthening, work hardening, solid solution strengthening, precipitation hardening, and martensitic transformation.

Physics

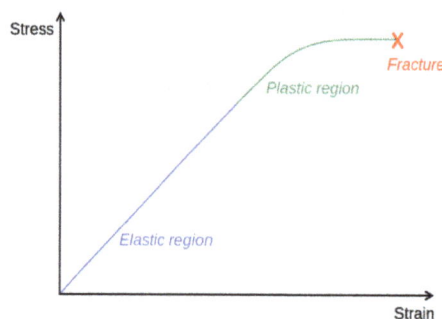

Diagram of a stress-strain curve, showing the relationship between stress (force applied per unit area) and strain or deformation of a ductile metal.

In solid mechanics, solids generally have three responses to force, depending on the amount of force and the type of material:

- They exhibit elasticity—the ability to temporarily change shape, but return to the original shape when the pressure is removed. "Hardness" in the elastic range—a small temporary change in shape for a given force—is known as stiffness in the case of a given object, or a high elastic modulus in the case of a material.

- They exhibit plasticity—the ability to permanently change shape in response to the force, but remain in one piece. The yield strength is the point at which elastic deformation gives way to plastic deformation. Deformation in the plastic range is non-linear, and is described by the stress-strain curve. This response produces the observed properties of scratch and indentation hardness, as described and measured in materials science. Some materials exhibit both elasticity and viscosity when undergoing plastic deformation; this is called viscoelasticity.

- They fracture—split into two or more pieces.

Strength is a measure of the extent of a material's elastic range, or elastic and plastic ranges together. This is quantified as compressive strength, shear strength, tensile strength depending on the direction of the forces involved. Ultimate strength is an engineering measure of the maximum load a part of a specific material and geometry can withstand.

Brittleness, in technical usage, is the tendency of a material to fracture with very little or no detectable plastic deformation beforehand. Thus in technical terms, a material can be both brittle and strong. In everyday usage "brittleness" usually refers to the tendency to fracture under a small amount of force, which exhibits both brittleness and a lack of strength (in the technical sense). For perfectly brittle materials, yield strength and ultimate strength are the same, because they do not experience detectable plastic deformation. The opposite of brittleness is ductility.

The toughness of a material is the maximum amount of energy it can absorb before fracturing, which is different from the amount of force that can be applied. Toughness tends to be small for brittle materials, because elastic and plastic deformations allow materials to absorb large amounts of energy.

Hardness increases with decreasing particle size. This is known as the Hall-Petch relationship. However, below a critical grain-size, hardness decreases with decreasing grain size. This is known as the inverse Hall-Petch effect.

Hardness of a material to deformation is dependent on its microdurability or small-scale shear modulus in any direction, not to any rigidity or stiffness properties such as its bulk modulus or Young's modulus. Stiffness is often confused for hardness. Some materials are stiffer than diamond (e.g. osmium) but are not harder, and are prone to spalling and flaking in squamose or acicular habits.

Mechanisms and Theory

The key to understanding the mechanism behind hardness is understanding the metallic microstructure, or the structure and arrangement of the atoms at the atomic level. In fact, most important metallic properties critical to the manufacturing of today's goods are determined by the microstructure of a material. At the atomic level, the atoms in a metal are arranged in an orderly three-dimensional array called a crystal lattice. In reality, however, a given specimen of a metal likely never contains a consistent single crystal lattice. A given sample of metal will contain many grains, with each grain having a fairly consistent array pattern. At an even smaller scale, each grain contains irregularities.

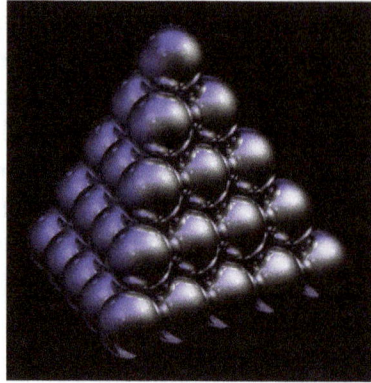

A representation of the crystal lattice showing the planes of atoms.

There are two types of irregularities at the grain level of the microstructure that are responsible for the hardness of the material. These irregularities are point defects and line defects. A point defect is an irregularity located at a single lattice site inside of the overall three-dimensional lattice of the grain. There are three main point defects. If there is an atom missing from the array, a vacancy defect is formed. If there is a different type of atom at the lattice site that should normally be occupied by a metal atom, a substitutional defect is formed. If there exists an atom in a site where there should normally not be, an interstitial defect is formed. This is possible because space exists between atoms in a crystal lattice. While point defects are irregularities at a single site in the crystal lattice, line defects are irregularities on a plane of atoms. Dislocations are a type of line defect involving the misalignment of these planes. In the case of an edge dislocation, a half plane of atoms is wedged between two planes of atoms. In the case of a screw dislocation two planes of atoms are offset with a helical array running between them.

In glasses, hardness seems to depend linearly on the number of topological constraints acting between the atoms of the network. Hence, the rigidity theory has allowed predicting hardness values with respect to composition.

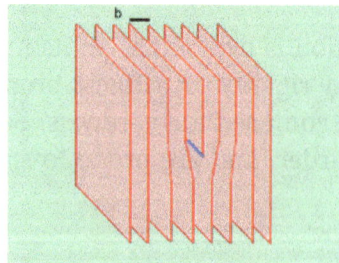

Planes of atoms split by an edge dislocation.

Dislocations provide a mechanism for planes of atoms to slip and thus a method for plastic or permanent deformation. Planes of atoms can flip from one side of the dislocation to the other effectively allowing the dislocation to traverse through the material and the material to deform permanently. The movement allowed by these dislocations causes a decrease in the material's hardness.

The way to inhibit the movement of planes of atoms, and thus make them harder, involves the interaction of dislocations with each other and interstitial atoms. When a dislocation intersects with a second dislocation, it can no longer traverse through the crystal lattice. The intersection of dislocations creates an anchor point and does not allow the planes of atoms to continue to slip over one another A dislocation can also be anchored by the interaction with interstitial atoms. If a dislocation comes in contact with two or more interstitial atoms, the slip of the planes will again be disrupted. The interstitial atoms create anchor points, or pinning points, in the same manner as intersecting dislocations.

By varying the presence of interstitial atoms and the density of dislocations, a particular metal's hardness can be controlled. Although seemingly counter-intuitive, as the density of dislocations increases, there are more intersections created and consequently more anchor points. Similarly, as more interstitial atoms are added, more pinning points that impede the movements of dislocations are formed. As a result, the more anchor points added, the harder the material will become.

Scales of Hardness

Rosiwal scale

The Rosiwal scale is a hardness scale, with its name given in memory of the Austrian geologist August Karl Rosiwal. The Rosiwal scale bases its measure on absolute values, unlike the Mohs scale whose values are relative values, its interest is relegated to the amateur or an approach that makes it useful in the research field ('in situ').

The Rosiwal method (also called Delesse and Rosiwal) is basically a method of petrographic analysis and led to the development of the stereograph.

Rosiwal Scale Values

Valor MOHS		ROSIWAL value							Chemical composition
Hardness	Mineral	1	10	100	1.000	10.000	100.000	1000000	
1	Talc	****							$Mg_3Si_4O_{10}(OH)_2$
2	Plaster	*****	*						$CaSO_4 \cdot 2H_2O$
3	Calcite	*****	*****						$CaCO_3$
4	Fluorite	*****	*****						CaF_2
5	Apatite	*****	******						$Ca_5(PO_4)_3(OH^-,Cl^-,F^-)$
6	Feldspar	*****	*****	****					$KAlSi_3O_8$
7	Quartz	*****	*****	*****	*				SiO_2
8	Topaz	*****	*****	*****	***				$Al_2SiO_4(OH^-,F^-)_2$
9	Corundum	*****	*****	*****	*****				Al_2O_3
10	Diamond	*****	*****	*****	*****	*****	*****	**	C

Mineral	Mohs scale Degree	Rosiwal hardness
Talc	1°	0.03
Plaster	2°	1.25
Calcite	3°	4.5
Fluorite	4°	5
Apatite	5°	5.5
Feldspar	6°	37
Quartz	7°	120
Topaz	8°	175
Corundum	9°	1,000
Diamond	10°	140,000

Measures in an absolute scale the hardness of minerals expressed as abrasion's resistance, measured at laboratory, starting with corundum with a base value of 1000.

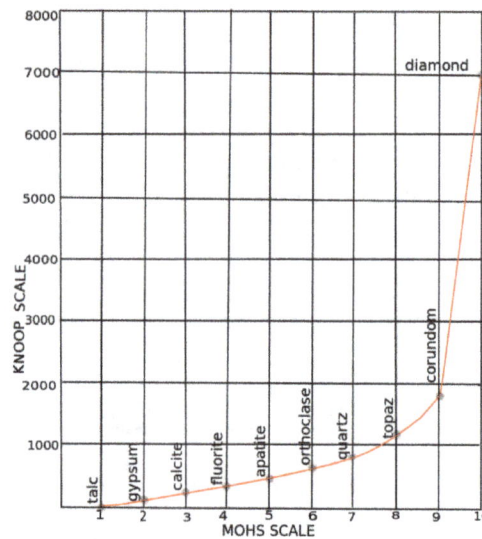

Comparison Mohs scale and Knoop.

Knoop Scale

The Rosiwal scale is used in mineralogy in the same way as Mohs scale and Knoop scale.

Mohs Scale of Mineral Hardness

The Mohs scale of mineral hardness is a qualitative ordinal scale that characterizes the scratch resistance of various minerals through the ability of a harder material to scratch a softer material. It was created in 1812 by the German geologist and mineralogist Friedrich Mohs and is one of several definitions of hardness in materials science, some of which are more quantitative. The method of comparing hardness by seeing which minerals can visibly scratch others, however, is of great antiquity, having been mentioned by Theophrastus in his treatise *On Stones*, c. 300 BC, followed by Pliny the Elder in his *Naturalis Historia*, c. 77 AD. While greatly facilitating the identification of minerals in the field, the Mohs scale does not show how well hard materials perform in an industrial setting.

Mohs hardness kit, containing one specimen of each mineral on the ten-point hardness scale

Usage

Despite its simplicity and lack of precision, the Mohs scale is highly relevant for field geologists, who use the scale to roughly identify minerals using scratch kits. The Mohs scale hardness of minerals can be commonly found in reference sheets. Reference materials may be expected to have a uniform Mohs hardness.

Minerals

The Mohs scale of mineral hardness is based on the ability of one natural sample of mineral to scratch another mineral visibly. The samples of matter used by Mohs are all different minerals. Minerals are pure substances found in nature. Rocks are made up of one or more minerals. As the hardest known naturally occurring substance when the scale was designed, diamonds are at the top of the scale. The hardness of a material is measured against the scale by finding the hardest material that the given material can scratch, and/or the softest material that can scratch the given material. For example, if some material is scratched by apatite but not by fluorite, its hardness on the Mohs scale would fall between 4 and 5. "Scratching" a material for the purposes of the Mohs scale means creating non-elastic dislocations visible to the naked eye. Frequently, materials that are lower on the Mohs scale can create microscopic, non-elastic dislocations on materials that have a higher Mohs number. While these microscopic dislocations are permanent and sometimes detrimental to the harder material's structural integrity, they are not considered "scratches" for the determination of a Mohs scale number.

The Mohs scale is a purely ordinal scale. For example, corundum (9) is twice as hard as topaz (8), but diamond (10) is four times as hard as corundum. The table below shows the comparison with the absolute hardness measured by a sclerometer, with pictorial examples.

Mohs hardness	Mineral	Chemical formula	Absolute hardness	Image
1	Talc	$Mg_3Si_4O_{10}(OH)_2$	1	
2	Gypsum	$CaSO_4 \cdot 2H_2O$	3	
3	Calcite	$CaCO_3$	9	
4	Fluorite	CaF_2	21	
5	Apatite	$Ca_5(PO_4)_3(OH^-,Cl^-,F^-)$	48	
6	Orthoclase feldspar	$KAlSi_3O_8$	72	
7	Quartz	SiO_2	100	
8	Topaz	$Al_2SiO_4(OH^-,F^-)_2$	200	

9	Corundum	Al_2O_3	400	
10	Diamond	C	1600	

On the Mohs scale, a streak plate (unglazed porcelain) has a hardness of 7.0. Using these ordinary materials of known hardness can be a simple way to approximate the position of a mineral on the scale.

Intermediate Hardness

The table below incorporates additional substances that may fall between levels:

Hardness	Substance or mineral
0.2–0.3	caesium, rubidium
0.5–0.6	lithium, sodium, potassium
1	talc
1.5	gallium, strontium, indium, tin, barium, thallium, lead, graphite, ice
2	hexagonal boron nitride, calcium, selenium, cadmium, sulfur, tellurium, bismuth, gypsum
2.5–3	gold, silver, aluminium, zinc, lanthanum, cerium, Jet (lignite)
3	calcite, copper, arsenic, antimony, thorium, dentin
3.5	platinum
4	fluorite, iron, nickel
4–4.5	steel
5	apatite (tooth enamel), zirconium, palladium, obsidian (volcanic glass)
5.5	beryllium, molybdenum, hafnium, glass, cobalt
6	orthoclase, titanium, manganese, germanium, niobium, rhodium, uranium
6–7	fused quartz, iron pyrite, silicon, ruthenium, iridium, tantalum, opal, peridot, tanzanite, jade
7	osmium, quartz, rhenium, vanadium
7.5–8	emerald, hardened steel, tungsten, spinel
8	topaz, cubic zirconia
8.5	chrysoberyl, chromium, silicon nitride, tantalum carbide
9	corundum, tungsten carbide, titanium nitride
9–9.5	silicon carbide (carborundum), titanium carbide
9.5–10	boron, boron nitride, rhenium diboride (a-axis), stishovite, titanium diboride
10	diamond, carbonado
>10	nanocrystalline diamond (hyperdiamond, ultrahard fullerite), rhenium diboride (c-axis)

Hardness (Vickers)

Comparison between Hardness (Mohs) and Hardness (Vickers):

Mineral name	Hardness (Mohs)	Hardness (Vickers) kg/mm²
Graphite	1–2	VHN_{10}=7–11
Tin	1½	VHN_{10}=7–9
Bismuth	2–2½	VHN_{100}=16–18
Gold	2½	VHN_{10}=30–34
Silver	2½	VHN_{100}=61–65
Chalcocite	2½–3	VHN_{100}=84–87
Copper	2½–3	VHN_{100}=77–99
Galena	2½	VHN_{100}=79–104
Sphalerite	3½–4	VHN_{100}=208–224
Heazlewoodite	4	VHN_{100}=230–254
Carrollite	4½–5½	VHN_{100}=507–586
Goethite	5–5½	VHN_{100}=667
Hematite	5–6	VHN_{100}=1,000–1,100
Chromite	5½	VHN_{100}=1,278–1,456
Anatase	5½–6	VHN_{100}=616–698
Rutile	6–6½	VHN_{100}=894–974
Pyrite	6–6½	VHN_{100}=1,505–1,520
Bowieite	7	VHN_{100}=858–1,288
Euclase	7½	VHN_{100}=1,310
Chromium	8½	VHN_{100}=1,875–2,000

Tenacity (Mineralogy)

In mineralogy, tenacity is a mineral's behavior when deformed or broken.

Common Terms

Brittle

The mineral breaks or powders easily. Most ionic-bonded minerals are brittle.

Malleable

The mineral may be pounded out into thin sheets. Metallic-bonded minerals may be malleable.

Ductile

The mineral may be drawn into a wire. Obviously not easy to test. Malleable materials also may be ductile.

Sectile

May be cut smoothly with a knife. Relatively few minerals are sectile.

Elastic

If bent, will spring back to its original position when the stress is released.

Plasticity

If bent, will NOT spring back to its original position when the stress is released. It stays bent. In contrast, flexibility is the ability of a material to deform elastically and return to its original shape when the applied stress is removed.

Cleavage (Crystal)

Green fluorite with prominent cleavage.

Biotite with basal cleavage.

Cleavage, in mineralogy, is the tendency of crystalline materials to split along definite crystallographic structural planes. These planes of relative weakness are a result of the regular locations of atoms and ions in the crystal, which create smooth repeating surfaces that are visible both in the microscope and to the naked eye.

Types of Cleavage

Miller indices {h k ℓ}

Cleavage forms parallel to crystallographic planes:

- Basal or pinacoidal cleavage occurs when there is only one cleavage plane. Graphite has basal cleavage. Mica (like muscovite or biotite) also has basal cleavage; this is why mica can be peeled into thin sheets.

- Cubic cleavage occurs on when there are three cleavage planes intersecting at 90 degrees. Halite (or salt) has cubic cleavage, and therefore, when halite crystals are broken, it will form more cubes.

- Octahedral cleavage occurs when there are four cleavage planes in a crystal. Fluorite exhibits perfect octahedral cleavage. Octahedral cleavage is common for semiconductors. Diamond also has octahedral cleavage.

- Rhombohedral cleavage occurs when there are three cleavage planes intersecting at angles that are not 90 degrees. Calcite had rhombohedral cleavage.

- Prismatic cleavage occurs when there are two cleavage planes in a crystal. Spodumene exhibits prismatic cleavage.

- Dodecahedral cleavage occurs when there are six cleavage planes in a crystal. Sphalerite has dodecahedral cleavage.

Parting

Crystal parting occurs when minerals break along planes of structural weakness due to external stress or along twin composition planes. Parting breaks are very similar in appearance to cleavage, but only occur due to stress. Examples include magnetite which shows octahedral parting, the rhombohedral parting of corundum and basal parting in pyroxenes.

Uses

Cleavage is a physical property traditionally used in mineral identification, both in hand specimen and microscopic examination of rock and mineral studies. As an example, the angles between the prismatic cleavage planes for the pyroxenes (88–92°) and the amphiboles (56–124°) are diagnostic.

Crystal cleavage is of technical importance in the electronics industry and in the cutting of gemstones.

Precious stones are generally cleaved by impact, as in diamond cutting.

Synthetic single crystals of semiconductor materials are generally sold as thin wafers which are much easier to cleave. Simply pressing a silicon wafer against a soft surface and scratching its edge with a diamond scribe is usually enough to cause cleavage; however, when dicing a wafer to form chips, a procedure of scoring and breaking is often followed for greater control. Elemental semiconductors (Si, Ge, and diamond) are diamond cubic, a space group for which octahedral cleavage is observed. This means that some orientations of wafer allow near-perfect rectangles to be cleaved. Most other commercial semiconductors (GaAs, InSb, etc.) can be made in the related zinc blende structure, with similar cleavage planes.

Fracture (Mineralogy)

In the field of mineralogy, fracture is the texture and shape of a rock's surface formed when a mineral is fractured. Minerals often have a highly distinctive fracture, making it a principal feature used in their identification.

Fracture differs from cleavage in that the latter involves clean splitting along the cleavage planes of the mineral's crystal structure, as opposed to more general breakage. All minerals exhibit fracture, but when very strong cleavage is present.

Terminology

Conchoidal Fracture

Conchoidal fracture breakage that resembles the concentric ripples of a mussel shell. It often occurs in amorphous or fine-grained minerals such as flint, opal or obsidian, but may also occur in crystalline minerals such as quartz. Subconchoidal fracture is similar to conchoidal fracture, but with less significant curvature. (Note that obsidian is an igneous rock, not a mineral, but it does illustrate conchoidal fracture well.)

Obsidian

Earthy Fracture

Limonite

Earthy fracture is reminiscent of freshly broken soil. It is frequently seen in relatively soft, loosely bound minerals, such as limonite, kaolinite and aluminite.

Hackly Fracture

Native copper

Hackly fracture (also known as jagged fracture) is jagged, sharp and not even. It occurs when metals are torn, and so is often encountered in native metals such as copper and silver.

Splintery Fracture

Chrysotile

Splintery fracture comprises sharp elongated points. It is particularly seen in fibrous minerals such as chrysotile, but may also occur in non-fibrous minerals such as kyanite.

Uneven Fracture

Uneven fracture is a rough surface or one with random irregularities. It occurs in a wide range of minerals including arsenopyrite, pyrite and magnetite.

Crystal Habit

This article is about the descriptive term used in mineralogy. For the addictive drug.

Pyrite sun (or dollar) in laminated shale matrix. Between tightly spaced layers of shale, the aggregate was forced to grow in a laterally compressed, radiating manner. Under normal conditions, pyrite would form cubes or pyritohedrons.

In mineralogy, crystal habit is the characteristic external shape of an individual crystal or crystal group. A single crystal's habit is a description of its general shape and its crystallographic forms, plus how well developed each are. Recognizing the habit may help in identifying a mineral. When the faces are well-developed due to uncrowded growth a crystal is called *euhedral*, one with partially developed faces is *subhedral*, and one with undeveloped crystal faces is called *anhedral*. The long axis of a euhedral quartz crystal typically has a six-sided prismatic habit with parallel opposite faces. Aggregates can be formed of individual crystals with euhedral to anhedral grains. The arrangement of crystals within the aggregate can be characteristic of certain minerals. For example, minerals used for asbestos insulation often grow in a fibrous habit, a mass of very fine fibers.

The terms used by mineralogists to report crystal habits describe the typical appearance of an ideal mineral. Recognizing the habit can aid in identification as some habits are characteristic. Most minerals, however, do not display ideal habits due to conditions during crystallization. Euhedral crystals formed in uncrowded conditions with no adjacent crystal grains are not common; more often faces are poorly formed or unformed against adjacent grains and the mineral's habit may not be easily recognized.

Goethite replacing pyrite cubes

Factors influencing habit include: a combination of two or more crystal forms; trace impurities present during growth; crystal twinning and growth conditions (i.e., heat, pressure, space); and specific growth tendencies such as growth striations. Minerals belonging to the same crystal system do not necessarily exhibit the same habit. Some habits of a mineral are unique to its variety and locality: For example, while most sapphires form elongate barrel-shaped crystals, those found in Montana form stout *tabular* crystals. Ordinarily, the latter habit is seen only in ruby. Sapphire and ruby are both varieties of the same mineral; corundum.

Some minerals may replace other existing minerals while preserving the original's habit: this process is called pseudomorphous replacement. A classic example is tiger's eye quartz, crocidolite asbestos replaced by silica. While quartz typically forms *prismatic* (elongate, prism-like) crystals, in tiger's eye the original *fibrous* habit of crocidolite is preserved.

The names of crystal habits are derived from:

Predominant crystal faces (prism – prismatic, pyramid – pyramidal and pinacoid – platy). Crystal forms (cubic, octahedral, dodecahedral). Aggregation of crystals or aggregates (fibrous, botryoidal, radiating, massive). Crystal appearance (foliated/lamellar (layered), dendritic, bladed, acicular, lenticular, tabular (tablet shaped)).

Lustre (Mineralogy)

Lustre or luster is the way light interacts with the surface of a crystal, rock, or mineral. The word traces its origins back to the latin *lux*, meaning "light", and generally implies radiance, gloss, or brilliance.

A range of terms are used to describe lustre, such as *earthy, metallic, greasy,* and *silky*. Similarly, the term *vitreous* (derived from the Latin for glass, *vitrum*) refers to a glassy lustre. A list of these terms is given below.

Lustre varies over a wide continuum, and so there are no rigid boundaries between the different types of lustre. (For this reason, different sources can often describe the same mineral differently. This ambiguity is further complicated by lustre's ability to vary widely within a particular mineral species.) The terms are frequently combined to describe intermediate types of lustre (for example, a "vitreous greasy" lustre).

Some minerals exhibit unusual optical phenomena, such as asterism (the display of a star-shaped luminous area) or chatoyancy (the display of luminous bands, which appear to move as the specimen is rotated). A list of such phenomena is given below.

Common Terms

Adamantine Lustre

Cut diamonds.

Adamantine minerals possess a superlative lustre, which is most notably seen in diamond. Such minerals are transparent or translucent, and have a high refractive index (of 1.9 or more). Minerals with a true adamantine lustre are uncommon, with examples being cerussite and Cubic zirconia.

Minerals with a lesser (but still relatively high) degree of lustre are referred to as subadamantine, with some examples being garnet and corundum.

Dull Lustre

Kaolinite

Dull (or earthy) minerals exhibit little to no lustre, due to coarse granulations which scatter light in all directions, approximating a Lambertian reflector. An example is kaolinite. A distinction is sometimes drawn between dull minerals and earthy minerals, with the latter being coarser, and having even less lustre.

Greasy Lustre

Greasy minerals resemble fat or grease. A greasy lustre often occurs in minerals containing a great abundance of microscopic inclusions, with examples including opal and cordierite. Many minerals with a greasy lustre also feel greasy to the touch.

Moss opal

Metallic (or splendent) minerals have the lustre of polished metal, and with ideal surfaces will work as a reflective surface. Examples include galena, pyrite and magnetite.

References

- Raymond Serway; John Jewett (2005), Principles of Physics: A Calculus-Based Text, Cengage Learning, p. 467, ISBN 0-534-49143-X

- Hugh D. Young; Roger A. Freedman. University Physics with Modern Physics. Addison-Wesley; 2012. ISBN 978-0-321-69686-1. p. 374.

- Mukherjee, Swapna (2012). Applied Mineralogy: Applications in Industry and Environment. Springer Science & Business Media. pp. 373–. ISBN 978-94-007-1162-4.

- Berger, Lev I. (1996). Semiconductor Materials (First ed.). Boca Raton, FL: CRC Press. p. 126. ISBN 978-0849389122.

- Wenk, Hans-Rudolph and Andrei Bulakh, 2004, Minerals: Their Constitution and Origin, Cambridge, first edition, ISBN 978-0521529587

- Bonewitz, Ronald Louis (2005). Rock and Gem. Dorling Kindersley. pp. 152–153. ISBN 0-7513-4400-1.

- Emsley, John (2001). Nature's Building Blocks: An A-Z Guide to the Elements. Oxford University Press. pp. 451–53. ISBN 0-19-850341-5.

- Leslie, W. C. (1981). The physical metallurgy of steels. Washington: Hempisphere Pub. Corp., New York: Mc-Graw-Hill, ISBN 0070377804.

- D H Maling (24 September 2013). Measurements from Maps: Principles and Methods of Cartometry. Elsevier. pp. 437–. ISBN 978-1-4832-5767-9.

- New carbon nanotube struructure aerographite is lightest material champ. Phys.org (July 13, 2012). Retrieved on July 14, 2012.

- Aerographit: Leichtestes Material der Welt entwickelt – SPIEGEL ONLINE. Spiegel.de (July 11, 2012). Retrieved on July 14, 2012.

Magnetic Properties of Minerals

6

Minerals differ in their response to magnetic fields. Based on their attraction or repulsion, minerals can be classified as possessing features like diamagnetism, paramagnetism, ferromagnetism, ferrimagnetism, antiferromagnetism and superparamagnetism. This chapter exclusively deals with the origins of magnetism and cites examples to deepen the understanding of each category of magnetic ability.

Diamagnetism

Diamagnetic materials create an induced magnetic field in a direction opposite to an externally applied magnetic field, and are repelled by the applied magnetic field. In contrast, the opposite behavior is exhibited by paramagnetic materials. Diamagnetism is a quantum mechanical effect that occurs in all materials; when it is the only contribution to the magnetism the material is called a *diamagnet*. Unlike a ferromagnet, a diamagnet is not a permanent magnet. Its magnetic permeability is less than μ_o, the permeability of vacuum. In most materials diamagnetism is a weak effect, but a superconductor repels the magnetic field entirely, apart from a thin layer at the surface.

Diamagnets were first discovered when Sebald Justinus Brugmans observed in 1778 that bismuth and antimony were repelled by magnetic fields. In 1845, Michael Faraday demonstrated that it was a property of matter and concluded that every material responded (in either a diamagnetic or paramagnetic way) to an applied magnetic field. He adopted the term *diamagnetism* after it was suggested to him by William Whewell.

Materials

Notable diamagnetic materials	
Material	χ_v ($\times 10^{-5}$)
Superconductor	-10^5
Pyrolytic carbon	−40.9
Bismuth	−16.6
Mercury	−2.9
Silver	−2.6
Carbon (diamond)	−2.1
Lead	−1.8
Carbon (graphite)	−1.6
Copper	−1.0

Notable diamagnetic materials	
Material	χ_v ($\times 10^{-5}$)
Water	−0.91

Diamagnetism, to a greater or lesser degree, is a property of all materials and always makes a weak contribution to the material's response to a magnetic field. For materials that show some other form of magnetism (such as ferromagnetism or paramagnetism), the diamagnetic contribution becomes negligible. Substances that mostly display diamagnetic behaviour are termed diamagnetic materials, or diamagnets. Materials called diamagnetic are those that laymen generally think of as *non-magnetic*, and include water, wood, most organic compounds such as petroleum and some plastics, and many metals including copper, particularly the heavy ones with many core electrons, such as mercury, gold and bismuth. The magnetic susceptibility values of various molecular fragments are called Pascal's constants.

Diamagnetic materials, like water, or water based materials, have a relative magnetic permeability that is less than or equal to 1, and therefore a magnetic susceptibility less than or equal to 0, since susceptibility is defined as $\chi_v = \mu_v - 1$. This means that diamagnetic materials are repelled by magnetic fields. However, since diamagnetism is such a weak property its effects are not observable in everyday life. For example, the magnetic susceptibility of diamagnets such as water is $\chi_v = -9.05 \times 10^{-6}$. The most strongly diamagnetic material is bismuth, $\chi_v = -1.66 \times 10^{-4}$, although pyrolytic carbon may have a susceptibility of $\chi_v = -4.00 \times 10^{-4}$ in one plane. Nevertheless, these values are orders of magnitude smaller than the magnetism exhibited by paramagnets and ferromagnets. Note that because χ_v is derived from the ratio of the internal magnetic field to the applied field, it is a dimensionless value.

All conductors exhibit an effective diamagnetism when they experience a changing magnetic field. The Lorentz force on electrons causes them to circulate around forming eddy currents. The eddy currents then produce an induced magnetic field opposite the applied field, resisting the conductor's motion.

Superconductors

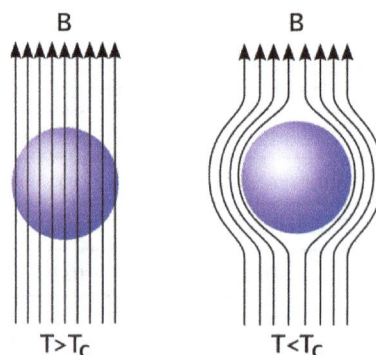

Transition from ordinary conductivity (left) to superconductivity (right). At the transition, the superconductor expels the magnetic field and then acts as a perfect diamagnet.

Superconductors may be considered perfect diamagnets ($\chi_v = -1$), because they expel all fields (except in a thin surface layer) due to the Meissner effect. However this effect is not due to eddy currents, as in ordinary diamagnetic materials.

Demonstrations

Curving Water Surfaces

If a powerful magnet (such as a supermagnet) is covered with a layer of water (that is thin compared to the diameter of the magnet) then the field of the magnet significantly repels the water. This causes a slight dimple in the water's surface that may be seen by its reflection.

Levitation

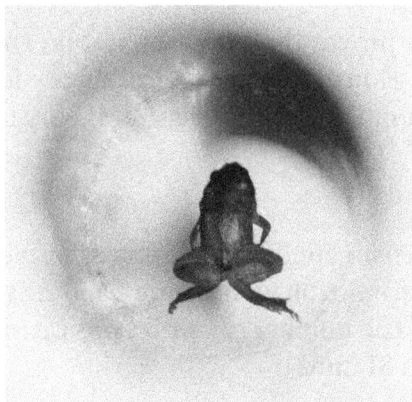

A live frog levitates inside a 32 mm (1.26 in) diameter vertical bore of a Bitter solenoid in a magnetic field of about 16 teslas at the Nijmegen High Field Magnet Laboratory.

Diamagnets may be levitated in stable equilibrium in a magnetic field, with no power consumption. Earnshaw's theorem seems to preclude the possibility of static magnetic levitation. However, Earnshaw's theorem only applies to objects with positive susceptibilities, such as ferromagnets (which have a permanent positive moment) and paramagnets (which induce a positive moment). These are attracted to field maxima, which do not exist in free space. Diamagnets (which induce a negative moment) are attracted to field minima, and there can be a field minimum in free space.

A thin slice of pyrolytic graphite, which is an unusually strong diamagnetic material, can be stably floated in a magnetic field, such as that from rare earth permanent magnets. This can be done with all components at room temperature, making a visually effective demonstration of diamagnetism.

The Radboud University Nijmegen, the Netherlands, has conducted experiments where water and other substances were successfully levitated. Most spectacularly, a live frog was levitated.

In September 2009, NASA's Jet Propulsion Laboratory in Pasadena, California announced they had successfully levitated mice using a superconducting magnet, an important step forward since mice are closer biologically to humans than frogs. They hope to perform experiments regarding the effects of microgravity on bone and muscle mass.

Recent experiments studying the growth of protein crystals have led to a technique using powerful magnets to allow growth in ways that counteract Earth's gravity.

A simple homemade device for demonstration can be constructed out of bismuth plates and a few permanent magnets that levitate a permanent magnet.

Theory

The electrons in a material generally circulate in orbitals, with effectively zero resistance and act like current loops. Thus it might be imagined that diamagnetism effects in general would be very, very common, since any applied magnetic field would generate currents in these loops that would oppose the change, in a similar way to superconductors, which are essentially perfect diamagnets. However, since the electrons are rigidly held in orbitals by the charge of the protons and are further constrained by the Pauli exclusion principle, many materials exhibit diamagnetism, but typically respond very little to the applied field.

The Bohr–van Leeuwen theorem proves that there cannot be any diamagnetism or paramagnetism in a purely classical system. However, the classical theory for Langevin diamagnetism gives the same prediction as the quantum theory. The classical theory is given below.

Langevin Diamagnetism

The Langevin theory of diamagnetism applies to materials containing atoms with closed shells. A field with intensity B, applied to an electron with charge e and mass m, gives rise to Larmor precession with frequency $\omega = eB / 2m$. The number of revolutions per unit time is $\omega / 2\pi$, so the current for an atom with Z electrons is (in SI units)

$$I = -\frac{Ze^2 B}{4\pi m}.$$

The magnetic moment of a current loop is equal to the current times the area of the loop. Suppose the field is aligned with the z axis. The average loop area can be given as $\pi \langle \rho^2 \rangle$, where $\langle \rho^2 \rangle$ is the mean square distance of the electrons perpendicular to the z axis. The magnetic moment is therefore

$$\mu = -\frac{Ze^2 B}{4m} \langle \rho^2 \rangle.$$

If the distribution of charge is spherically symmetric, we can suppose that the distribution of x, y, z coordinates are independent and identically distributed. Then $\langle x^2 \rangle = \langle y^2 \rangle = \langle z^2 \rangle = \frac{1}{3} \langle r^2 \rangle$, where $\langle r^2 \rangle$ is the mean square distance of the electrons from the nucleus. Therefore, $\langle \rho^2 \rangle = \langle x^2 \rangle + \langle y^2 \rangle = \frac{2}{3} \langle r^2 \rangle$. If N is the number of atoms per unit volume, the diamagnetic susceptibility in SI units is

$$\chi = \frac{\mu_0 N \mu}{B} = -\frac{\mu_0 N Z e^2}{6m} \langle r^2 \rangle.$$

In Metals

The Langevin theory does not apply to metals because they have non-localized electrons. The theory for the diamagnetism of a free electron gas is called Landau diamagnetism, and instead considers the weak counter-acting field that forms when their trajectories are curved due to the Lorentz force. Landau diamagnetism, however, should be contrasted with Pauli paramagnetism, an effect associated with the polarization of delocalized electrons' spins.

Paramagnetism

Paramagnetism is a form of magnetism whereby certain materials are attracted by an externally applied magnetic field, and form internal, induced magnetic fields in the direction of the applied magnetic field. In contrast with this behavior, diamagnetic materials are repelled by magnetic fields and form induced magnetic fields in the direction opposite to that of the applied magnetic field. Paramagnetic materials include most chemical elements and some compounds; they have a relative magnetic permeability greater than or equal to 1 (i.e., a positive magnetic susceptibility) and hence are attracted to magnetic fields. The magnetic moment induced by the applied field is linear in the field strength and rather weak. It typically requires a sensitive analytical balance to detect the effect and modern measurements on paramagnetic materials are often conducted with a SQUID magnetometer.

Paramagnetic materials have a small, positive susceptibility to magnetic fields. These materials are slightly attracted by a magnetic field and the material does not retain the magnetic properties when the external field is removed. Paramagnetic properties are due to the presence of some unpaired electrons, and from the realignment of the electron paths caused by the external magnetic field. Paramagnetic materials include magnesium, molybdenum, lithium, and tantalum.

Unlike ferromagnets, paramagnets do not retain any magnetization in the absence of an externally applied magnetic field because thermal motion randomizes the spin orientations. (Some paramagnetic materials retain spin disorder even at absolute zero, meaning they are paramagnetic in the ground state, i.e. in the absence of thermal motion.) Thus the total magnetization drops to zero when the applied field is removed. Even in the presence of the field there is only a small induced magnetization because only a small fraction of the spins will be oriented by the field. This fraction is proportional to the field strength and this explains the linear dependency. The attraction experienced by ferromagnetic materials is non-linear and much stronger, so that it is easily observed, for instance, in the attraction between a refrigerator magnet and the iron of the refrigerator itself.

Relation to Electron Spins

Constituent atoms or molecules of paramagnetic materials have permanent magnetic moments (dipoles), even in the absence of an applied field. The permanent moment generally is due to the spin of unpaired electrons in atomic or molecular electron orbitals. In pure paramagnetism, the dipoles do not interact with one another and are randomly oriented in the absence of an external field due to thermal agitation, resulting in zero net magnetic moment. When a magnetic field is applied, the dipoles will tend to align with the applied field, resulting in a net magnetic moment in the direction of the applied field. In the classical description, this alignment can be understood to occur due to a torque being provided on the magnetic moments by an applied field, which tries to align the dipoles parallel to the applied field. However, the true origins of the alignment can only be understood via the quantum-mechanical properties of spin and angular momentum.

If there is sufficient energy exchange between neighbouring dipoles they will interact, and may spontaneously align or anti-align and form magnetic domains, resulting in ferromagnetism (permanent magnets) or antiferromagnetism, respectively. Paramagnetic behavior can also be observed in ferromagnetic materials that are above their Curie temperature, and in antiferromagnets above their Néel temperature. At these temperatures, the available thermal energy simply overcomes the interaction energy between the spins.

In general, paramagnetic effects are quite small: the magnetic susceptibility is of the order of 10^{-3} to 10^{-5} for most paramagnets, but may be as high as 10^{-1} for synthetic paramagnets such as ferrofluids.

Delocalization

Selected Pauli-paramagnetic metals	
Material	Magnetic susceptibility, $[10^{-5}]$
Tungsten	6.8
Cesium	5.1
Aluminium	2.2
Lithium	1.4
Magnesium	1.2
Sodium	0.72

In conductive materials the electrons are delocalized, that is, they travel through the solid more or less as free electrons. Conductivity can be understood in a band structure picture as arising from the incomplete filling of energy bands. In an ordinary nonmagnetic conductor the conduction band is identical for both spin-up and spin-down electrons. When a magnetic field is applied, the conduction band splits apart into a spin-up and a spin-down band due to the difference in magnetic potential energy for spin-up and spin-down electrons. Since the Fermi level must be identical for both bands, this means that there will be a small surplus of the type of spin in the band that moved downwards. This effect is a weak form of paramagnetism known as *Pauli paramagnetism*.

The effect always competes with a diamagnetic response of opposite sign due to all the core electrons of the atoms. Stronger forms of magnetism usually require localized rather than itinerant electrons. However, in some cases a band structure can result in which there are two delocalized sub-bands with states of opposite spins that have different energies. If one subband is preferentially filled over the other, one can have itinerant ferromagnetic order. This situation usually only occurs in relatively narrow (d-)bands, which are poorly delocalized.

S and P Electrons

Generally, strong delocalization in a solid due to large overlap with neighboring wave functions means that there will be a large Fermi velocity; this means that the number of electrons in a band is less sensitive to shifts in that band's energy, implying a weak magnetism. This is why s- and p-type metals are typically either Pauli-paramagnetic or as in the case of gold even diamagnetic. In the latter case the diamagnetic contribution from the closed shell inner electrons simply wins from the weak paramagnetic term of the almost free electrons.

D and F Electrons

Stronger magnetic effects are typically only observed when d or f electrons are involved. Particularly the latter are usually strongly localized. Moreover, the size of the magnetic moment on a lanthanide atom can be quite large as it can carry up to 7 unpaired electrons in the case of gadolinium(III) (hence its use in MRI). The high magnetic moments associated with lanthanides is one reason why superstrong magnets are typically based on elements like neodymium or samarium.

Molecular Localization

Of course the above picture is a *generalization* as it pertains to materials with an extended lattice rather than a molecular structure. Molecular structure can also lead to localization of electrons. Although there are usually energetic reasons why a molecular structure results such that it does not exhibit partly filled orbitals (i.e. unpaired spins), some non-closed shell moieties do occur in nature. Molecular oxygen is a good example. Even in the frozen solid it contains di-radical molecules resulting in paramagnetic behavior. The unpaired spins reside in orbitals derived from oxygen p wave functions, but the overlap is limited to the one neighbor in the O_2 molecules. The distances to other oxygen atoms in the lattice remain too large to lead to delocalization and the magnetic moments remain unpaired.

Curie's Law

For low levels of magnetization, the magnetization of paramagnets follows what is known as Curie's law, at least approximately. This law indicates that the susceptibility, , of paramagnetic materials is inversely proportional to their temperature, i.e. that materials become more magnetic at lower temperatures. The mathematical expression is:

$$\mathbf{M} = \chi\mathbf{H} = \frac{C}{T}\mathbf{H}$$

where:

$\mathbf{M}$ is the resulting magnetization

χ is the magnetic susceptibility

$\mathbf{H}$ is the auxiliary magnetic field, measured in amperes/meter

T is absolute temperature, measured in kelvins

C is a material-specific Curie constant

Curie's law is valid under the commonly encountered conditions of low magnetization ($\mu_B H \lesssim k_B T$), but does not apply in the high-field/low-temperature regime where saturation of magnetization occurs ($\mu_B H \gtrsim k_B T$) and magnetic dipoles are all aligned with the applied field. When the dipoles are aligned, increasing the external field will not increase the total magnetization since there can be no further alignment.

For a paramagnetic ion with noninteracting magnetic moments with angular momentum J, the Curie constant is related the individual ions' magnetic moments,

$$C = \frac{N_A}{3k_B} \mu_{eff}^2 \text{ where } \mu_{eff} = g_J \mu_B \sqrt{J(J+1)}.$$

The parameter μ_{eff} is interpreted as the effective magnetic moment per paramagnetic ion. If one uses a classical treatment with molecular magnetic moments represented as discrete magnetic dipoles, μ, a Curie Law expression of the same form will emerge with μ appearing in place of μ_{eff}.

When orbital angular momentum contributions to the magnetic moment are small, as occurs for most organic radicals or for octahedral transition metal complexes with d^3 or high-spin d^5 configurations, the effective magnetic moment takes the form ($g_e = 2.0023... \approx 2$),

$\mu_{eff} \simeq 2\sqrt{S(S+1)}\mu_B = \sqrt{n(n+2)}\mu_B$, where n is the number of unpaired electrons. In other transition metal complexes this yields a useful, if somewhat cruder, estimate.

Examples of Paramagnets

Materials that are called "paramagnets" are most often those that exhibit, at least over an appreciable temperature range, magnetic susceptibilities that adhere to the Curie or Curie–Weiss laws. In principle any system that contains atoms, ions, or molecules with unpaired spins can be called a paramagnet, but the interactions between them need to be carefully considered.

Systems with Minimal Interactions

The narrowest definition would be: a system with unpaired spins that *do not interact* with each other. In this narrowest sense, the only pure paramagnet is a dilute gas of monatomic hydrogen atoms. Each atom has one non-interacting unpaired electron. Of course, the latter could be said about a gas of lithium atoms but these already possess two paired core electrons that produce a diamagnetic response of opposite sign. Strictly speaking Li is a mixed system therefore, although admittedly the diamagnetic component is weak and often neglected. In the case of heavier elements the diamagnetic contribution becomes more important and in the case of metallic gold it dominates the properties. Of course, the element hydrogen is virtually never called 'paramagnetic' because the monatomic gas is stable only at extremely high temperature; H atoms combine to form molecular H_2 and in so doing, the magnetic moments are lost (*quenched*), because the spins pair. Hydrogen is therefore *diamagnetic* and the same holds true for many other elements. Although the electronic configuration of the individual atoms (and ions) of most elements contain unpaired spins, they are not necessarily paramagnetic, because at ambient temperature quenching is very much the rule rather than the exception. The quenching tendency is weakest for f-electrons because f (especially $4f$) orbitals are radially contracted and they overlap only weakly with orbitals on adjacent atoms. Consequently, the lanthanide elements with incompletely filled 4f-orbitals are paramagnetic or magnetically ordered.

μ_{eff} values for typical d^3 and d^5 transition metal complexes.	
Material	μ_{eff}/μ_B
$[Cr(NH_3)_6]Br_3$	3.77
$K_3[Cr(CN)_6]$	3.87
$K_3[MoCl_6]$	3.79
$K_4[V(CN)_6]$	3.78
$[Mn(NH_3)_6]Cl_2$	5.92
$(NH_4)_2[Mn(SO_4)_2]\cdot 6H_2O$	5.92
$NH_4[Fe(SO_4)_2]\cdot 12H_2O$	5.89

Thus, condensed phase paramagnets are only possible if the interactions of the spins that lead either to quenching or to ordering are kept at bay by structural isolation of the magnetic centers. There are two classes of materials for which this holds:

- Molecular materials with a (isolated) paramagnetic center.

  - Good examples are coordination complexes of d- or f-metals or proteins with such centers, e.g. myoglobin. In such materials the organic part of the molecule acts as an envelope shielding the spins from their neighbors.

  - Small molecules can be stable in radical form, oxygen O_2 is a good example. Such systems are quite rare because they tend to be rather reactive.

- Dilute systems.

  - Dissolving a paramagnetic species in a diamagnetic lattice at small concentrations, e.g. Nd^{3+} in $CaCl_2$ will separate the neodymium ions at large enough distances that they do not interact. Such systems are of prime importance for what can be considered the most sensitive method to study paramagnetic systems: EPR.

Systems with Interactions

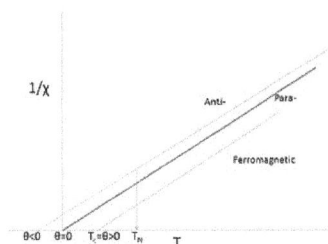

Idealized Curie–Weiss behavior; N.B. $T_C = \theta$, but T_N is not θ. Paramagnetic regimes are denoted by solid lines. Close to T_N or T_C the behavior usually deviates from ideal.

As stated above, many materials that contain d- or f-elements do retain unquenched spins. Salts of such elements often show paramagnetic behavior but at low enough temperatures the magnetic moments may order. It is not uncommon to call such materials 'paramagnets', when referring to their paramagnetic behavior above their Curie or Néel-points, particularly if such temperatures are very low or have never been properly measured. Even for iron it is not uncommon to say that *iron becomes a paramagnet* above its relatively high Curie-point. In that case the Curie-point is seen as a phase transition between a ferromagnet and a 'paramagnet'. The word paramagnet now merely refers to the linear response of the system to an applied field, the temperature dependence of which requires an amended version of Curie's law, known as the Curie–Weiss law:

$$M = \frac{C}{T - \theta} H$$

This amended law includes a term θ that describes the exchange interaction that is present albeit overcome by thermal motion. The sign of θ depends on whether ferro- or antiferromagnetic interactions dominate and it is seldom exactly zero, except in the dilute, isolated cases mentioned above.

Obviously, the paramagnetic Curie–Weiss description above T_N or T_C is a rather different interpretation of the word "paramagnet" as it does *not* imply the *absence* of interactions, but rather that the magnetic structure is random in the absence of an external field at these sufficiently high temperatures. Even if θ is close to zero this does not mean that there are no interactions, just that the aligning ferro- and the anti-aligning antiferromagnetic ones cancel. An additional complication is that the interactions are often different in different directions of the crystalline lattice (anisotropy), leading to complicated magnetic structures once ordered.

Randomness of the structure also applies to the many metals that show a net paramagnetic response over a broad temperature range. They do not follow a Curie type law as function of temperature however, often they are more or less temperature independent. This type of behavior is of an itinerant nature and better called Pauli-paramagnetism, but it is not unusual to see e.g. the metal aluminium called a "paramagnet", even though interactions are strong enough to give this element very good electrical conductivity.

Superparamagnets

Some materials show induced magnetic behavior that follows a Curie type law but with exceptionally large values for the Curie constants. These materials are known as superparamagnets. They are characterized by a strong ferromagnetic or ferrimagnetic type of coupling into domains of a limited size that behave independently from one another. The bulk properties of such a system resembles that of a paramagnet, but on a microscopic level they are ordered. The materials do show an ordering temperature above which the behavior reverts to ordinary paramagnetism (with interaction). Ferrofluids are a good example, but the phenomenon can also occur inside solids, e.g., when dilute paramagnetic centers are introduced in a strong itinerant medium of ferromagnetic coupling such as when Fe is substituted in $TlCu_2Se_2$ or the alloy AuFe. Such systems contain ferromagnetically coupled clusters that freeze out at lower temperatures. They are also called mictomagnets.

Ferromagnetism

Ferromagnetism is the basic mechanism by which certain materials (such as iron) form permanent magnets, or are attracted to magnets. In physics, several different types of magnetism are distinguished. Ferromagnetism (including ferrimagnetism) is the strongest type: it is the only one that typically creates forces strong enough to be felt, and is responsible for the common phenomena of magnetism in magnets encountered in everyday life. Substances respond weakly to magnetic fields with three other types of magnetism, paramagnetism, diamagnetism, and antiferromagnetism, but the forces are usually so weak that they can only be detected by sensitive instruments in a laboratory. An everyday example of ferromagnetism is a refrigerator magnet used to hold notes on a refrigerator door. The attraction between a magnet and ferromagnetic material is "the quality of magnetism first apparent to the ancient world, and to us today".

Permanent magnets (materials that can be magnetized by an external magnetic field and remain magnetized after the external field is removed) are either ferromagnetic or ferrimagnetic, as are the materials that are noticeably attracted to them. Only a few substances are ferromagnetic. The common ones are iron, nickel, cobalt and most of their alloys, some compounds of rare earth metals, and a few naturally-occurring minerals such as lodestone.

Ferromagnetism is very important in industry and modern technology, and is the basis for many electrical and electromechanical devices such as electromagnets, electric motors, generators, transformers, and magnetic storage such as tape recorders, and hard disks.

History and Distinction from Ferrimagnetism

Historically, the term *ferromagnetism* was used for any material that could exhibit spontaneous magnetization: a net magnetic moment in the absence of an external magnetic field. This general definition is still in common use. More recently, however, different classes of spontaneous magnetization have been identified when there is more than one magnetic ion per primitive cell of the material, leading to a stricter definition of "ferromagnetism" that is often used to distinguish it from ferrimagnetism. In particular,

- a material is "ferromagnetic" in this narrower sense only if *all* of its magnetic ions add a positive contribution to the net magnetization.

- If some of the magnetic ions *subtract* from the net magnetization (if they are partially *anti*-aligned), then the material is "ferrimagnetic".

- If the moments of the aligned and anti-aligned ions balance completely so as to have zero net magnetization, despite the magnetic ordering, then it is an antiferromagnet.

These alignment effects only occur at temperatures below a certain critical temperature, called the Curie temperature (for ferromagnets and ferrimagnets) or the Néel temperature (for antiferromagnets).

Among the first investigations of ferromagnetism are the pioneering works of Aleksandr Stoletov on measurement of the magnetic permeability of ferromagnetics, known as the Stoletov curve.

Ferromagnetic Materials

Curie temperatures for some crystalline ferromagnetic (* = ferrimagnetic) materials	
Material	Curie temp. (K)
Co	1388
Fe	1043
Fe_2O_3*	948
$FeOFe_2O_3$*	858
$NiOFe_2O_3$*	858
$CuOFe_2O_3$*	728
$MgOFe_2O_3$*	713
MnBi	630
Ni	627
MnSb	587
$MnOFe_2O_3$*	573
$Y_3Fe_5O_{12}$*	560
CrO_2	386
MnAs	318
Gd	292
Tb	219
Dy	88
EuO	69

The table on the right lists a selection of ferromagnetic and ferrimagnetic compounds, along with the temperature above which they cease to exhibit spontaneous magnetization.

Ferromagnetism is a property not just of the chemical make-up of a material, but of its crystalline structure and microstructure. There are ferromagnetic metal alloys whose constituents are not themselves ferromagnetic, called Heusler alloys, named after Fritz Heusler. Conversely there are non-magnetic alloys, such as types of stainless steel, composed almost exclusively of ferromagnetic metals.

Amorphous (non-crystalline) ferromagnetic metallic alloys can be made by very rapid quenching (cooling) of a liquid alloy. These have the advantage that their properties are nearly isotropic (not aligned along a crystal axis); this results in low coercivity, low hysteresis loss, high permeability, and high electrical resistivity. One such typical material is a transition metal-metalloid alloy, made from about 80% transition metal (usually Fe, Co, or Ni) and a metalloid component (B, C, Si, P, or Al) that lowers the melting point.

A relatively new class of exceptionally strong ferromagnetic materials are the rare-earth magnets. They contain lanthanide elements that are known for their ability to carry large magnetic moments in well-localized f-orbitals.

Actinide Ferromagnets

A number of actinide compounds are ferromagnets at room temperature or exhibit ferromagnetism upon cooling. PuP is a paramagnet with cubic symmetry at room temperature, but which undergoes a structural transition into a tetragonal state with ferromagnetic order when cooled below its T_c = 125 K. In its ferromagnetic state, PuP's easy axis is in the <100> direction.

In $NpFe_2$ the easy axis is <111>. Above T_c ≈ 500 K $NpFe_2$ is also paramagnetic and cubic. Cooling below the Curie temperature produces a rhombohedral distortion wherein the rhombohedral angle changes from 60° (cubic phase) to 60.53°. An alternate description of this distortion is to consider the length c along the unique trigonal axis (after the distortion has begun) and a as the distance in the plane perpendicular to c. In the cubic phase this reduces to c/a = 1.00. Below the Curie temperature

$$\frac{c}{a} - 1 = -(120 \pm 5) \times 10^{-4}$$

which is the largest strain in any actinide compound. $NpNi_2$ undergoes a similar lattice distortion below T_c = 32 K, with a strain of $(43 \pm 5) \times 10^{-4}$. $NpCo_2$ is a ferrimagnet below 15 K.

Lithium Gas

In 2009, a team of MIT physicists demonstrated that a lithium gas cooled to less than one kelvin can exhibit ferromagnetism. The team cooled fermionic lithium-6 to less than 150 billionths of one kelvin above absolute zero using infrared laser cooling. This demonstration is the first time that ferromagnetism has been demonstrated in a gas.

Explanation

The Bohr–van Leeuwen theorem, discovered in the 1910s, showed that classical physics theories are unable to account for any form of magnetism, including ferromagnetism. Magnetism is now regarded as a purely quantum mechanical effect. Ferromagnetism arises due to two effects from quantum mechanics: spin and the Pauli exclusion principle.

Origin of Magnetism

One of the fundamental properties of an electron (besides that it carries charge) is that it has a magnetic dipole moment, i.e., it behaves like a tiny magnet. This dipole moment comes from the more fundamental property of the electron that it has quantum mechanical spin. Due to its quantum nature, the spin of the electron can be in one of only two states; with the magnetic field either pointing "up" or "down" (for any choice of up and down). The spin of the electrons in atoms is the main source of ferromagnetism, although there is also a contribution from the orbital angular momentum of the electron about the nucleus. When these magnetic dipoles in a piece of matter are aligned, (point in the same direction) their individually tiny magnetic fields add together to create a much larger macroscopic field.

However, materials made of atoms with filled electron shells have a total dipole moment of zero, because every electron's magnetic moment is cancelled by the opposite moment of the second electron in the pair. Only atoms with partially filled shells (i.e., unpaired spins) can have a net magnetic moment, so ferromagnetism only occurs in materials with partially filled shells. Because of Hund's rules, the first few electrons in a shell tend to have the same spin, thereby increasing the total dipole moment.

These unpaired dipoles (often called simply "spins" even though they also generally include angular momentum) tend to align in parallel to an external magnetic field, an effect called paramagnetism. Ferromagnetism involves an additional phenomenon, however: The dipoles tend to align spontaneously, giving rise to a spontaneous magnetization, even when there is no applied field.

Exchange Interaction

When two nearby atoms have unpaired electrons, whether the electron spins are parallel or antiparallel affects whether the electrons can share the same orbit as a result of the quantum mechanical effect called the exchange interaction. This in turn affects the electron location and the Coulomb (electrostatic) interaction and thus the energy difference between these states.

The exchange interaction is related to the Pauli exclusion principle, which says that two electrons with the same spin cannot also have the same "position". Therefore, under certain conditions, when the orbitals of the unpaired outer valence electrons from adjacent atoms overlap, the distributions of their electric charge in space are farther apart when the electrons have parallel spins than when they have opposite spins. This reduces the electrostatic energy of the electrons when their spins are parallel compared to their energy when the spins are anti-parallel, so the parallel-spin state is more stable. In simple terms, the electrons, which repel one another, can move "further apart" by aligning their spins, so the spins of these electrons tend to line up. This difference in energy is called the exchange energy.

This energy difference can be orders of magnitude larger than the energy differences associated with the magnetic dipole-dipole interaction due to dipole orientation, which tends to align the dipoles antiparallel. In certain doped semiconductor oxides RKKY interactions have been shown to bring about periodic longer-range magnetic interactions, a phenomenon of significance in the study of spintronic materials.

The materials in which the exchange interaction is much stronger than the competing dipole-dipole interaction are frequently called *magnetic materials*. For instance, in iron (Fe) the exchange force is about 1000 times stronger than the dipole interaction. Therefore, below the Curie temperature virtually all of the dipoles in a ferromagnetic material will be aligned. In addition to ferromagnetism, the exchange interaction is also responsible for the other types of spontaneous ordering of atomic magnetic moments occurring in magnetic solids, antiferromagnetism and ferrimagnetism. There are different exchange interaction mechanisms which create the magnetism in different ferromagnetic, ferrimagnetic, and antiferromagnetic substances. These mechanisms include direct exchange, RKKY exchange, double exchange, and superexchange.

Magnetic Anisotropy

Although the exchange interaction keeps spins aligned, it does not align them in a particular direction. Without magnetic anisotropy, the spins in a magnet randomly change direction in response to thermal fluctuations and the magnet is superparamagnetic. There are several kinds of magnetic anisotropy, the most common of which is magnetocrystalline anisotropy. This is a dependence of the energy on the direction of magnetization relative to the crystallographic lattice. Another common source of anisotropy, inverse magnetostriction, is induced by internal strains. Single-domain magnets also can have a *shape anisotropy* due to the magnetostatic effects of the particle shape. As the temperature of a magnet increases, the anisotropy tends to decrease, and there is often a blocking temperature at which a transition to superparamagnetism occurs.

Magnetic Domains

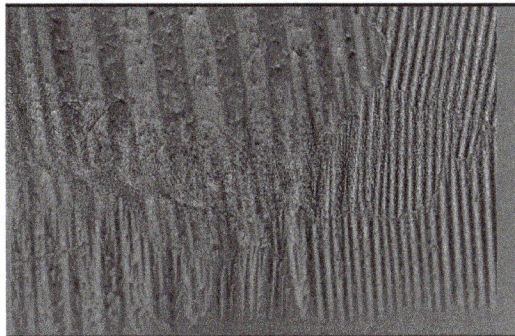

Electromagnetic dynamic magnetic domain motion of grain oriented electrical silicon steel

Kerr micrograph of metal surface showing magnetic domains, with red and green stripes denoting opposite magnetization directions.

The above would seem to suggest that every piece of ferromagnetic material should have a strong magnetic field, since all the spins are aligned, yet iron and other ferromagnets are often found in an "unmagnetized" state. The reason for this is that a bulk piece of ferromagnetic material is divided into tiny regions called *magnetic domains* (also known as *Weiss domains*). Within each domain, the spins are aligned, but (if the bulk material is in its lowest energy configuration, i.e. *unmagnetized*), the spins of separate domains point in different directions and their magnetic fields cancel out, so the object has no net large scale magnetic field.

Ferromagnetic materials spontaneously divide into magnetic domains because the *exchange interaction* is a short-range force, so over long distances of many atoms the tendency of the magnetic dipoles to reduce their energy by orienting in opposite directions wins out. If all the dipoles in a piece of ferromagnetic material are aligned parallel, it creates a large magnetic field extending into the space around it. This contains a lot of magnetostatic energy. The material can reduce this energy by splitting into many domains pointing in different directions, so the magnetic field is confined to small local fields in the material, reducing the volume of the field. The domains are separated by thin domain walls a number of molecules thick, in which the direction of magnetization of the dipoles rotates smoothly from one domain's direction to the other.

Magnetized Materials

Thus, a piece of iron in its lowest energy state ("unmagnetized") generally has little or no net magnetic field. However, if it is placed in a strong enough external magnetic field, the domain walls will move, reorienting the domains so more of the dipoles are aligned with the external field. The domains will remain aligned when the external field is removed, creating a magnetic field of their own extending into the space around the material, thus creating a "permanent" magnet. The domains do not go back to their original minimum energy configuration when the field is removed because the domain walls tend to become 'pinned' or 'snagged' on defects in the crystal lattice, preserving their parallel orientation. This is shown by the Barkhausen effect: as the magnetizing field is changed, the magnetization changes in thousands of tiny discontinuous jumps as the domain walls suddenly "snap" past defects.

This magnetization as a function of the external field is described by a hysteresis curve. Although this state of aligned domains found in a piece of magnetized ferromagnetic material is not a minimal-energy configuration, it is metastable, and can persist for long periods, as shown by samples of magnetite from the sea floor which have maintained their magnetization for millions of years.

Heating and then cooling (annealing) a magnetized material, subjecting it to vibration by hammering it, or applying a rapidly oscillating magnetic field from a degaussing coil tends to release the domain walls from their pinned state, and the domain boundaries tend to move back to a lower energy configuration with less external magnetic field, thus *demagnetizing* the material.

Commercial magnets are made of "hard" magnetic materials with very large magnetic anisotropy, such as alnico and hard ferrites, with a very strong tendency for the magnetization to be pointed along one axis of the crystal, the "easy axis". During manufacture the materials are subjected to various metallurgical processes in a powerful magnetic field, which aligns the crystal grains so their "easy" axes of magnetization all point in the same direction. Thus the magnetization, and the resulting magnetic field, is "built in" to the crystal structure of the material, making it very difficult to demagnetize.

Curie Temperature

As the temperature increases, thermal motion, or entropy, competes with the ferromagnetic tendency for dipoles to align. When the temperature rises beyond a certain point, called the Curie temperature, there is a second-order phase transition and the system can no longer maintain a spontaneous magnetization, so its ability to be magnetized or attracted to a magnet disappears, although it still responds paramagnetically to an external field. Below that temperature, there is a spontaneous symmetry breaking and magnetic moments become aligned with their neighbors. The Curie temperature itself is a critical point, where the magnetic susceptibility is theoretically infinite and, although there is no net magnetization, domain-like spin correlations fluctuate at all length scales.

The study of ferromagnetic phase transitions, especially via the simplified Ising spin model, had an important impact on the development of statistical physics. There, it was first clearly shown that mean field theory approaches failed to predict the correct behavior at the critical point (which was found to fall under a *universality class* that includes many other systems, such as liquid-gas transitions), and had to be replaced by renormalization group theory.

Ferrimagnetism

In physics, a ferrimagnetic material is one that has populations of atoms with opposing magnetic moments, as in antiferromagnetism; however, in ferrimagnetic materials, the opposing moments are unequal and a spontaneous magnetization remains. This happens when the populations consist of different materials or ions (such as Fe^{2+} and Fe^{3+}).

Ferrimagnetism is exhibited by ferrites and magnetic garnets. The oldest known magnetic substance, magnetite (iron(II,III) oxide; Fe_{3O_4}), is a ferrimagnet; it was originally classified as a ferromagnet before Néel's discovery of ferrimagnetism and antiferromagnetism in 1948.

Some ferrimagnetic materials are YIG (yttrium iron garnet), cubic ferrites composed of iron oxides and other elements such as aluminum, cobalt, nickel, manganese and zinc, hexagonal ferrites such as $PbFe_{12}O_{19}$ and $BaFe_{12}O_{19}$, and pyrrhotite, $Fe_{1-x}S$.

Effects of Temperature

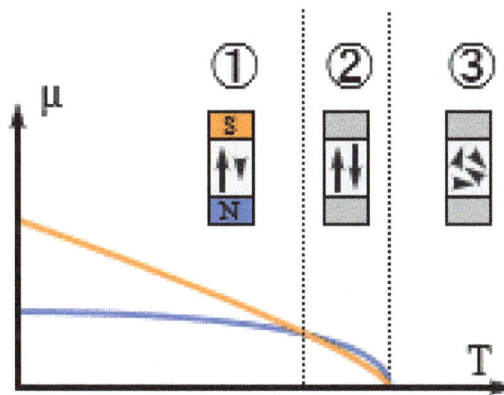

Below the magnetization compensation point, ferrimagnetic material is magnetic. ② At the compensation point, the magnetic components cancel each other and the total magnetic moment is zero. ③ Above the Curie point, the material loses magnetism.

Ferrimagnetic materials are like ferromagnets in that they hold a spontaneous magnetization below the Curie temperature, and show no magnetic order (are paramagnetic) above this temperature. However, there is sometimes a temperature *below* the Curie temperature at which the two opposing moments are equal, resulting in a net magnetic moment of zero; this is called the *magnetization compensation point*. This compensation point is observed easily in garnets and rare earth-transition metal alloys (RE-TM). Furthermore, ferrimagnets may also have an *angular momentum compensation point* at which the net angular momentum vanishes. This compensation point is a crucial point for achieving high speed magnetization reversal in magnetic memory devices.

Properties

Ferrimagnetic materials have high resistivity and have anisotropic properties. The anisotropy is actually induced by an external applied field. When this applied field aligns with the magnetic dipoles it causes a net magnetic dipole moment and causes the magnetic dipoles to precess at a frequency controlled by the applied field, called *Larmor* or *precession frequency*. As a particular example, a microwave signal circularly polarized in the same direction as this precession strongly interacts with the magnetic dipole moments; when it is polarized in the opposite direction the interaction is very low. When the interaction is strong, the microwave signal can pass through the material. This directional property is used in the construction of microwave devices like isolators, circulators and gyrators. Ferrimagnetic materials are also used to produce optical isolators and circulators. Ferrimagnetic minerals in various rock types are used to study ancient geomagnetic properties of Earth and other planets. That field of study is known as paleomagnetism.

Molecular Ferrimagnets

Ferrimagnetism can also occur in molecular magnets. A classic example is a dodecanuclear manganese molecule with an effective spin of S = 10 derived from antiferromagnetic interaction on Mn(IV) metal centres with Mn(III) and Mn(II) metal centres.

Antiferromagnetism

In materials that exhibit antiferromagnetism, the magnetic moments of atoms or molecules, usually related to the spins of electrons, align in a regular pattern with neighboring spins (on different sublattices) pointing in opposite directions. This is, like ferromagnetism and ferrimagnetism, a manifestation of ordered magnetism. Generally, antiferromagnetic order may exist at sufficiently low temperatures, vanishing at and above a certain temperature, the Néel temperature (named after Louis Néel, who had first identified this type of magnetic ordering). Above the Néel temperature, the material is typically paramagnetic.

Measurement

When no external field is applied, the antiferromagnetic structure corresponds to a vanishing total magnetization. In an external magnetic field, a kind of ferrimagnetic behavior may be displayed in the antiferromagnetic phase, with the absolute value of one of the sublattice magnetizations differing from that of the other sublattice, resulting in a nonzero net magnetization. Although the net magnetization should be zero at a temperature of absolute zero, the effect of spin canting often causes a small net magnetization to develop, as seen for example in hematite.

The magnetic susceptibility of an antiferromagnetic material typically shows a maximum at the Néel temperature. In contrast, at the transition between the ferromagnetic to the paramagnetic phases the susceptibility will diverge. In the antiferromagnetic case, a divergence is observed in the *staggered susceptibility*.

Various microscopic (exchange) interactions between the magnetic moments or spins may lead to antiferromagnetic structures. In the simplest case, one may consider an Ising model on an bipartite lattice, e.g. the simple cubic lattice, with couplings between spins at nearest neighbor sites. Depending on the sign of that interaction, ferromagnetic or antiferromagnetic order will result. Geometrical frustration or competing ferro- and antiferromagnetic interactions may lead to different and, perhaps, more complicated magnetic structures.

Antiferromagnetic Materials

Antiferromagnetic materials occur commonly among transition metal compounds, especially oxides. Examples include hematite, metals such as chromium, alloys such as iron manganese (FeMn), and oxides such as nickel oxide (NiO). There are also numerous examples among high nuclearity metal clusters. Organic molecules can also exhibit antiferromagnetic coupling under rare circumstances, as seen in radicals such as 5-dehydro-m-xylylene.

Antiferromagnets can couple to ferromagnets, for instance, through a mechanism known as exchange bias, in which the ferromagnetic film is either grown upon the antiferromagnet or annealed in an aligning magnetic field, causing the surface atoms of the ferromagnet to align with the surface atoms of the antiferromagnet. This provides the ability to "pin" the orientation of a ferromagnetic film, which provides one of the main uses in so-called spin valves, which are the basis of magnetic sensors including modern hard drive read heads. The temperature at or above which an antiferromagnetic layer loses its ability to "pin" the magnetization direction of an adjacent ferromagnetic layer is called the blocking temperature of that layer and is usually lower than the Néel temperature.

Geometric Frustration

Unlike ferromagnetism, anti-ferromagnetic interactions can lead to multiple optimal states (ground states—states of minimal energy). In one dimension, the anti-ferromagnetic ground state is an alternating series of spins: up, down, up, down, etc. Yet in two dimensions, multiple ground states can occur.

Consider an equilateral triangle with three spins, one on each vertex. If each spin can take on only two values (up or down), there are $2^3 = 8$ possible states of the system, six of which are ground states. The two situations which are not ground states are when all three spins are up or are all down. In any of the other six states, there will be two favorable interactions and one unfavorable one. This illustrates frustration: the inability of the system to find a single ground state. This type of magnetic behavior has been found in minerals that have a crystal stacking structure such as a Kagome lattice or hexagonal lattice.

Other Properties

Synthetic antiferromagnets (often abbreviated by SAF) are artificial antiferromagnets consisting of two or more thin ferromagnetic layers separated by a nonmagnetic layer. Due to dipole coupling of the ferromagnetic layers results in antiparallel alignment of the magnetization of the ferromagnets.

Antiferromagnetism plays a crucial role in giant magnetoresistance, as had been discovered in 1988 by the Nobel prize winners Albert Fert and Peter Grünberg (awarded in 2007) using synthetic antiferromagnets.

There are also examples of disordered materials (such as iron phosphate glasses) that become antiferromagnetic below their Néel temperature. These disordered networks 'frustrate' the antiparallelism of adjacent spins; i.e. it is not possible to construct a network where each spin is surrounded by opposite neighbour spins. It can only be determined that the average correlation of neighbour spins is antiferromagnetic. This type of magnetism is sometimes called *speromagnetism*.

Superparamagnetism

Superparamagnetism is a form of magnetism, which appears in small ferromagnetic or ferrimagnetic nanoparticles. In sufficiently small nanoparticles, magnetization can randomly flip direction under the influence of temperature. The typical time between two flips is called the Néel relaxation time. In the absence of an external magnetic field, when the time used to measure the magnetization of the nanoparticles is much longer than the *Néel relaxation time*, their magnetization appears to be in average zero: they are said to be in the superparamagnetic state. In this state, an external magnetic field is able to magnetize the nanoparticles, similarly to a paramagnet. However, their magnetic susceptibility is much larger than that of paramagnets.

The Néel Relaxation in The Absence of Magnetic Field

Normally, any ferromagnetic or ferrimagnetic material undergoes a transition to a paramagnetic state above its Curie temperature. Superparamagnetism is different from this standard transition since it occurs below the Curie temperature of the material.

Superparamagnetism occurs in nanoparticles which are single-domain, i.e. composed of a single magnetic domain. This is possible when their diameter is below 3–50 nm, depending on the materials. In this condition, it is considered that the magnetization of the nanoparticles is a single giant magnetic moment, sum of all the individual magnetic moments carried by the atoms of the nanoparticle. Those in the field of superparamagnetism call this "macro-spin approximation".

Because of the nanoparticle's magnetic anisotropy, the magnetic moment has usually only two stable orientations antiparallel to each other, separated by an energy barrier. The stable orientations define the nanoparticle's so called "easy axis". At finite temperature, there is a finite probability for the magnetization to flip and reverse its direction. The mean time between two flips is called the Néel relaxation time τ_N and is given by the following Néel-Arrhenius equation:

$$\tau_N = \tau_0 \exp\left(\frac{KV}{k_B T}\right),$$

where:

- τ_N is thus the average length of time that it takes for the nanoparticle's magnetization to randomly flip as a result of thermal fluctuations.

- τ_0 is a length of time, characteristic of the material, called the *attempt time* or *attempt period* (its reciprocal is called the *attempt frequency*); its typical value is 10^{-9}–10^{-10} second.

- K is the nanoparticle's magnetic anisotropy energy density and V its volume. KV is therefore the energy barrier associated with the magnetization moving from its initial easy axis direction, through a "hard plane", to the other easy axis direction.

- k_B is the Boltzmann constant.

- T is the temperature.

This length of time can be anywhere from a few nanoseconds to years or much longer. In particular, it can be seen that the Néel relaxation time is an exponential function of the grain volume, which explains why the flipping probability becomes rapidly negligible for bulk materials or large nanoparticles.

Blocking Temperature

Let us imagine that the magnetization of a single superparamagnetic nanoparticle is measured and let us define τ_m as the measurement time. If $\tau_m \gg \tau_N$, the nanoparticle magnetization will flip several times during the measurement, then the measured magnetization will average to zero. If $\tau_m \ll \tau_N$, the magnetization will not flip during the measurement, so the measured magnetization will be what the instantaneous magnetization was at the beginning of the measurement. In the former case, the nanoparticle will appear to be in the superparamagnetic state whereas in the latter case it will appear to be "blocked" in its initial state. The state of the nanoparticle (superparamagnetic or blocked) depends on the measurement time. A transition between superparamagnetism and blocked state occurs when $\tau_m = \tau_N$ In several experiments, the measurement time is kept constant but the temperature is varied, so the transition between superparamagnetism and blocked state is seen as a function of the temperature. The temperature for which $\tau_m = \tau_N$ is called the blocking temperature:

$$T_B = \frac{KV}{k_B \ln\left(\dfrac{\tau_m}{\tau_0}\right)}$$

For typical laboratory measurements, the value of the logarithm in the previous equation is in the order of 20–25.

Effect of a Magnetic Field

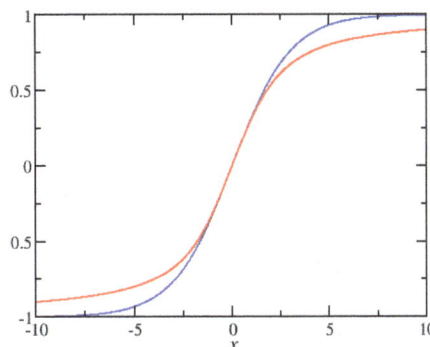

Langevin function (red line), compared with (blue line).

When an external magnetic field is applied to an assembly of superparamagnetic nanoparticles, their magnetic moments tend to align along the applied field, leading to a net magnetization. The magnetization curve of the assembly, i.e. the magnetization as a function of the applied field, is a reversible S-shaped increasing function. This function is quite complicated but for some simple cases:

1. If all the particles are identical (same energy barrier and same magnetic moment), their easy axes are all oriented parallel to the applied field and the temperature is low enough ($T_B < T \lesssim KV/(10\, k_B)$), then the magnetization of the assembly is

$$M(H) \approx n\mu \tanh\left(\frac{\mu_0 H \mu}{k_B T}\right).$$

2. If all the particles are identical and the temperature is high enough ($T \gtrsim KV/k_B$), then, irrespective of the orientations of the easy axes:

$$M(H) \approx n\mu L\left(\frac{\mu_0 H \mu}{k_B T}\right)$$

In the above equations:

- n in the density of nanoparticles in the sample
- μ_0 is the magnetic permeability of vacuum
- μ is the magnetic moment of a nanoparticle
- $L(x) = 1/\tanh(x) - 1/x$ is the Langevin function

The initial slope of the $M(H)$ function is the magnetic susceptibility of the sample χ :

$$\chi = \frac{n\mu_0 \mu^2}{k_B T} \text{ in the first case}$$

$$\chi = \frac{n\mu_0 \mu^2}{3k_B T} \text{ in the second case.}$$

The later susceptibility is also valid for all temperatures $T > T_B$ if the easy axes of the nanoparticles are randomly oriented.

It can be seen from these equations that large nanoparticles have a larger μ and so a larger susceptibility. This explains why superparamagnetic nanoparticles have a much larger susceptibility than standard paramagnets: they behave exactly as a paramagnet with a huge magnetic moment.

Time Dependence of The Magnetization

There is no time-dependence of the magnetization when the nanoparticles are either completely blocked ($T \ll T_B$) or completely superparamagnetic ($T \gg T_B$). There is, however, a narrow window around T_B where the measurement time and the relaxation time have comparable magnitude. In this case, a frequency-dependence of the susceptibility can be observed. For a randomly oriented sample, the complex susceptibility is:

$$\chi(\omega) = \frac{\chi_{sp} + i\omega\tau\chi_b}{1 + i\omega\tau}$$

where

$\frac{\omega}{2\pi}$ is the frequency of the applied field

$\chi_{sp} = \frac{n\mu_0\mu^2}{3k_BT}$ is the susceptibility in the superparamagnetic state

$\chi_b = \frac{n\mu_0\mu^2}{3KV}$ is the susceptibility in the blocked state

$\tau = \frac{\tau_N}{2}$ is the relaxation time of the assembly

From this frequency-dependent susceptibility, the time-dependence of the magnetization for low-fields can be derived:

$$\tau\frac{\mathrm{d}M}{\mathrm{d}t} + M = \tau\chi_b\frac{\mathrm{d}H}{\mathrm{d}t} + \chi_{sp}H$$

Measurements

A superparamagnetic system can be measured with AC susceptibility measurements, where an applied magnetic field varies in time, and the magnetic response of the system is measured. A superparamagnetic system will show a characteristic frequency dependence: When the frequency is much higher than $1/\tau_N$, there will be a different magnetic response than when the frequency is much lower than $1/\tau_N$, since in the latter case, but not the former, the ferromagnetic clusters will have time to respond to the field by flipping their magnetization. The precise dependence can be calculated from the Néel-Arrhenius equation, assuming that the neighboring clusters behave independently of one another (if clusters interact, their behavior becomes more complicated). It is also possible to perform magneto-optical AC susceptibility measurements with magneto-optically active superparamagnetic materials such as iron oxide nanoparticles in the visible wavelength range.

Effect on Hard Drives

Superparamagnetism sets a limit on the storage density of hard disk drives due to the minimum size of particles that can be used. This limit is known as the superparamagnetic limit.

Older hard disk technology uses longitudinal recording. It has an estimated limit of 100 to 200 Gbit/in²

Current hard disk technology uses perpendicular recording. As of August 2010 drives with densities of 667Gb/in² are available commercially. Perpendicular recording is predicted to allow information densities of up to around 1 Tbit/in² (1024 Gbit/in²).

Future hard disk technologies currently in development include: heat-assisted magnetic recording

(HAMR), which use materials that are stable at much smaller sizes. They require heating before the magnetic orientation of a bit can be changed; and bit-patterned recording (BPR).

Applications of Superparamagnetism

General Applications

- Ferrofluid: tunable viscosity

Biomedical Applications

- Imaging: Contrast agents in Magnetic Resonance Imaging (MRI)

- Magnetic separation: cell-, DNA-, protein- separation, RNA fishing

- Treatments: targeted drug delivery, magnetic hyperthermia, magnetofection

References

- Poole, Jr., Charles P. (2007). Superconductivity (2nd ed.). Amsterdam: Academic Press. p. 23. ISBN 9780080550480.

- Kittel, Charles (1986). Introduction to Solid State Physics (6th ed.). John Wiley & Sons. pp. 299–302. ISBN 0-471-87474-4.

- Miessler, G. L. and Tarr, D. A. (2010) Inorganic Chemistry 3rd ed., Pearson/Prentice Hall publisher, ISBN 0-13-035471-6.

- Chikazumi, Sōshin (2009). Physics of ferromagnetism. English edition prepared with the assistance of C.D. Graham, Jr (2nd ed.). Oxford: Oxford University Press. p. 118. ISBN 9780199564811.

- Bozorth, Richard M. Ferromagnetism, first published 1951, reprinted 1993 by IEEE Press, New York as a "Classic Reissue." ISBN 0-7803-1032-2.

- Chikazumi, Sōshin (2009). Physics of ferromagnetism. English edition prepared with the assistance of C.D. Graham, Jr (2nd ed.). Oxford: Oxford University Press. pp. 129–130. ISBN 9780199564811.

- Feynman, Richard P.; Robert B. Leighton; Matthew Sands (1963). The Feynman Lectures on Physics, Vol. I. USA: California Inst. of Technology. pp. 37.5–37.6. ISBN 0465024939.

- Spaldin, Nicola A. (2010). "9. Ferrimagnetism". Magnetic materials : fundamentals and applications (2nd ed.). Cambridge: Cambridge University Press. pp. 113–129. ISBN 9780521886697.

- Jackson, Roland (21 July 2014). "John Tyndall and the Early History of Diamagnetism". Annals of Science: 4. doi:10.1080/00033790.2014.929743. Retrieved 28 October 2014.

- Drakos, Nikos; Moore, Ross; Young, Peter (2002). "Landau diamagnetism". Electrons in a magnetic field. Retrieved 27 November 2012.

- "Hitachi achieves nanotechnology milestone for quadrupling terabyte hard drive" (Press release). Hitachi. October 15, 2007. Retrieved September 2011.

- "Fun with diamagnetic levitation". ForceField. 02-12-2008. Archived from the original on February 12, 2008. Retrieved September 2011.

- Beatty, Bill (2005). "Neodymium supermagnets: Some demonstrations—Diamagnetic water". Science Hobbyist. Retrieved September 2011.

Significant Topics of Mineraloid

Mineraloids are mineral like substances that do not exhibit crystalline nature. Most mineraloids are amorphous and do not possess an ordered atomic structure. The chapter introduces the reader to this category of minerals, lists the characteristic properties with suitable examples. Some examples explored include pearl, jet, obsidian, amber, ebonite etc.

Mineraloid

A mineraloid is a mineral-like substance that does not demonstrate crystallinity. Mineraloids possess chemical compositions that vary beyond the generally accepted ranges for specific minerals. For example, obsidian is an amorphous glass and not a crystal. Jet is derived from decaying wood under extreme pressure. Opal is another mineraloid because of its non-crystalline nature. Pearl, considered by some to be a mineral because of the presence of calcium carbonate crystals within its structure, would be better considered a mineraloid because the crystals are bonded by an organic material, and there is no definite proportion of the components.

Common Mineraloids

- Amber, non-crystalline structure, organic
- Ebonite, vulcanized natural or synthetic rubber (organic); lacks a crystalline structure
- Jet, non-crystalline nature, organic (very compact coal)
- Lechatelierite, nearly pure silica glass
- Limonite, a mixture of oxides and hydroxides of iron
- Mercury, liquid (IMA/CNMNC valid mineral name)
- Obsidian, volcanic glass - non-crystalline structure, a glass and quartz mixture
- Opal, non-crystalline silicon dioxide, a mix of minerals (IMA/CNMNC valid mineral name)
- Pearl, organically produced carbonate
- Petroleum, liquid, organic
- Pyrobitumen, amorphous fossilized petroleum (noncrystalline, organic)
- Tektites, meteoritic silica glass

Amber

Amber pendants made of modified amber. The oval pendant is 52 by 32 mm (2.0 by 1.3 in).

An ant inside Baltic amber

Unpolished amber stones

Wood resin, the source of amber

Extracting Baltic amber from Holocene deposits, Gdansk, Poland

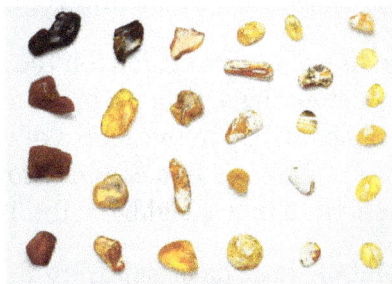

Unique colors of Baltic amber. Polished stones.

Fishing for amber on the coast of Baltic Sea. Winter storms throw out amber nuggets. Close to Gdansk, Poland.

Amber is fossilized tree resin (not sap), which has been appreciated for its color and natural beauty since Neolithic times. Much valued from antiquity to the present as a gemstone, amber is made into a variety of decorative objects. Amber is used as an ingredient in perfumes, as a healing agent in folk medicine, and as jewelry.

There are five classes of amber, defined on the basis of their chemical constituents. Because it originates as a soft, sticky tree resin, amber sometimes contains animal and plant material as inclusions. Amber occurring in coal seams is also called resinite, and the term ambrite is applied to that found specifically within New Zealand coal seams.

History and Names

The English word *amber* derives from Arabic '*anbar* عنبر (cognate with Middle Persian *ambar*) via Middle Latin *ambar* and Middle French *ambre*. The word was adopted in Middle English in the 14th century as referring to what is now known as *ambergris* (*ambre gris* or "grey amber"), a solid waxy substance derived from the sperm whale. In the Romance languages, the sense of the word had come to be extended to Baltic amber (fossil resin) from as early as the late 13th century. At first called white or yellow amber (*ambre jaune*), this meaning was adopted in English by the early 15th century. As the use of ambergris waned, this became the main sense of the word.

The two substances ("yellow amber" and "grey amber") conceivably became associated or confused because they both were found washed up on beaches. Ambergris is less dense than water and floats, whereas amber is too dense to float, though less dense than stone.

According to myth, when Phaëton son of Helios (the Sun) was killed, his mourning sisters became poplar trees, and their tears became *elektron*, amber.

Amber is discussed by Theophrastus in the 4th century BC, and again by Pytheas (c. 330 BC) whose work "On the Ocean" is lost, but was referenced by Pliny the Elder, according to whose *The Natural History* (in what is also the earliest known mention of the name *Germania*):

Pytheas says that the Gutones, a people of Germany, inhabit the shores of an estuary of the Ocean called Mentonomon, their territory extending a distance of six thousand stadia; that, at one day's sail from this territory, is the Isle of Abalus, upon the shores of which, amber is thrown up by the waves in spring, it being an excretion of the sea in a concrete form; as, also, that the inhabitants use this amber by way of fuel, and sell it to their neighbors, the Teutones.

Earlier Pliny says that a large island of three days' sail from the Scythian coast called Balcia by Xenophon of Lampsacus, author of a fanciful travel book in Greek, is called Basilia by Pytheas. It is generally understood to be the same as Abalus. Based on the amber, the island could have been Heligoland, Zealand, the shores of Bay of Gdansk, the Sambia Peninsula or the Curonian Lagoon, which were historically the richest sources of amber in northern Europe. It is assumed that there were well-established trade routes for amber connecting the Baltic with the Mediterranean (known as the "Amber Road"). Pliny states explicitly that the Germans export amber to Pannonia, from where it was traded further abroad by the Veneti. The ancient Italic peoples of southern Italy were working amber, the most important examples are on display at the National Archaeological Museum of Siritide to Matera. Amber used in antiquity as at Mycenae and in the prehistory of the Mediterranean comes from deposits of Sicily.

Pliny also cites the opinion of Nicias, according to whom amber "is a liquid produced by the rays of the sun; and that these rays, at the moment of the sun's setting, striking with the greatest force upon the surface of the soil, leave upon it an unctuous sweat, which is carried off by the tides of the Ocean, and thrown up upon the shores of Germany." Besides the fanciful explanations according to which amber is "produced by the Sun", Pliny cites opinions that are well aware of its origin in tree resin, citing the native Latin name of *succinum* (*sūcinum*, from *sucus* "juice"). "Amber is produced from a marrow discharged by trees belonging to the pine genus, like gum from the cherry, and resin from the ordinary pine. It is a liquid at first, which issues forth in considerable quantities, and is gradually hardened [...] Our forefathers, too, were of opinion that it is the juice of a tree, and for this reason gave it the name of 'succinum' and one great proof that it is the produce of a tree of the pine genus, is the fact that it emits a pine-like smell when rubbed, and that it burns, when ignited, with the odour and appearance of torch-pine wood."

He also states that amber is also found in Egypt and in India, and he even refers to the electrostatic properties of amber, by saying that "in Syria the women make the whorls of their spindles of this substance, and give it the name of *harpax* from the circumstance that it attracts leaves towards it, chaff, and the light fringe of tissues."

Pliny says that the German name of amber was *glæsum*, "for which reason the Romans, when Germanicus Cæsar commanded the fleet in those parts, gave to one of these islands the name of Glæsaria, which by the barbarians was known as Austeravia". This is confirmed by the recorded Old High German *glas* and Old English *glær* for "amber" (c.f. *glass*). In Middle Low German, amber was known as *berne-, barn-, börnstēn*. The Low German term became dominant also in High German by the 18th century, thus modern German *Bernstein* besides Dutch Dutch *barn-steen*.

The Baltic Lithuanian term for amber is *gintaras* and Latvian *dzintars*. They, and the Slavic *jantar* or Hungarian *gyanta* ('resin'), are thought to originate from Phoenician *jainitar* ("sea-resin").

Early in the nineteenth century, the first reports of amber from North America came from discoveries in New Jersey along Crosswicks Creek near Trenton, at Camden, and near Woodbury.

Legends

The origins of Baltic amber are associated with the Lithuanian legend about Juratė, the queen of the sea, who fell in love with Kastytis, a fisherman. According to one of the versions, her jealous father punished his daughter by destroying her amber palace and changing her into sea foam. The pieces of the Juratė's palace can still be found on the Baltic shore.

Composition and Formation

Amber is heterogeneous in composition, but consists of several resinous bodies more or less soluble in alcohol, ether and chloroform, associated with an insoluble bituminous substance. Amber is a macromolecule by free radical polymerization of several precursors in the labdane family, e.g. communic acid, cummunol, and biformene. These labdanes are diterpenes ($C_{20}H_{32}$) and trienes, equipping the organic skeleton with three alkene groups for polymerization. As amber matures over the years, more polymerization takes place as well as isomerization reactions, crosslinking and cyclization.

Heated above 200 °C (392 °F), amber suffers decomposition, yielding an oil of amber, and leaving a black residue which is known as "amber colophony", or "amber pitch"; when dissolved in oil of turpentine or in linseed oil this forms "amber varnish" or "amber lac".

Formation

Molecular polymerization, resulting from high pressures and temperatures produced by overlying sediment, transforms the resin first into copal. Sustained heat and pressure drives off terpenes and results in the formation of amber.

For this to happen, the resin must be resistant to decay. Many trees produce resin, but in the majority of cases this deposit is broken down by physical and biological processes. Exposure to sunlight, rain, microorganisms (such as bacteria and fungi), and extreme temperatures tends to disintegrate resin. For resin to survive long enough to become amber, it must be resistant to such forces or be produced under conditions that exclude them.

Botanical Origin

Fossil resins from Europe fall into two categories, the famous Baltic ambers and another that resembles the *Agathis* group. Fossil resins from the Americas and Africa are closely related to the modern genus *Hymenaea*, while Baltic ambers are thought to be fossil resins from Sciadopityaceae family plants that used to live in north Europe.

Inclusions

Baltic amber with inclusions

The abnormal development of resin in living trees (*succinosis*) can result in the formation of amber. Impurities are quite often present, especially when the resin dropped onto the ground, so the material may be useless except for varnish-making. Such impure amber is called *firniss*.

Such inclusion of other substances can cause amber to have an unexpected color. Pyrites may give a bluish color. *Bony amber* owes its cloudy opacity to numerous tiny bubbles inside the resin. However, so-called *black amber* is really only a kind of jet.

In darkly clouded and even opaque amber, inclusions can be imaged using high-energy, high-contrast, high-resolution X-rays.

Extraction and Processing

Distribution and Mining

Amber mine "Primorskoje" in Jantarny, Kaliningrad Oblast,Russia

Amber is globally distributed, mainly in rocks of Cretaceous age or younger. Historically, the Samland coast west of Königsberg in Prussia was the world's leading source of amber. About 90% of the world's extractable amber is still located in that area, which became the Kaliningrad Oblast of Russia in 1946.

Pieces of amber torn from the seafloor are cast up by the waves, and collected by hand, dredging, or diving. Elsewhere, amber is mined, both in open works and underground galleries. Then nodules of *blue earth* have to be removed and an opaque crust must be cleaned off, which can be done in revolving barrels containing sand and water. Erosion removes this crust from sea-worn amber.

Blue amber from Dominican Republic

Caribbean amber, especially Dominican blue amber, is mined through bell pitting, which is dangerous due to the risk of tunnel collapse.

Treatment

The Vienna amber factories, which use pale amber to manufacture pipes and other smoking tools, turn it on a lathe and polish it with whitening and water or with rotten stone and oil. The final luster is given by friction with flannel.

When gradually heated in an oil-bath, amber becomes soft and flexible. Two pieces of amber may be united by smearing the surfaces with linseed oil, heating them, and then pressing them together while hot. Cloudy amber may be clarified in an oil-bath, as the oil fills the numerous pores to which the turbidity is due. Small fragments, formerly thrown away or used only for varnish, are now used on a large scale in the formation of "ambroid" or "pressed amber".

The pieces are carefully heated with exclusion of air and then compressed into a uniform mass by intense hydraulic pressure, the softened amber being forced through holes in a metal plate. The product is extensively used for the production of cheap jewelry and articles for smoking. This pressed amber yields brilliant interference colors in polarized light. Amber has often been imitated by other resins like copal and kauri gum, as well as by celluloid and even glass. Baltic amber is sometimes colored artificially, but also called "true amber".

Appearance

Amber occurs in a range of different colors. As well as the usual yellow-orange-brown that is associated with the color "amber", amber itself can range from a whitish color through a pale lemon yellow, to brown and almost black. Other uncommon colors include red amber (sometimes known as "cherry amber"), green amber, and even blue amber, which is rare and highly sought after.

Yellow amber is a hard, translucent, yellow, orange, or brown fossil resin from evergreen trees. Known to the Iranians by the Pahlavi compound word kah-ruba (from kah "straw" plus rubay "attract, snatch," referring to its electrical properties), which entered Arabic as kahraba' or kahraba (which later became the Arabic word for electricity, كهرباء *kahrabā*'), it too was called amber in

Europe (Old French and Middle English ambre). Found along the southern shore of the Baltic Sea, yellow amber reached the Middle East and western Europe via trade. Its coastal acquisition may have been one reason yellow amber came to be designated by the same term as ambergris. Moreover, like ambergris, the resin could be burned as an incense. The resin's most popular use was, however, for ornamentation—easily cut and polished, it could be transformed into beautiful jewelry. Much of the most highly prized amber is transparent, in contrast to the very common cloudy amber and opaque amber. Opaque amber contains numerous minute bubbles. This kind of amber is known as "bony amber".

Although all Dominican amber is fluorescent, the rarest Dominican amber is blue amber. It turns blue in natural sunlight and any other partially or wholly ultraviolet light source. In long-wave UV light it has a very strong reflection, almost white. Only about 100 kg (220 lb) is found per year, which makes it valuable and expensive.

Sometimes amber retains the form of drops and stalactites, just as it exuded from the ducts and receptacles of the injured trees. It is thought that, in addition to exuding onto the surface of the tree, amber resin also originally flowed into hollow cavities or cracks within trees, thereby leading to the development of large lumps of amber of irregular form.

Classification

Amber can be classified into several forms. Most fundamentally, there are two types of plant resin with the potential for fossilization. Terpenoids, produced by conifers and angiosperms, consist of ring structures formed of isoprene (C_5H_8) units. Phenolic resins are today only produced by angiosperms, and tend to serve functional uses. The extinct medullosans produced a third type of resin, which is often found as amber within their veins. The composition of resins is highly variable; each species produces a unique blend of chemicals which can be identified by the use of pyrolysis–gas chromatography–mass spectrometry. The overall chemical and structural composition is used to divide ambers into five classes. There is also a separate classification of amber gemstones, according to the way of production.

Class I

This class is by far the most abundant. It comprises labdatriene carboxylic acids such as communic or ozic acids. It is further split into three sub-classes. Classes Ia and Ib utilize regular labdanoid diterpenes (e.g. communic acid, communol, biformenes), while Ic uses *enantio* labdanoids (ozic acid, ozol, *enantio* biformenes).

Ia

Includes *Succinite* (= 'normal' Baltic amber) and *Glessite*. Have a communic acid base. They also include much succinic acid.

Baltic amber yields on dry distillation succinic acid, the proportion varying from about 3% to 8%, and being greatest in the pale opaque or *bony* varieties. The aromatic and irritating fumes emitted by burning amber are mainly due to this acid. Baltic amber is distinguished by its yield of succinic acid, hence the name *succinite*. Succinite has a hardness between 2 and 3, which is rather greater than that of many other fossil resins. Its specific gravity varies from 1.05 to 1.10. It can be distinguished from other ambers via IR spectroscopy due to a specific carbonyl absorption peak. IR spectroscopy can detect the relative age of an amber sample. Succinic acid may not be an original component of amber, but rather a degradation product of abietic acid.

Ib

Like class Ia ambers, these are based on communic acid; however, they lack succinic acid.

Ic

This class is mainly based on *enantio*-labdatrienonic acids, such as ozic and zanzibaric acids. Its most familiar representative is Dominican amber.

Dominican amber differentiates itself from Baltic amber by being mostly transparent and often containing a higher number of fossil inclusions. This has enabled the detailed reconstruction of the ecosystem of a long-vanished tropical forest. Resin from the extinct species *Hymenaea protera* is the source of Dominican amber and probably of most amber found in the tropics. It is not "succinite" but "retinite".

Class II

These ambers are formed from resins with a sesquiterpenoid base, such as cadinene.

Class III

These ambers are polystyrenes.

Class IV

Class IV is something of a wastebasket; its ambers are not polymerized, but mainly consist of cedrene-based sesquiterpenoids.

Class V

Class V resins are considered to be produced by a pine or pine relative. They comprise a mixture of diterpinoid resins and *n*-alkyl compounds. Their type mineral is *highgate copalite*.

Geological Record

The oldest amber recovered dates to the Upper Carboniferous period (320 million years ago). Its chemical composition makes it difficult to match the amber to its producers – it is most similar to the resins produced by flowering plants; however, there are no flowering plant fossils until the Cretaceous, and they were not common until the Upper Cretaceous. Amber becomes abundant long after the Carboniferous, in the Early Cretaceous, 150 million years ago, when it is found in association with insects. The oldest amber with arthropod inclusions comes from the Levant, from Lebanon and Jordan. This amber, roughly 125–135 million years old, is considered of high scientific value, providing evidence of some of the oldest sampled ecosystems.

Typical amber specimen with a number of indistinct inclusions

In Lebanon more than 450 outcrops of Lower Cretaceous amber were discovered by Dany Azar a Lebanese paleontologist and entomologist. Among these outcrops 20 have yielded biological inclusions comprising the oldest representatives of several recent families of terrestrial arthropods. Even older, Jurassic amber has been found recently in Lebanon as well. Many remarkable insects and spiders were recently discovered in the amber of Jordan including the oldest zorapterans, clerid beetles, umenocoleid roaches, and achiliid planthoppers.

Baltic amber or succinite (historically documented as Prussian amber) is found as irregular nodules in marine glauconitic sand, known as *blue earth*, occurring in the Lower Oligocene strata of Sambia in Prussia (in historical sources also referred to as *Glaesaria*). After 1945 this territory around Königsberg was turned into Kaliningrad Oblast, Russia, where amber is now systematically mined.

It appears, however, to have been partly derived from older Eocene deposits and it occurs also as a derivative phase in later formations, such as glacial drift. Relics of an abundant flora occur as inclusions trapped within the amber while the resin was yet fresh, suggesting relations with the flora of Eastern Asia and the southern part of North America. Heinrich Göppert named the common amber-yielding pine of the Baltic forests *Pinites succiniter*, but as the wood does not seem to differ from that of the existing genus it has been also called *Pinus succinifera*. It is improbable, however, that the production of amber was limited to a single species; and indeed a large number of conifers belonging to different genera are represented in the amber-flora.

Paleontological Significance

Amber is a unique preservational mode, preserving otherwise unfossilizable parts of organisms; as such it is helpful in the reconstruction of ecosystems as well as organisms; the chemical composition of the resin, however, is of limited utility in reconstructing the phylogenetic affinity of the resin producer.

Amber sometimes contains animals or plant matter that became caught in the resin as it was secreted. Insects, spiders and even their webs, annelids, frogs, crustaceans, bacteria and amoebae, marine microfossils, wood, flowers and fruit, hair, feathers and other small organisms have been recovered in ambers dating to 130 million years ago.

In August 2012, two mites preserved in amber were determined to be the oldest animals ever to have been found in the substance; the mites are 230 million years old and were discovered in north-eastern Italy.

Use

Solutrean of Altamira – MHNT

Amber has been used since prehistory (Solutrean) in the manufacture of jewelry and ornaments, and also in folk medicine. Amber also forms the flavoring for akvavit liquor. Amber has been used as an ingredient in perfumes.

Jewelry

Amber has been used since the stone age, from 13,000 years ago. Amber ornaments have been found in Mycenaean tombs and elsewhere across Europe. To this day it is used in the manufacture of smoking and glassblowing mouthpieces. Amber's place in culture and tradition lends it a tourism value; Palanga Amber Museum is dedicated to the fossilized resin.

Amber jewelry from Dominican Republic

Historic Medicinal Uses

Amber has long been used in folk medicine for its purported healing properties. Amber and extracts were used from the time of Hippocrates in ancient Greece for a wide variety of treatments through the Middle Ages and up until the early twentieth century.

Scent of Amber and Amber Perfumery

In ancient China it was customary to burn amber during large festivities. If amber is heated under the right conditions, oil of amber is produced, and in past times this was combined carefully with nitric acid to create "artificial musk" – a resin with a peculiar musky odor. Although when burned, amber does give off a characteristic "pinewood" fragrance, modern products, such as perfume, do not normally use actual amber due to the fact that fossilized amber produces very little scent. In perfumery, scents referred to as "amber" are often created and patented to emulate the opulent golden warmth of the fossil.

Lithuanian amber jewelry

The modern name for amber is thought to come from the Arabic word, ambar, meaning ambergris. Ambergris is the waxy aromatic substance created in the intestines of sperm whales and was used in making perfumes both in ancient times as well as modern.

The scent of amber was originally derived from emulating the scent of ambergris and/or labdanum but due to the endangered species status of the sperm whale the scent of amber is now largely derived from labdanum. The term "amber" is loosely used to describe a scent that is warm, musky, rich and honey-like, and also somewhat oriental and earthy. It can be synthetically created or derived from natural resins. When derived from natural resins it is most often created out of labdanum. Benzoin is usually part of the recipe. Vanilla and cloves are sometimes used to enhance the aroma.

"Amber" perfumes may be created using combinations of labdanum, benzoin resin, copal (itself a type of tree resin used in incense manufacture), vanilla, Dammara resin and/or synthetic materials.

Imitation

Imitation Made in Natural Resins

Young resins, these are used as imitations:

- Kauri resin from trees *Agathis australis*, New Zealand.

- The copals (subfossil resins). The African and American (Colombia) copals from *Leguminosae* trees family (genus *Hymenaea*). Amber of the Dominican or Mexican type (Class I of fossil resins). Copals from Manilia (Indonesia) and from New Zealand from trees of the genus *Agathis* (*Araucariaceae* family)

- Other fossil resins: burmite in Burma, rumenite in Romania, simetite in Sicilia.

- Other natural resins — cellulose or chitin, etc.

Imitations Made of Plastics

Plastics, these are used as imitations:

- Stained glass (inorganic material) and other ceramic materials

- Celluloid

- Cellulose nitrate (obtain first time in 1833) — a product of treatment of cellulose with nitration mixture.

- Acetylcellulose (not in the use at present)

- Galalith or «artificial horn» (condensation product of casein and formaldehyde), other trade names: Alladinite, Erinoid, Lactoid.

- Casein — a conjugated protein forming from the casein precursor – caseinogen.

- Resolane (phenolic resins or phenoplasts, not in the use at present)

- Bakelite resine (resol, phenolic resins), product from Africa are known under the misleading name: «African amber».

- Carbamide resins — melamine, formaldehyde and urea-formaldehyde resins.

- Epoxy novolac (phenolic resins), non officially name: «antique amber», not in the use at present

- Polyesters (Polish amber imitation) with styrene. Ex.: unsaturated polyester resins (polymals) are produced by Chemical Industrial Works «Organika» in Sarzyna, Poland; estomal are produced by Laminopol firm. Polybern or sticked amber is artificial resins the curled chips are obtained, whereas in the case of amber – small scraps. «African amber» (polyester, synacryl is then probably other name of the same resine) are produced by Reichhold firm; Styresol trade mark or alkid resin (used in Russia, Reichhold, Inc. patent, 1948.

- Polyethylene

- Eepoxide resins

- Polystyrene and polystyrene-like polymers (vinyl polymers).

- The resins of acrylic type (vinyl polymers), especially polymethyl methacrylate PMMA (trade mark Plexiglass, metaplex).

Ebonite

Ebonite applications from the 19th century

Ebonite is a brand name for very hard rubber first obtained by Charles Goodyear by vulcanizing natural rubber for prolonged periods. For vulcanizing natural rubber he received patent number 3633 from the United States Patent Office on June 15, 1844. Besides natural rubber, Ebonite contains about 25% to 80% sulfur and linseed oil. Its name comes from its intended use as an artificial substitute for ebony wood.

The material is known generically as hard rubber and has formerly been called "vulcanite", although that name now refers to the mineral vulcanite.

History

Charles Goodyear's brother Nelson Goodyear experimented with the chemistry of ebonite composites. In 1851 he used zinc oxide as a filler.

Properties

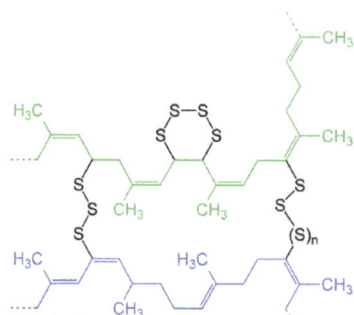

Schematic presentation of two strands (**blue** and **green**) of natural rubber after vulcanization with elemental sulfur

The sulfur percentage and the applied temperatures and duration during vulcanizing are the main variables that determine the technical properties of the hard rubber polysulfide elastomer. The occurring reaction is basically addition of sulfur at the double bonds, forming intramoleculer ring structures, so a large portion of the sulfur is highly cross-linked in the form of intramoleculer addition. High sulfur content up to 40% may be used for greatest resistance to swelling and minimal dielectric loss. The strongest mechanical properties and greatest heat resistance is obtained with sulfur contents around 35%. Best impact strength is obtained with somewhat lower sulfur contents around 30%. The rigidity of hard rubber at room temperature is attributed to the van der Waals forces between the intramoleculer sulfur atoms. Raising the temperature gradually increases the molecular vibrations that overcome the van der Waals forces making it elastic. Hard rubber has a content mixture dependent density around 1.1 to 1.2. When reheated hard rubber exhibits shape-memory effect and can be fairly easy reshaped within certain limits. Depending on the sulfur percentage hard rubber has a thermoplastic transition or softening temperature of 70 to 80 °C (158 to 176 °F).

The material is brittle, which produces problems in its use in battery cases for example, where the integrity of the case is vital to prevent leakage of sulfuric acid. It has now been generally replaced by carbon black -filled polypropylene.

Under the influence of the ultraviolet portion in daylight hard rubber oxidizes and exposure to moisture bonds water with free sulfur on the surface creating sulfates and sulfuric acid at the surface that are very hygroscopic. The sulfates condense water from the air, forming a hydrophilic film with favorable wettability characteristics on the surface. These aging processes will gradually discolor the surface grayish green to brown and cause rapid deterioration of electric surface resistivity.

Contamination

Ebonite contamination was problematic when it was used for electronics. The ebonite was rolled between metal foil sheets, which were peeled off, leaving traces of metal behind. For electronic use the surface was ground to remove metal particles.

Applications

Green/black rippled ebonite fountain pen made in 2014

Hard rubber was used in early 20th century bowling balls, however was phased out in favor of other materials (the Ebonite name remains as a trade name for one of the major manufacturers of polymer balls). It has been used in electric plugs, smoking pipe mouthpieces (in competition with lucite), fountain pen bodies and nib feeds, saxophone and clarinet mouthpieces as well as complete humidity-stable clarinets . Hard rubber is often seen as the wheel material in casters. It is also commonly used in physics classrooms to demonstrate static electricity because it is at or near the negative end of the triboelectric series.

Hard rubber was used in the cases of automobile batteries for years, thus establishing black as their traditional colour even long after stronger modern plastics like polypropylene were substituted. It was used famously for many decades in hair combs made by Ace, now part of Newell Rubbermaid, although the current models are known to be produced solely with plastics. (An easy way to identify a hard rubber comb is to rub part of its surface vigorously, then immediately smell the comb. Hard rubber's scent, resulting from the sulfur in the ebonite, can usually be detected temporarily. The same effect can often be produced by running the comb under hot tap water.) Ebonite is used as an anticorrosive lining for various (mainly storage) vessels that contain hydrochloric acid. It forms bubbles when storing hydrofluoric acid at temperatures above room temperature, or for prolonged durations.

Jet (Lignite)

Jet is a type of lignite, a precursor to coal, and is considered to be a minor gemstone. Jet is not considered a true mineral, but rather a mineraloid as it has an organic origin, being derived from decaying wood under extreme pressure.

Sample of jet

Pendant in Jet, Magdalenian, Marsoulas MHNT

Hallstatt culture bracelets made from jet and bronze, unearthed at Magdalenenberg

Mourning jewellery: jet brooch, 19th century

The English noun "jet" derives from the French word for the same material: *jaiet*. Jet is either black or dark brown, but may contain pyrite inclusions, which are of brassy colour and metallic lustre. The adjective "jet-black", meaning as dark a black as possible, derives from this material.

Origin

Jet is a product of high pressure decomposition of wood from millions of years ago, commonly the wood of trees of the family Araucariaceae. Jet is found in two forms, hard and soft. Hard jet is the result of the carbon compression and salt water; soft jet is the result of the carbon compression and fresh water.

The jet found at Whitby, in England, is of Early Jurassic (Toarcian) age, approximately 182 million years old. Whitby Jet is the fossilized wood from species similar to the extant Chile pine or Monkey Puzzle tree (*Araucaria araucana*).

History

Jet has been used in Britain since the Neolithic period, but the earliest known object is a 10,000 BC model of a botfly larva, from Baden-Württemberg, Germany. It continued in use in Britain through the Bronze Age where it was used for necklace beads. During the Iron Age jet went out of fashion until the early third century AD in Roman Britain. The End of Roman Britain marked the end of jet's ancient popularity until, despite sporadic use in the Anglo-Saxon and Viking periods, the later Medieval period. Jet saw a massive resurgence during the Victorian era.

Roman Use

Whitby jet was a popular material for jewellery in Roman Britain from the third century onward. It was used in rings, hair pins, beads, bracelets, bangles, necklaces and pendants; many of which are visible in the Yorkshire Museum. There is no evidence for Roman jet working in Whitby itself, rather it was transferred to Eboracum (modern York) where considerable evidence for jet production has been found. The collection of jet at this time was based on beachcombing rather than quarrying.

Jet cameo depicting a Medusa in the Yorkshire Museum

In the Roman period it saw use as a magical material, frequently used in amulets and pendants because of its supposed protective qualities and ability to deflect the gaze of the evil eye. Pliny the Elder suggests that "*the kindling of jet drives off snakes and relieves suffocation of the uterus. Its fumes detect attempts to stimulate a disabling illness or a state of virginity.*" and has been referenced by other Ancient writers including Solinus and Galen.

Jet objects were exported from Eboracum all over Roman Britain and into Europe.

Around the Rhine some jet bracelets from the period have been found that feature grooves with gold inserts.

Victorian Use

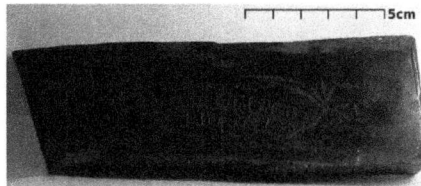
A large piece of jet from Whitby

Jet as a gemstone was fashionable during the reign of Queen Victoria, during which the Queen wore Whitby jet as part of her mourning dress, mourning the death of Prince Albert. Jet was associated with mourning jewellery in the 19th century because of its sombre colour and modest appearance, and it has been traditionally fashioned into rosaries for monks.

In some jewellery designs of the period jet was combined with cut steel.

20th Century

In the United States, long necklaces of jet beads were very popular during the Roaring Twenties, when women and young flappers would wear multiple strands of jet beads stretching from the neckline to the waistline. In these necklaces, the jet was strung using heavy cotton thread; small knots were made on either side of each bead to keep the beads spaced evenly, much in the same way that fine pearl necklaces are made. Jet has also been known as black amber, as it may induce an electric charge like that of amber when rubbed.

Latticework San Miguel de Lillo, Asturias, Asturian jet of Oles, Villaviciosa.

Properties

Jet is very easy to carve but it is difficult to create fine details without breaking so it takes an experienced lapidary to execute more elaborate carvings.

Jet has a Mohs hardness ranging between 2.5 and 4 and a specific gravity of 1.30 to 1.34. The refractive index of jet is approximately 1.66. The touch of a red-hot needle should cause jet to emit an odour similar to coal.

Authenticating Jet

Although now much less popular than in the past, authentic jet jewels are valued by collectors.

Unlike black glass, which is cool to the touch, jet is not cool, due to its lower thermal conductivity. Glass was used as a jet substitute during the peak of jet's popularity. When it was used in this way it was known as French jet or Vauxhall glass. Ebonite was also used as a jet substitute and initially looks very similar to jet but it fades over time. In some cases jet offcuts were mixed with glue and molded into jewellery.

Anthracite (hard coal) is superficially similar to fine jet, and has been used to imitate it. This imitation are not always easy to distinguish from real jet. When rubbed against unglazed porcelain, true jet will leave a chocolate brown streak.

The microstructure of jet, which strongly resembles the original wood, can be seen under 120× or greater magnification.

Opal

Opal is a hydrated amorphous form of silica ($SiO_2 \cdot nH_2O$); its water content may range from 3 to 21% by weight, but is usually between 6 and 10%. Because of its amorphous character, it is classed as a mineraloid, unlike crystalline forms of silica, which are classed as minerals. It is deposited at a relatively low temperature and may occur in the fissures of almost any kind of rock, being most commonly found with limonite, sandstone, rhyolite, marl, and basalt. Opal is the national gemstone of Australia.

The internal structure of precious opal makes it diffract light; depending on the conditions in which it formed, it can take on many colors. Precious opal ranges from clear through white, gray, red, orange, yellow, green, blue, magenta, rose, pink, slate, olive, brown, and black. Of these hues, the black opals are the most rare, whereas white and greens are the most common. It varies in optical density from opaque to semitransparent.

Precious Opal

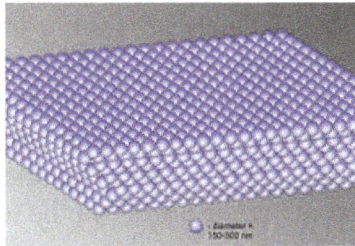

Precious opal consists of spheres of silicon dioxide molecules arranged in regular, closely packed planes. (Idealized diagram)

Multicolor rough crystal opal from Coober Pedy, South Australia, expressing nearly every color of the visible spectrum

Precious opal replacing ichthyosaur backbone; display specimen, South Australian Museum

Precious opal shows a variable interplay of internal colors, and though it is a mineraloid, it has an internal structure. At microscopic scales, precious opal is composed of silica spheres some 150 to 300 nm in diameter in a hexagonal or cubic close-packed lattice. It was shown by J. V. Sanders in the mid-1960s, that these ordered silica spheres produce the internal colors by causing the interference and diffraction of light passing through the microstructure of the opal. The regularity of the sizes and the packing of these spheres determines the quality of precious opal. Where the distance between the regularly packed planes of spheres is around half the wavelength of a component of visible light, the light of that wavelength may be subject to diffraction from the grating created by the stacked planes. The colors that are observed are determined by the spacing between the planes and the orientation of planes with respect to the incident light. The process can be described by Bragg's law of diffraction.

Visible light of diffracted wavelengths cannot pass through large thicknesses of the opal. This is the basis of the optical band gap in a photonic crystal. The notion that opals are photonic crystals for visible light was expressed in 1995 by Vasily Astratov's group. In addition, microfractures may be filled with secondary silica and form thin lamellae inside the opal during solidification. The term opalescence is commonly and erroneously used to describe this unique and beautiful phenomenon, which is correctly termed play of color. Contrarily, opalescence is correctly applied to the milky, turbid appearance of common or potch opal. Potch does not show a play of color.

For gemstone use, most opal is cut and polished to form a cabochon. "Solid" opal refers to polished stones consisting wholly of precious opal. Opals too thin to produce a "solid" may be combined with other materials to form attractive gems. An opal doublet consists of a relatively thin layer of precious opal, backed by a layer of dark-colored material, most commonly ironstone, dark or black common opal (potch), onyx, or obsidian. The darker backing emphasizes the play of color, and results in a more attractive display than a lighter potch. An opal triplet is similar to a doublet, but has a third layer, a domed cap of clear quartz or plastic on the top. The cap takes a high polish and acts as a protective layer for the opal. The top layer also acts as a magnifier, to emphasize the play of color of the opal beneath, which is often of lower quality. Triplet opals therefore have a more artificial appearance, and are not classed as precious opal. Jewelry applications of precious opal can be somewhat limited by opal's sensitivity to heat due primarily to its relatively high water content and predisposition to scratching.

Combined with modern techniques of polishing, doublet opal produces a similar effect to black or boulder opal at a fraction of the price. Doublet opal also has the added benefit of having genuine opal as the top visible and touchable layer, unlike triplet opals.

Common Opal

Besides the gemstone varieties that show a play of color, the other kinds of common opal include the milk opal, milky bluish to greenish (which can sometimes be of gemstone quality); resin opal, which is honey-yellow with a resinous luster; wood opal, which is caused by the replacement of the organic material in wood with opal; menilite, which is brown or grey; hyalite, a colorless glass-clear opal sometimes called Muller's glass; geyserite, also called siliceous sinter, deposited around hot springs or geysers; and diatomite or diatomaceous earth, the accumulations of diatom shells or tests.

A piece of milky raw opal from Andamooka, South Australia

An opal "triplet" from Andamooka showing blue and green fire

Fire opal from Mexico

A rock showing striations of opal throughout

Other Varieties of Opal

Brightness of the fire in opal ranges on a scale of 1 to 5 (5 being the brightest)

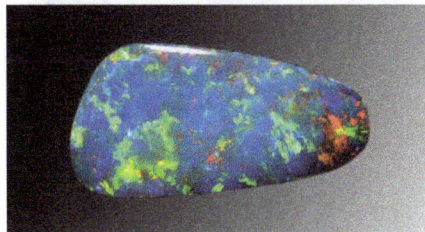

Australian opal doublet, a slice of precious opal with a backing of ironstone

Faceted yellow fire opal, 4.76 ct, Mexico

Fire opal is a transparent to translucent opal, with warm body colors of yellow to orange to red. Although it does not usually show any play of color, occasionally a stone will exhibit bright green flashes. The most famous source of fire opals is the state of Querétaro in Mexico; these opals are commonly called Mexican fire opals. Fire opals that do not show play of color are sometimes referred to as jelly opals. Mexican opals are sometimes cut in their ryholitic host material if it is hard enough to allow cutting and polishing. This type of Mexican opal is referred to as a Cantera opal. Also, a type of opal from Mexico, referred to as Mexican water opal, is a colorless opal which exhibits either a bluish or golden internal sheen.

Girasol opal is a term sometimes mistakenly and improperly used to refer to fire opals, as well as a type of transparent to semitransparent type milky quartz from Madagascar which displays an asterism, or star effect, when cut properly. However, the true girasol opal is a type of hyalite opal that exhibits a bluish glow or sheen that follows the light source around. It is not a play of color as seen in precious opal, but rather an effect from microscopic inclusions. It is also sometimes referred to as water opal, too, when it is from Mexico. The two most notable locations of this type of opal are Oregon and Mexico.

Peruvian opal (also called blue opal) is a semiopaque to opaque blue-green stone found in Peru, which is often cut to include the matrix in the more opaque stones. It does not display pleochroism. Blue opal also comes from Oregon in the Owyhee region, as well as from Nevada around Virgin Valley.

Sources of Opal

Multicolored solid black opal cabochon from Lightning Ridge, NSW

Boulder opal, Carisbrooke Station near Winton, Queensland

Australian opal has often been cited as accounting for 95–97% of the world's supply of precious opal, with the state of South Australia accounting for 80% of the world's supply. Recent data suggests that the world supply of precious opal may have changed. In 2012, Ethiopian opal production was estimated to be 14,000 kg (31,000 lb) by the United States Geological Survey. USGS data from the same period (2012), reveals that Australian opal production to be $41 million. Because of the units of measurement, it is not possible to directly compare Australian and Ethiopian opal production, but these data and others suggest that the traditional percentages given for Australian opal production may be overstated. Yet, the validity of data in the USGS report appears to conflict with that of Laurs and others and Mesfin, who estimated the 2012 Ethiopian opal output (from Wegal Tena) to be only 750 kg (1,650 lb).

Australian Opal

The town of Coober Pedy in South Australia is a major source of opal. The world's largest and most valuable gem opal "Olympic Australis" was found in August 1956 at the "Eight Mile" opal field in Coober Pedy. It weighs 17,000 carats (3450 g) and is 11 in (280 mm) long, with a height of 4.75 in (121 mm) and a width of 4.5 in (110 mm). The Mintabie Opal Field located about 250 km (160 mi) north west of Coober Pedy has also produced large quantities of crystal opal and the rarer black opal. Over the years, it has been sold overseas incorrectly as Coober Pedy opal. The black opal is said to be some of the best examples found in Australia. Andamooka in South Australia is also a major producer of matrix opal, crystal opal, and black opal. Another Australian town, Lightning Ridge in New South Wales, is the main source of black opal, opal containing a predominantly dark background (dark-gray to blue-black displaying the play of color). Boulder opal consists of concretions and fracture fillings in a dark siliceous ironstone matrix. It is found sporadically in western Queensland, from Kynuna in the north, to Yowah and Koroit in the south. Its largest quantities are found around Jundah and Quilpie (known as the "home of the boulder opal") in South West Queensland. Australia also has opalised fossil remains, including dinosaur bones in New South Wales, and marine creatures in South Australia. The rarest type of Australian opal is "pipe" opal, closely related to boulder opal, which forms in sandstone with some iron oxide content, usually as fossilized tree roots.

Ethiopian Opal

Gem-grade precious Ethiopian Welo opal pendant

Although it has been reported that Northern African opal was used to make tools as early as 4000 BC, the first published report of gem opal from Ethiopia appeared in 1994, with the discovery of precious opal in the Menz Gishe District, North Shewa Province. The opal, found mostly in the form of nodules, was of volcanic origin and was found predominantly within weathered layers of rhyolite. This Shewa Province opal was mostly dark brown in color and had a tendency to crack. These qualities made it unpopular in the gem trade. In 2008, a new opal deposit was found near the town of Wegel Tena, in Ethiopia's Wollo Province. The Wollo Province opal was different from the previous Ethiopian opal finds in that it more closely resembled the sedimentary opals of Australia and Brazil, with a light background and often vivid play-of-color. Wollo Province opal, more commonly referred to as "Welo" or "Wello" opal, has become the dominant Ethiopian opal in the gem trade.

Virgin Valley, Nevada

Multicolored rough opal specimen from Virgin Valley, Nevada, US

The Virgin Valley opal fields of Humboldt County in northern Nevada produce a wide variety of precious black, crystal, white, fire, and lemon opal. The black fire opal is the official gemstone of Nevada. Most of the precious opal is partial wood replacement. The precious opal is hosted and found within a subsurface horizon or zone of bentonite in-place which is considered a "lode" deposit. Opals which have weathered out of the in-place deposits are alluvial and considered placer deposits. Miocene-age opalised teeth, bones, fish, and a snake head have been found. Some of the opal has high water content and may desiccate and crack when dried. The largest producing mines of Virgin Valley have been the famous Rainbow Ridge, Royal Peacock, Bonanza, Opal Queen, and WRT Stonetree/Black Beauty Mines. The largest unpolished black opal in the Smithsonian Institution, known as the "Roebling opal", came out of the tunneled portion of the Rainbow Ridge Mine in 1917, and weighs 2,585 carats. The largest polished black opal in the Smithsonian Institution comes from the Royal Peacock opal mine in the Virgin Valley, weighing 160 carats, known as the "Black Peacock".

Other Locations

Another source of white base opal or creamy opal in the United States is Spencer, Idaho. A high percentage of the opal found there occurs in thin layers.

Other significant deposits of precious opal around the world can be found in the Czech Republic, Slovakia, Hungary, Turkey, Indonesia, Brazil (in Pedro II, Piauí), Honduras (more precisely in Erandique), Guatemala and Nicaragua.

In late 2008, NASA announced it had discovered opal deposits on Mars.

Synthetic Opal

Opals of all varieties have been synthesized experimentally and commercially. The discovery of the ordered sphere structure of precious opal led to its synthesis by Pierre Gilson in 1974. The resulting material is distinguishable from natural opal by its regularity; under magnification, the patches of color are seen to be arranged in a "lizard skin" or "chicken wire" pattern. Furthermore, synthetic opals do not fluoresce under ultraviolet light. Synthetics are also generally lower in density and are often highly porous.

Two notable producers of synthetic opal are Kyocera and Inamori of Japan. Most so-called synthetics, however, are more correctly termed "imitation opal", as they contain substances not found in natural opal (e.g., plastic stabilizers). The imitation opals seen in vintage jewelry are often foiled glass, glass-based "Slocum stone", or later plastic materials.

Other research in macroporous structures have yielded highly ordered materials that have similar optical properties to opals and have been used in cosmetics.

Local Atomic Structure of Opals

The lattice of spheres of opal that cause the interference with light are several hundred times larger than the fundamental structure of crystalline silica. As a mineraloid, no unit cell describes the structure of opal. Nevertheless, opals can be roughly divided into those that show no signs of crystalline order (amorphous opal) and those that show signs of the beginning of crystalline order, commonly termed cryptocrystalline or microcrystalline opal. Dehydration experiments and infrared spectroscopy have shown that most of the H_2O in the formula of $SiO_2 \cdot nH_2O$ of opals is present in the familiar form of clusters of molecular water. Isolated water molecules, and silanols, structures such as SiOH, generally form a lesser proportion of the total and can reside near the surface or in defects inside the opal.

The crystal structure of crystalline α-cristobalite. Locally, the structures of some opals, opal-C, are similar to this.

The structure of low-pressure polymorphs of anhydrous silica consist of frameworks of fully corner bonded tetrahedra of SiO_4. The higher temperature polymorphs of silica cristobalite and tridymite are frequently the first to crystallize from amorphous anhydrous silica, and the local structures of microcrystalline opals also appear to be closer to that of cristobalite and tridymite than to quartz. The structures of tridymite and cristobalite are closely related and can be described as hexagonal and cubic close-packed layers. It is therefore possible to have intermediate structures in which the layers are not regularly stacked.

Microcrystalline Opal

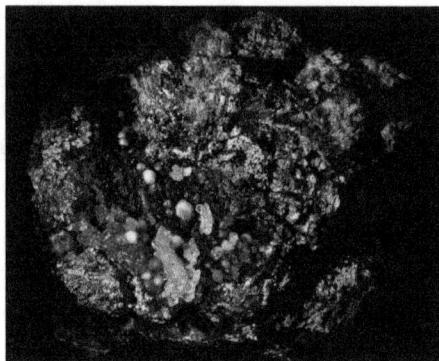

Lussatite (Opal-CT)

Opal-CT has been interpreted as consisting of clusters of stacking of cristobalite and tridymite over very short length scales. The spheres of opal in opal-CT are themselves made up of tiny microcrystalline blades of cristobalite and tridymite. Opal-CT has occasionally been further subdivided in the literature. Water content may be as high as 10 wt%. Lussatite is a synonym. Opal-C, also called lussatine, is interpreted as consisting of localized order of -cristobalite with a lot of stacking disorder. Typical water content is about 1.5wt%.

Noncrystalline Opal

Two broad categories of noncrystalline opals, sometimes just referred to as "opal-A", have been proposed. The first of these is opal-AG consisting of aggregated spheres of silica, with water filling the space in between. Precious opal and potch opal are generally varieties of this, the difference being in the regularity of the sizes of the spheres and their packing. The second "opal-A" is opal-AN or water-containing amorphous silica-glass. Hyalite is another name for this.

Noncrystalline silica in siliceous sediments is reported to gradually transform to opal-CT and then opal-C as a result of diagenesis, due to the increasing overburden pressure in sedimentary rocks, as some of the stacking disorder is removed.

Naming

The word 'opal' is adapted from the Roman term *opalus*, but the origin of this word is a matter of debate. However, most modern references suggest it is adapted from the Sanskrit word *úpala*.

References to the gem are made by Pliny the Elder. It is suggested to have been adapted from Ops, the wife of Saturn and goddess of fertility. The portion of Saturnalia devoted to Ops was "Opalia", similar to *opalus*.

Another common claim that the term is adapted from the Greek word, *opallios*. This word has two meanings, one is related to "seeing" and forms the basis of the English words like "opaque"; the other is "other" as in "alias" and "alter". It is claimed that *opalus* combined these uses, meaning "to see a change in color". However, historians have noted the first appearances of *opallios* do not occur until after the Romans had taken over the Greek states in 180 BC, and they had previously used the term *paederos*.

However, the argument for the Sanskrit origin is strong. The term first appears in Roman references around 250 BC, at a time when the opal was valued above all other gems. The opals were supplied by traders from the Bosporus, who claimed the gems were being supplied from India. Before this the stone was referred to by a variety of names, but these fell from use after 250 BC.

Historical Superstitions

In the Middle Ages, opal was considered a stone that could provide great luck because it was believed to possess all the virtues of each gemstone whose color was represented in the color spectrum of the opal. It was also said to confer the power of invisibility if wrapped in a fresh bay leaf and held in the hand. Following the publication of Sir Walter Scott's *Anne of Geierstein* in 1829, opal acquired a less auspicious reputation. In Scott's novel, the Baroness of Arnheim wears an opal talisman with supernatural powers. When a drop of holy water falls on the talisman, the opal turns into a colorless stone and the Baroness dies soon thereafter. Due to the popularity of Scott's novel, people began to associate opals with bad luck and death. Within a year of the publishing of Scott's novel in April 1829, the sale of opals in Europe dropped by 50%, and remained low for the next 20 years or so.

Even as recently as the beginning of the 20th century, it was believed that when a Russian saw an opal among other goods offered for sale, he or she should not buy anything more, as the opal was believed to embody the evil eye.

Opal is considered the birthstone for people born in October or under the signs of Scorpio and Libra.

Obsidian

Obsidian is a naturally occurring volcanic glass formed as an extrusive igneous rock.

It is produced when felsic lava extruded from a volcano cools rapidly with minimal crystal growth. Obsidian is commonly found within the margins of rhyolitic lava flows known as obsidian flows, where the chemical composition (high silica content) induces a high viscosity and polymerization degree of the lava. The inhibition of atomic diffusion through this highly viscous and polymerized lava explains the lack of crystal growth. Obsidian is hard and brittle; it therefore fractures with very sharp edges, which were used in the past in cutting and piercing tools, and it has been used experimentally as surgical scalpel blades.

Origin and Properties

Obsidian talus at Obsidian Dome, California

Polished snowflake obsidian, formed through the inclusion of cristobalite crystals

... among the various forms of glass we may reckon Obsian glass, a substance very similar to the stone found by Obsius in Ethiopia.

The translation into English of *Natural History* written by Pliny the Elder of Rome shows a few sentences on the subject of a volcanic glass called Obsian, so named from its resemblance to a stone (*obsiānus lapis*) found in Ethiopia by Obsius, a Roman explorer.

Obsidian is the rock formed as a result of quickly cooled lava, which is the parent material. Tektites were once thought by many to be obsidian produced by lunar volcanic eruptions, though few scientists now adhere to this hypothesis.

Obsidian is mineral-like, but not a true mineral because as a glass it is not crystalline; in addition, its composition is too complex to comprise a single mineral. It is sometimes classified as a mineraloid. Though obsidian is usually dark in color similar to mafic rocks such as basalt, obsidian's composition is extremely felsic. Obsidian consists mainly of SiO_2 (silicon dioxide), usually 70% or more. Crystalline rocks with obsidian's composition include granite and rhyolite. Because obsidian is metastable at the Earth's surface (over time the glass becomes fine-grained mineral crystals), no obsidian has been found that is older than Cretaceous age. This breakdown of obsidian is accelerated by the presence of water. Having a low water content when newly formed, typically less than 1% water by weight, obsidian becomes progressively hydrated when exposed to groundwater, forming perlite.

Pure obsidian is usually dark in appearance, though the color varies depending on the presence of impurities. Iron and magnesium typically give the obsidian a dark brown to black color. Very few samples are nearly colorless. In some stones, the inclusion of small, white, radially clustered crystals of cristobalite in the black glass produce a blotchy or snowflake pattern (*snowflake obsidian*). Obsidian may contain patterns of gas bubbles remaining from the lava flow, aligned along layers created as the molten rock was flowing before being cooled. These bubbles can produce interesting effects such as a golden sheen (*sheen obsidian*). An iridescent, rainbow-like sheen (*rainbow obsidian*) is caused by inclusions of magnetite nanoparticles.

Occurrence

Obsidian can be found in locations which have experienced rhyolitic eruptions. It can be found in Argentina, Armenia, Azerbaijan, Australia, Canada, Chile, Georgia, Greece, El Salvador, Guatemala, Iceland, Italy, Japan, Kenya, Mexico, New Zealand, Papua New Guinea, Peru, Scotland,

Turkey and the United States. Obsidian flows which may be hiked on are found within the calderas of Newberry Volcano and Medicine Lake Volcano in the Cascade Range of western North America, and at Inyo Craters east of the Sierra Nevada in California. Yellowstone National Park has a mountainside containing obsidian located between Mammoth Hot Springs and the Norris Geyser Basin, and deposits can be found in many other western U.S. states including Arizona, Colorado, New Mexico, Texas, Utah, Washington, Oregon and Idaho. Obsidian can also be found in the eastern U.S. states of Virginia, as well as Pennsylvania and North Carolina.

Glass Mountain, a large obsidian flow at Medicine Lake Volcano

There are only four major deposit areas in the central Mediterranean: Lipari, Pantelleria, Palmarola and Monte Arci.

Ancient sources in the Aegean were Melos and Giali.

Acigöl town and the Göllü Dağ volcano were the most important sources in central Anatolia, one of the more important source areas in the prehistoric Near East.

Historical Use

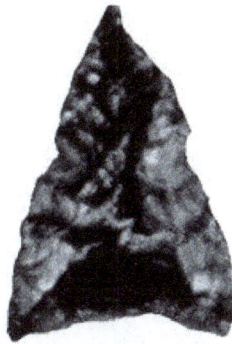

Obsidian arrowhead

The first known archaeological evidence of usage was in Kariandusi and other sites of the Acheulian age (beginning 1.5 million years BP) dated 700,000 BC, although the number of objects found at these sites were very low relative to the Neolithic. Use of obsidian in pottery of the Neolithic in the area around Lipari was found to be significantly less at a distance representing two weeks journeying. Anatolian sources of obsidian are known to have been the material used in the Levant and modern-day Iraqi Kurdistan from a time beginning sometime about 12,500 BC. The first attested civilized use is from excavations at Tell Brak dated the late fifth millennia. Obsidian was valued in Stone Age cultures because, like flint, it could be fractured to produce sharp blades or arrowheads. Like all glass and some other types of naturally occurring rocks, obsidian breaks with a characteristic conchoidal fracture. It was also polished to create early mirrors. Modern archaeologists have developed a relative dating system, obsidian hydration dating, to calculate the age of obsidian artifacts.

Middle East

Obsidian tools from Tilkitepe, Turkey, 5th millennium BC. Museum of Anatolian Civilizations

In the Ubaid in the 5th millennium BC, blades were manufactured from obsidian extracted from outcrops located in modern-day Turkey. Ancient Egyptians used obsidian imported from the eastern Mediterranean and southern Red Sea regions. Obsidian was also used in ritual circumcisions because of its deftness and sharpness. In the eastern Mediterranean area the material was used to make tools, mirrors and decorative objects.

Obsidian has also been found in Gilat, a site in the western Negev in Israel. Eight obsidian artifacts dating to the Chalcolithic Age found at this site were traced to obsidian sources in Anatolia. Neutron activation analysis (NAA) on the obsidian found at this site helped to reveal trade routes and exchange networks previously unknown.

Americas

Obsidian worked into plates and other wares by Victor Lopez Pelcastre of Nopalillo, Epazoyucan, Hidalgo. On display at the Museo de Arte Popular, Mexico City.

Lithic analysis can be instrumental in understanding prehispanic groups in Mesoamerica. A careful analysis of obsidian in a culture or place can be of considerable use to reconstruct commerce, production, distribution and thereby understand economic, social and political aspects of a civilization. This is the case in Yaxchilán, a Maya city where even warfare implications have been studied linked with obsidian use and its debris. Another example is the archeological recovery at coastal Chumash sites in California indicating considerable trade with the distant site of Casa Diablo, California in the Sierra Nevada Mountains.

Pre-Columbian Mesoamericans' use of obsidian was extensive and sophisticated; including carved and worked obsidian for tools and decorative objects. Mesoamericans also made a type of sword with obsidian blades mounted in a wooden body. Called a *macuahuitl*, the weapon was capable of inflicting terrible injuries, combining the sharp cutting edge of an obsidian blade with the ragged cut of a serrated weapon.

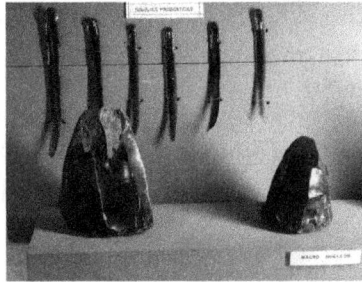

Raw obsidian and obsidian blades from the Mayan site of Takalik Abaj

Native American people traded obsidian throughout the Americas. Each volcano and in some cases each volcanic eruption produces a distinguishable type of obsidian, making it possible for archaeologists to trace the origins of a particular artifact. Similar tracing techniques have allowed obsidian to be identified in Greece also as coming from Melos, Nisyros or Yiali, islands in the Aegean Sea. Obsidian cores and blades were traded great distances inland from the coast.

In Chile obsidian tools from Chaitén Volcano have been found as far away as in Chan-Chan 400 km (250 mi) north of the volcano and also in sites 400 km south of it.

Easter Island

Obsidian was also used on Rapa Nui (Easter Island) for edged tools such as *Mataia* and the pupils of the eyes of their Moai (statues).

Current Use

Obsidian can be used to make extremely sharp knives, and obsidian blades are a type of glass knife made using naturally occurring obsidian instead of manufactured glass. Obsidian is used by some surgeons for scalpel blades, although this is not approved by the US Food and Drug Administration (FDA) for use on humans. Well-crafted obsidian blades, as with any glass knife, can have a cutting edge many times sharper than high-quality steel surgical scalpels, the cutting edge of the blade being only about 3 nanometers thick. Even the sharpest metal knife has a jagged, irregular blade when viewed under a strong enough microscope; when examined even under an electron microscope an obsidian blade is still smooth and even. One study found that obsidian incisions produced fewer inflammatory cells and less granulation tissue at seven days, in a group of rats, although no differences were found after 21 days. Don Crabtree produced obsidian blades for surgery and other purposes, and has written articles on the subject. Obsidian scalpels may currently be purchased for surgical use on research animals.

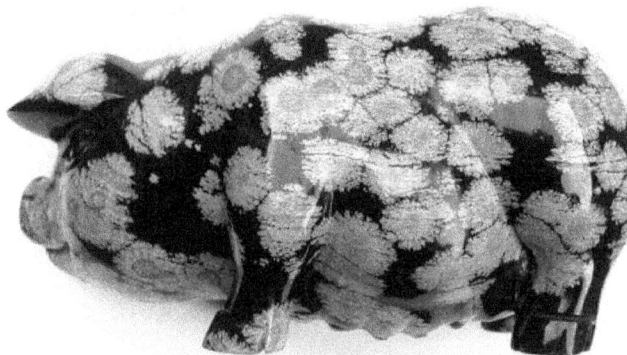

Pig carved in snowflake obsidian, 10 centimeters (4 in) long. The markings are spherulites.

Obsidian is also used for ornamental purposes and as a gemstone. It presents a different appearance depending on how it is cut: in one direction it is jet black, while in another it is glistening gray. "Apache tears" are small rounded obsidian nuggets often embedded within a grayish-white perlite matrix.

Plinths for audio turntables have been made of obsidian since the 1970s; e.g. the grayish-black SH-10B3 plinth by Technics.

References

- "Amber" (2004). In Maxine N. Lurie and Marc Mappen (eds.) Encyclopedia of New Jersey, Rutgers University Press, ISBN 0813533252.

- "Amber". (1999). In G. W. Bowersock, Peter Brown, Oleg Grabar (eds.) Late Antiquity: A Guide to the Postclassical World, Harvard University Press, ISBN 0674511735.

- George Poinar, Jr. and Roberta Poinar, 1999. The Amber Forest: A Reconstruction of a Vanished World, (Princeton University Press) ISBN 0-691-02888-5

- Poinar, P.O., Jr., and R.K. Milki (2001) Lebanese Amber: The Oldest Insect Ecosystem in Fossilized Resin. Oregon State University Press, Corvallis. ISBN 0-87071-533-X.

- Langenheim, Jean (2003). Plant Resins: Chemistry, Evolution, Ecology, and Ethnobotany. Timber Press Inc. ISBN 0-88192-574-8.

- Johns, Catherine (1996). The Jewellery of Roman Britain Celtic and classical Traditions. Routledge. pp. 120–121. ISBN 9780415516129.

- Anderson, Ken B. (1996). "New Evidence Concerning the Structure, Composition, and Maturation of Class I (Polylabdanoid) Resinites". Amber, Resinite, and Fossil Resins. ACS Symposium Series. 617. pp. 105–129. doi:10.1021/bk-1995-0617.ch006. ISBN 0-8412-3336-5.

- Gribble, C. D. (1988). "Tektosilicates (framework silicates)". Rutley's Elements of Mineralogy (27th ed.). London: Unwin Hyman. p. 431. ISBN 0-04-549011-2.

- Peter Roger Stuart Moorey (1999). Ancient mesopotamian materials and industries: the archaeological evidence. Eisenbrauns. pp. 108–. ISBN 978-1-57506-042-2.

- Pliny the Elder (translated by John Bostock, Henry Thomas Riley) (1857). The natural history of Pliny. 6. H G Bohn. ISBN 1851099301.

- M Martini; M Milazzo; M Piacentini; Società Italiana di Fisica (2004). Physics Methods in Archaeometry. 154. IOS Press. ISBN 1586034243.

- P R S Moorey. Ancient Mesopotamian Materials and Industries: The Archaeological Evidence Eisenbrauns,

1999 ISBN 1575060426.

- L Romano. 6 ICAANE, Otto Harrassowitz Verlag, 2010 Volume 3 of Proceedings of the 6th International Congress of the Archaeology of the Ancient Near East: 5–10 May 2009 ISBN 3447062177.

- D Schmandt-Besserat. Volume 3 of Invited lectures on the Middle East at the University of Texas at Austin, Undena Publications, 1979, ISBN 0890030316

- J. D. Fage. The Cambridge history of Africa: From c. 1600 to c. 1790, Part 1050, Cambridge University Press, 1979 ISBN 0521215927

Permissions

Index

www.ingramcontent.com/pod-product-compliance
Lightning Source LLC
Chambersburg PA
CBHW061311190326
41458CB00011B/3779